TECHNICAL EDITING

THE ALLYN AND BACON SERIES IN TECHNICAL COMMUNICATION

Series Editor: Sam Dragga, Texas Tech University

Thomas T. Barker
*Writing Software Documentation:
A Task-Oriented Approach,* Second Edition

Carol M. Barnum
Usability Testing and Research

Deborah S. Bosley
*Global Contexts: Case Studies in International
Technical Communication*

Melody Bowdon and Blake Scott
*Service Learning in Technical and
Professional Communication*

R. Stanley Dicks
*Management Principles and Practices
for Technical Communicators*

Paul M. Dombrowski
Ethics in Technical Communication

David K. Farkas
Principles of Web Design

Laura Gurak
Oral Presentations for Technical Communication

Sandra Harner and Tom Zimmerman
Technical Marketing Communication

Barbara A. Heifferon
Writing in the Health Professions

TyAnna K. Herrington
A Legal Primer for the Digital Age

Richard Johnson-Sheehan
Writing Proposals

Dan Jones
Technical Writing Style

**Charles Kostelnick and
David D. Roberts**
*Designing Visual Strategies for
Professional Communities*

**Victoria Mikelonis,
Signe T. Betsinger,
and Constance E. Kampf**
Grant-Seeking in an Electronic Age

Ann M. Penrose and Steven B. Katz
*Writing in the Sciences: Exploring
Conventions of Scientific Discourse,*
Second Edition

Gerald J. Savage and Dale L. Sullivan
*Writing a Professional Life: Stories of
Technical Communicators On and Off the Job*

TECHNICAL EDITING

Fourth Edition

Carolyn D. Rude
Virginia Tech

Contributing Authors

David Dayton
Towson University

Bruce Maylath
University of Wisconsin–Stout

PEARSON
Longman

New York   San Francisco   Boston
London   Toronto   Sydney   Tokyo   Singapore   Madrid
Mexico City   Munich   Paris   Cape Town   Hong Kong   Montreal

To my students—past, present, and future.
You are the reason I teach and write.

Senior Sponsoring Editor: Virginia L. Blanford
Senior Marketing Manager: Melanie Craig
Senior Supplements Editor: Donna Campion
Media Supplements Editor: Jenna Egan
Managing Editor: Valerie Zaborski
Project Coordination, Text Design, and Electronic Page Makeup: Nesbitt Graphics, Inc.
Cover Designer/Manager: John Callahan
Manufacturing Buyer: Lucy Hebard

Library of Congress Cataloging-in-Publication Data

Rude, Carolyn D.
  Technical editing / Carolyn D. Rude; contributing authors, David Dayton,
Bruce Maylath.—4th ed.
      p. cm
  Includes bibliographical references and index.
  ISBN 0-321-33082-X
  1. Technical editing. I. Dayton, David. II. Maylath, Bruce. III. Title.

T11.4.R83 2006
808'.02—dc22   2005044260

Please visit our website at http://www.ablongman.com

ISBN 0-321-33082-X

  6 7 8 9 10—CRS—12 11 10 09

CONTENTS

PART
2   METHODS AND TOOLS   47

FOREWORD
By the Series Editor

The Allyn & Bacon Series in Technical Communication is designed for the growing number of students enrolled in undergraduate and graduate programs in technical communication. Such programs offer a wide variety of courses beyond the introductory technical writing course—advanced courses for which fully satisfactory and appropriately focused textbooks have often been impossible to locate. This series will also serve the continuing education needs of professional technical communicators, both those who desire to upgrade or update their own communication abilities as well as those who train or supervise writers, editors, and artists within their organization.

The chief characteristic of the books in this series is their consistent effort to integrate theory and practice. The books offer both research-based and experience-based instruction, describing not only what to do and how to do it but explaining why. The instructors who teach advanced courses and the students who enroll in these courses are looking for more than rigid rules and ad hoc guidelines. They want books that demonstrate theoretical sophistication and a solid foundation in the research of the field as well as pragmatic advice and perceptive applications. Instructors and students will also find these books filled with activities and assignments adaptable to the classroom and to the self-guided learning processes of professional technical communicators.

To operate effectively in the field of technical communication, today's students require extensive training in the creation, analysis, and design of information for both domestic and international audiences, for both paper and electronic environments. The books in the Allyn & Bacon Series address those subjects that are most frequently taught at the undergraduate and graduate levels as a direct response to both the educational needs of students and the practical demands of business and industry. Additional books will be developed for the series in order to satisfy or anticipate changes in writing technologies, academic curricula, and the profession of technical communication.

Sam Dragga
Texas Tech University

PREFACE

Concepts and Goals

Much workplace writing might more accurately be described as editing: compiling, selecting, reorganizing, updating, developing, and redesigning work that already exists in some form. This work may consist of drafts prepared by subject matter experts, related company documents, an existing version that requires new information, or information in secondary sources. Regarded in this sense, editors are information managers, people with a breadth of vision to perceive the possibilities for uses of information, both in the present and in the future. They have the ability to look beyond the existing versions to consider more effective possibilities; the ability to gather necessary information through interviews, secondary sources, databases, or field methods; the expertise in language use to evaluate not just sentences and paragraphs but also the whole document or document set; the awareness of documents in context and in use; and the ability to collaborate with others in developing information products.

This view of editing contrasts sharply with the traditional view of editors as grammar janitors, people who clean up mistakes after someone else has written an otherwise effective document. It is an ambitious view that demands a sophisticated sense of the use of information in the workplace and how this information becomes available for use through thoughtful choices of media, content, structure, and design. Editors remain expert on grammar and mechanics, but they offer so much more: analysis, evaluation, imagination, and good judgment applied to information design and management.

 This edition of *Technical Editing*, like its predecessors, aims to prepare students comprehensively for editing tasks in technical and other professional environments. One goal is to develop in students a rich appreciation (1) for the ways in which documents enable solutions to problems; (2) for the variety of options that editors can choose for documents; and (3) for the complex nature of language and communication and the expertise required to use information well. Students should be able to make choices about documents that are well grounded in the understanding of how users will respond. They also understand and can use the principles of organization, visual design, and style that are an editor's design tools. Although *Technical Editing* does not merely rehearse rules, it does review principles of grammar, punctuation, style, organization, and visual design so that editors will understand how all of these text features may be used to clarify meaning, to persuade, to make information easy to locate, or to encourage the accurate completion of a task. The book also offers a process for imagining varied audiences, including international users, reuse of information in multiple versions, and consequences of choice—all concepts that take the editor's eyes from the page and into various contexts of use.

Perhaps subtly, the themes of professionalism and respect permeate this textbook. Professionalism means taking responsibility for doing well: knowing one's area of specialization, including its vocabulary, and not just guessing or editing by preference; working well with colleagues; and doing quality work on time. Professionalism also means responsibility to users and to ethical standards. A professional person uses critical judgment and does not merely accept what he is given. A professional person offers respect to others: to users, to colleagues, to subject matter experts, to writers, to support staff. Self-respect matters too: it is based on the editor's confidence that she has something of value to contribute to a project.

Revisions for the Fourth Edition

Exciting changes in technology continue to change editorial procedures. Technology offers new options for publication, specifically the World Wide Web, corporate intranets, and online help. "Print" is becoming a dated concept in technical communication. Technology makes possible the generation and updating of documents in multiple forms from databases of information. It has changed the process of production from offset lithography offered by commercial vendors to electronic publishing offered by desktop operations and digital printers.

The main substantive changes to this edition relate to technology. The chapters on marking digital copy (Chapter 5) and electronic editing (Chapter 6) are substantially updated to reflect available tools, including XML and cascading style sheets, as they relate to websites and single-sourcing. Chapter 21 on editing websites replaces a chapter that discussed editing online documents in a more general way. Chapter 24, on project management, introduces new tools for managing publications development and production, including configuration management tools. A new chapter on client projects concludes the book, sending apprentice editors out into the world to apply their knowledge.

To increase the pedagogical effectiveness of the book, this edition introduces brief summaries of key concepts at the beginning of each chapter. Instead of ending with summaries, the chapters now end with sections titled Using Your Knowledge, which point students to applications of the chapter material. The tone of the book will always be serious, but I have tried to make it less solemn by cutting some of the words. Suggested materials for further reading and online resources have been updated.

Although technology and genres change the field almost daily, the principles of effective communication remain stable. These principles relate to developing content according to the needs of users, organizing it to make it understandable, structuring sentences for ease of comprehension, and using visual information when it works better than words to enhance comprehension and usability. In addition, technical communicators in general, and editors in particular, think beyond the page to users, to consequences of the textual choices, and to relationships of ideas. Understanding these principles and processes will always mark the value that editors bring to a communication task.

Audience, Pedagogical Methods, and Structure

This book is written for students who have completed at least one college course in technical communication and for practicing editors with some experience in technical genres. It presumes that readers have been introduced previously to such terms as *style, noun, line graph,* and *instruction manual,* and it presumes some competence in technical writing. Chapters on spelling, grammar, and punctuation review concepts readers have learned before and do not substitute for a basic textbook or handbook. The chapters refresh students' vocabularies so they can talk about their editorial choices using the language of the profession. Chapter activities and assignments at the Companion Website (www.ablongman.com/rude) support teaching in computer classrooms and online editing as well as providing documents to download for editing on paper. The glossary reinforces the premise that professional technical communicators master the vocabulary of their discipline.

Scenarios, examples, illustrations, discussion questions, and applications complement explanations to enhance learning. Tables summarizing key points in chapters on copyediting enable ready reference. An accompanying instructor's manual and website provide workplace documents correlated to chapters so that students can apply the principles that the chapters discuss without the distraction of other editorial needs.

The book is organized to parallel the typical career path of editors. Just as editors prove themselves as copyeditors before they accept responsibility for comprehensive editing, so does this book teach copyediting before the more complex and less structured principles of comprehensive editing and management. This structure also facilitates the use of the book in two quarters or semesters with the first term devoted to copyediting and the second to comprehensive editing. The modular character of the book, however, enables a top-down approach, with issues of comprehensive editing preceding the review of copyediting.

Part 1, "People and Purposes," includes introductory material. The first chapter illustrates the breadth and diversity of editorial responsibilities through scenarios and discussion. The second chapter explains what readers and users do with documents—the basis for editorial choices beyond adherence to rules. Chapter 3 positions the editor in a working relationship with the writer or subject matter expert.

Part 2, "Methods and Tools," shows students how to mark both paper and digital copy. It also describes the procedures and tools available for electronic editing.

Part 3, "Basic Copyediting," covers editorial choices that make a document conform to language standards, including grammar, punctuation, and consistency. It explains those standards in the context of a reader's need to understand, locate, and act.

Part 4, "Comprehensive Editing," offers an analytical process and principles for evaluating style, organization, and the visual features of a print or online text. Editors learn to look at whole documents and imagine their use by readers so that they can guide decisions about these high-level features of documents. Teachers who prefer the comprehensive approach to editing or whose students are skilled in basic copyediting can assign Part 4 directly after Parts 1 and 2, using Part 3 for reference as needed.

Part 5, "Management and Production," takes students into the workplace to consider legal and ethical issues of publication and methods of production and management. Chapter 25 offers advice about client projects and interacting with clients.

Instructor's Manual and Companion Website

Ancillary materials—an Instructor's Manual and a Companion Website—support instructors in using the textbook for a course.

The Instructor's Manual includes four sections:

- *Discussion of the philosophy and pedagogy of the editing course.* This section will help you conceptualize your own course and the methods you will use.
- *Options for a course syllabus and statement of objectives.* This section will help you plan the specific structure and requirements of your course.
- *Chapter notes.* This section offers suggestions for approaching each chapter. It also includes possible responses to the Discussion and Application activities following each chapter.
- *Worksheets and transparency masters.* This section includes materials that can be photocopied for a course workbook. Marked documents can be made into transparencies for discussion in class.

To request a copy of the Instructor's Manual, call 1-800-852-8024, or contact your Allyn & Bacon/Longman sales representative. (Go to www.ablongman.com/replocator/ to find contact information for your representative.)

The Companion Website (www.ablongman.com/rude) provides additional supplementary materials not included in the Instructor's Manual and materials that support electronic editing as well as completion of some activities at the computer. The website includes these sections:

- *Discussion and Application worksheets.* These documents replicate the end-of-chapter activities in the textbook, but they can be reproduced for a workbook or downloaded to computer workstations so that students do not have to write in their books. They are available both in a print version (for a workbook) and an online version (for completion at the computer).
- *Supplementary instructional materials.* Principles of grammar and style are reviewed in slide shows in more detail than they are discussed in the text.

For example, one show discusses dangling modifiers, and another distinguishes "although, but, and however" as connectors with similar meaning but different effects on sentence structure and punctuation. Two tutorials on advanced features of Microsoft Word help students develop skills for electronic editing.

- *Documents for editing practice.* Documents that are longer than those in the Instructor's Manual and suitable for comprehensive editing are available for download. A sample exam testing knowledge of grammar and punctuation is also available.
- *Internet resources for editors.* A list of sites with descriptions will help editors keep up-to-date on technology and locate reference materials.

Acknowledgments

Many people have helped make the fourth edition a better edition than its predecessors, but I especially would like to thank students for their suggestions. My own students have helped me understand the users of this textbook both by their responses to it and by their explicit suggestions. Nancy Allen and her students in an editing class at Eastern Michigan University reviewed the book and made numerous helpful suggestions. The students include David Brandt, Kristen Bretti, Katherine Caines, Matthew Carter, Cheryl Clark, Sarah Gorski, Candise Green, Tracey Kuffel, Erik Simmons, and Benjamin Simpson.

I am grateful to reviewers of the third edition, who gave me helpful advice about the fourth; Nancy Allen, Eastern Michigan University; Patricia Egan, University of California–Berkeley; Helen M. Grady, Mercer University; Kenneth Price, University of Alaska–Anchorage; Gerald Savage, Illinois State University; and Lee S. Tesdell, Minnesota State University–Mankato.

I am grateful for the revisions that David Dayton and Bruce Maylath provided for their chapters (6 and 20, respectively). Paula Green helped to update Chapter 1, Carlos Evia reviewed and contributed to Chapter 5, Ron Hampton reviewed and corrected sections of Chapter 12, Laura Palmer contributed to Chapter 23, and Heather Eisenbraun wrote some sections for Chapter 24. Tara L. Masih edited with an uncanny gift for recognizing inconsistencies even hundreds of pages apart and with an unfailing ear for style. Clair James proofread and created the index.

I continue to benefit from the assistance of Kathy Klimpel and Charlene Strickland, who provided insight into editorial practice in the scenarios for Chapter 1, beginning with the second edition. I continue to use examples provided for previous editions by Steven Auerbach, Kae Hentges, Ken Morgan, Lane Mayon, Carlos Orozco-Castillo, Ellen Peffley, Tony Santangelo, and William Stolgitis.

Love and appreciation to my husband, Don, who patiently supported me while the book once again took priority for time.

C. D. R.
Virginia Tech, Blacksburg
Carolyn.Rude@VT.edu

Longman Resources for Technical Communication Instructors and Students

Longman offers a number of resources for technical communication instructors and students. Instructors can check with their Longman sales representatives to order any of the following materials packaged free with this text:

- **Resources for Technical Communication** (ISBN 0-321-27870-4). This handy resource contains 50+ sample documents and 10 case studies. Organized by genre for easy reference, it also includes introductory materials and activities.
- **Visual Communication, 2/e** (ISBN 0-321-09981-8). Susan Hilligoss's popular text introduces document design principles and features practical discussions of space, type, organization, pattern, graphic elements, and visuals.
- **MyTechCommLab (www.mytechcommlab.com).** MyTechCommLab offers the best multimedia resources for technical writing in one easy-to-use place. Students will find guidelines, tutorials, and exercises for writing, grammar, and research, as well as Exchange, Longman's new online peer and instructor review program. MyTechCommLab is appropriate for any technical writing course when instructors want to give their students additional resources in technical writing, grammar, and research or want to do online peer and instructor review of papers.

People and Purposes

Editing: The Big Picture

If you are coming to the study of editing without prior experience, you may think of it as cleanup work after a document is written. Editors correct errors in spelling and punctuation.

Editing does require high standards of language use, but you will find that cleanup is a small part of what technical editors do. Technical editing, like writing, requires information design—creating documents that work for the people who will use them. Functional documents require more than correctness. Editors who help to create these documents must be able to imagine documents in use by particular readers, to use good judgment as well as grammar handbooks, to manage long-term projects, and to collaborate with others.

In this chapter, you will see specific editing responsibilities in the context of the whole process of conceiving, writing, reviewing, and publishing documents—the big picture of editing. Let us begin with questions: What does an editor do and why? How does a *technical* editor differ from any other editor?

To help answer these questions, we will review the work of two editors in different settings. One editor, Kathy Klimpel, works for a software company and edits software documentation. Her materials are developed in-house—within the company that needs the documentation to help customers use its software. The second editor, Charlene Strickland, works for a consulting company that contracts for documentation projects from other companies. She edited an online tutorial on computer security. These two settings represent typical employment for technical editors.

The editors' roles in both projects were comprehensive:

- helping to define the need, purpose, and scope for the document
- working with writers and subject matter experts
- reviewing the text for completeness, accuracy, visual design, and overall effectiveness

Both editors are experienced, not beginners, and represent advanced job responsibilities. Their methods differ somewhat because of their work situations.

Key Concepts

Basic copyediting, making the document correct and consistent, is just one editorial task. Editors are also information designers: they advise writers about the best choices for content, organization, and visual design, considering users and document purposes. They also help prepare documents for publication whether in print or online.

Scenario One: Print Document, In-House Editor

Kathy Klimpel works as an editor for BMC Software, Inc., in Houston. BMC Software provides software solutions that manage enterprise systems, applications, databases, and services. Kathy works with mainframe IMS products. (IMS is IBM's Information Management System, a database management system used in many major data processing centers around the world.) Before a BMC Software product can be sold, it must have documentation: all of the technical information that an end user needs to install and use the product. The products are so complex that a whole documentation set is needed to support them. A documentation set can consist of an installation guide, a getting started guide, a user guide, a reference manual, a message manual, a quick reference, an administrator guide, and other types of paper documents, as well as Help systems that are delivered with a variety of mechanisms. In addition, a documentation set often covers a range of related products rather than a single product. BMC Software does not provide on-site training, so the documentation must be complete and clear. Customers may call a toll-free number to talk with a product support representative, but BMC Software hopes clear documentation will minimize the number of calls. The scope of the documentation set can vary depending on how complicated the product is, how much background information users need, and how much time is available for developing the documentation.

The Product Team

At BMC Software, the Research and Development (R&D) division develops and supports products through a team approach. Technicians on the product team specialize in designing, coding, testing, and supporting the products. They report to product line managers or to development, quality assurance, or customer support managers. Other team members specialize in project management, usability, or marketing.

Information developers (technical writers) are members of product teams and work directly with other team members to create and maintain technical documentation for products. Depending on workloads, a writer may support multiple product teams, and a product team may require the support of multiple writers. The writer is assigned permanently to the product team, and a new product usually has an assigned writer from the beginning. Most writers are in daily contact with the technicians on the team, but they have regular contact with other team members too. Writers report to information development managers within R&D.

Editors also report to information development managers. They work directly with the writers, usually from the beginning of new projects, and are acknowledged as fully contributing members of the team. Editors work with many writers simultaneously on documentation for different products.

In some companies, information development is a separate activity that is performed in isolation from product development. By integrating writers and editors with technicians, the team approach at BMC Software helps writers and editors learn how to use a product and how to participate in its development.

Project Definition and Planning

Kathy works with Paula Green, a technical writer. When a product idea is approved, the entire product team meets a few times to discuss the scope of the project and to decide how long it will take to complete. Kathy and Paula attend the meetings and learn what expectations the team has for the product and the documentation.

When Paula writes a reference manual, she first writes an outline. Kathy reviews the outline and suggests revisions, mostly about content and organization. They develop a schedule for completing the manual so that it will be ready when the product is ready to ship. All products change during the course of development, and Paula and Kathy make corresponding changes to the structure of the manual and to the schedule.

Writing and Editing

As the product team begins developing the product, Paula begins writing. She uses FrameMaker templates that BMC Software has developed to give the documentation for all of their products a similar structure and appearance. (FrameMaker is publishing software, and a template is a set of instructions for display of different types of text, such as headings and paragraphs.) The templates save the writers and editors time by providing a standard page design for the documentation and by establishing some standard structural elements, such as chapter introductions. The templates also help users of multiple BMC Software products by providing some consistency in the various pieces of the document set. Paula also refers to the comprehensive BMC Software style guide that establishes BMC preferences for details such as capitalization of terms and use of numbers.

Paula submits each chapter to one or two members of the product team for a technical review of content as well as to Kathy. Kathy reviews the chapters comprehensively. Her overall goal is to help Paula produce readable and usable documentation. "Readable" and "usable" define whether readers can find and understand what they need to know in order to use the product. Kathy also reviews the drafts for these qualities:

- adherence to BMC Software standards as defined in the comprehensive style guide
- content and organization: completeness, logic; use of examples to explain concepts; logical arrangement of sections
- style: short but not choppy sentences; conciseness; ability to be translated into a foreign language

- visual features: graphics; page design; appropriate use of the templates
- grammar and punctuation
- mechanics and consistency: spelling, capitalization, hyphenation

An example illustrates the way Kathy may influence the content of the documentation and its usability. On one reference manual, Kathy and Paula changed the outline after Kathy reviewed it. The manual described three related products that allow a user to restart computer programs that have failed. The three products accomplish the same goal, but each one works in a different environment.

Paula wanted to write about all of the cross-product tasks together (for example, all implementation information in one chapter with subheadings for each environment). Kathy, who tries to take the reader's point of view, suggested that the book might be more usable if the subjects were grouped by environment rather than by task. If a customer buys the product for the IMS environment, that customer probably doesn't care about the other two environments the product works in. Therefore, it will probably be easier for customers to use a book if all the information they need is in one place and is not cluttered by other, irrelevant information.

Paula rearranged the topics according to environment. Information that applied to the product in more than one environment was centralized, and Paula used cross-references to point to information as needed.

When she edits, Kathy marks paper copy of the drafts because users see the book on paper. She prefers a purple pen, avoiding red because she doesn't want to suggest to Paula that she is grading her draft, like a teacher, instead of editing.

Kathy tries to return an edited chapter to Paula in a day or two. Paula incorporates the edits, but if she questions them or if Kathy has written "let's discuss," they talk together about alternatives. Often they arrive at a better solution together.

After Kathy edits a chapter, Paula sends the chapter to the entire product team for a technical review (for accuracy and completeness of content). This process of multiple reviews during the course of product development achieves technical accuracy as well as effectiveness of the writing. It also keeps the writer and editor in touch with the programmers in case there are changes in the original plan for the product. Because they use the product as they write about it, Kathy and Paula may also advise the programmers on how to modify the product to increase its efficiency or ease of use. If the writing for a procedure seems complex, the reason may be an unnecessarily complex program. Team members all benefit from the interaction throughout product development.

After she reviews each chapter individually, Kathy edits the entire manual, including all front and back matter (such as the title page and appendixes). She tries to finish this part of the editing task in two to five days. The chapters and overall structure of the manual are established by now. She mainly edits for completeness and the smooth connection of parts.

Developing a product and its documentation can take six to nine months. During this time the writer can get caught up in the details. The editor tries to keep the perspective of the user and the big picture.

Publication

BMC Software produces most of its documentation in electronic form and distributes it over the Web. Before distribution, multiple departments must sign off on the product and its documentation, using a detailed checklist to ensure that the products pass the various reviews. It also uses a document management system to organize and manage its electronic documents. This system is a huge database that enables BMC to catalog, store, and locate documents that are related to its products.

Collaboration of Writer and Editor

Because Paula and Kathy work together from the product idea to completion of the manual, they have a close working relationship. Paula has more technical knowledge than Kathy and has taught her how IMS works. When Kathy inadvertently changes the content as she edits, Paula explains why a different revision may make more sense. Kathy helps Paula tune specific writing skills.

Kathy says, "Paula and I work together very well, mainly because we have the same goal in mind—clear, correct, concise documentation—and because we have mutual respect. I enjoy working with Paula because she teaches me so much. She has helped me become a better editor."

Paula says, "I trust Kathy to see our documents from the reader's perspective, from the big picture to the tiny details. I rely on her intelligence, common sense, sharp eyes, and sound advice to perfect my work. She makes the difference between a good document and an excellent one."

Scenario Two: Online Tutorial, Contract Company

Charlene Strickland works in Information Technology for Science Applications International Corporation (SAIC) in Albuquerque. This firm works on contract for companies that need computer programs and documentation but do not have the staff or the time to complete the work themselves. The companies can achieve high quality at a fair price by hiring specialists to do the work.

Charlene was editor on a project to develop a computer security tutorial on an intranet within the World Wide Web for a national laboratory. (Access is limited on an intranet to people within a company.) Like typical workplace documents, the tutorial was created because a problem had to be solved. The laboratory is required by U.S. law to offer computer security training every year to each of its 7,000 employees. Each employee has to be recertified every year by passing a test that assesses his or her knowledge of the material. Classroom training was expensive and time-consuming. It was a logistical nightmare as well to train 7,000 employees with different levels of experience in computer use and knowledge of security procedures. The training was further complicated by the fact that three different types of computer workstations were used in various parts of the laboratory.

The laboratory contracted with Charlene's company to produce the training in an online version. Online training could solve the scheduling, computer platform, and expertise problems while reducing the cost of training. Users with different types of computers can use the same website. The design could incorporate testing as well as instruction. Employees could complete the tutorial at their own pace. They would study the tutorial on their own computers, answer questions and get feedback, and then take a test to earn certification in computer security.

The Product Team

The team that produced the tutorial included thirteen people with different types of expertise and responsibilities. Several of them were working on multiple projects simultaneously and did not devote all their time to this project.

- The project manager met with the client and was responsible for ensuring that the project met the client's specifications and was delivered on schedule.
- Three writers provided expertise in computer security and created the content for the tutorial.
- Four instructional designers organized information for maximum comprehension and retention.
- Three programmers made the pages work technically. They coded the text and graphics for each page and created the links between subjects and modules.
- One graphic artist created the color schemes, illustrations, and screen design for the instructional modules.
- One editor reviewed content development and sentence structure and maintained coherence in the parts produced by multiple writers and instructional designers.

All of these team members had specialized roles, but the editor was a generalist in addition to being a specialist. Charlene was the one person on the team with whom all the other team members interacted. Thus, she was responsible not just for the effectiveness of language but also for coordinating the efforts of the other team members and for ensuring that the information produced collaboratively through division of labor was complete, consistent, and appropriate for the purpose.

Project Definition and Planning: Content, Structure

The division of labor, in which different specialists were working simultaneously on different parts of the final product, required initial planning. The team used a software development process of planning, analysis, design, development, and implementation. Analysis required the team to learn the customer's expectations and to become familiar with the content of the training. Because the laboratory had previously delivered the computer security training in a classroom setting, some teaching materials were available to SAIC for developing the content. These materials, called *legacy* documents, were an important source of information for the training. Yet the materials contained inconsistencies and needed to be updated. The client provided this updated information in response to queries from the writers and editor.

The team planned six instructional modules plus a test. This plan for the content to be covered in each training module was necessary before the writers and programmers could begin working on the modules assigned to them. They displayed the plans on a *storyboard*, a poster-sized visual and verbal outline of the structure and contents of the tutorial. The storyboard included *scripts* for the components of the tutorial. Scripts contained both visual and textual elements, chunked by the sequence of screens that comprised each module. A template for the scripts, written as a Microsoft Word file, contained the scripts' elements in a three-column table format. Columns included the title of the electronic file, the action of the user, and the text that would appear as a result of the action.

The editor completed these tasks:

- contributed to information gathering and planning
- developed the template for the scripts
- drafted a style sheet of terms
- compared the scripts to the source document for accuracy

Planning for Design and Production

Design and production decisions affect writing and editing decisions from the start. The team had to find out what kind of equipment and software the laboratory would use for the tutorial. Software and hardware affect what the screen will display and how fast the text and images will load. The team planned screen design to encourage reading, comprehension, and ease of moving among the screens (navigation). A consistent layout of screens helps users to navigate because they can find similar types of information in the same place on every screen. Menu bars let users choose lessons. The team had to decide whether screens would require users to scroll or whether each one would be complete and how to create transitions from one screen to the next. They decided as well to use a minimalist style, with as few words as possible, to avoid filling the screens with endless blocks of text. They planned as much interaction as possible within the tutorial so that the laboratory employees would not merely read. These plans followed from analysis of the laboratory's needs and available options for meeting the needs as well as from principles of instructional design and good writing.

Planning for visual design was part of the planning for content and organization because design needed to support the learning goals, not substitute for them. The SAIC team wanted to avoid a trendy, eye-catching tutorial. Instead they aimed for one that would accomplish the goal of teaching computer security. Each design decision had to support the instructional goals.

All of these plans at the beginning of the project gave the editor some measures for reviewing the drafts of the modules. They functioned in the same way that the comprehensive style guide and the initial manual outline function at BMC Software. The editor's participation in planning meant that she would not come in at the end of the project with a different idea and try to make substantial changes. She and other team members pursued the same goals from the start.

Editorial Review

Charlene used the legacy materials plus the team's plans for content and design as guidelines for editing. She edited each module separately. Her top priority in editing was to make the content accurate, complete, and readable.

For example, she reviewed the grammar. She made changes in subject-verb agreement and the use of conjunctions (such as confusing *since* with *because* and *between* with *among*). Because online documents are often more informal in voice than material on the printed page, she also used contractions and did not spell out any numbers.

Because the modules were written by different people, Charlene edited to make the modules read as if written by a single author. Editing the tutorial as a whole was a necessary step to prevent inconsistency, contradiction, and redundancy. She eliminated arbitrary variations such as changing from second to third person. She also made terms consistent across the modules with regard to spelling, abbreviations, and capitalization. She referred throughout to the laboratory by its abbreviation, not the first word in its name.

To create consistency in bulleted lists, Charlene capitalized the first letter of each item and ended each item with no punctuation. She edited items in each list for parallel structure and used a colon to end the phrase preceding each list entry.

The writers submitted their drafts as computer files, and Charlene edited them on the computer, without printing. Editing on the computer helped her see the screens as the laboratory employees would and thus get a reader's perspective. Because she knew the subject matter and was in close contact with the writers, she could edit sentences or reorder paragraphs with confidence that she was not distorting the meaning. Still, it is easy to introduce new errors while editing, and the work required continual proofreading. Sharing files required careful version control, so that writers and editor would always work on the most recent version. Each writer or editor who modified a file included the date of revision in the file name.

Charlene used the tutorial as a laboratory employee would, choosing the different lessons through the menus, responding to questions, and navigating through the screens. She was conducting an informal usability test in conjunction with editing, completing tasks according to directions, not just reviewing the words on the page. Unclear directions or unexpected outcomes pointed to needs for editing the text.

Client Review

The clients reviewed printed pages of each module. Then, they viewed each module screen by screen, with a programmer and the editor presenting and discussing each screen. The clients made suggestions, and the programmer changed the screen so the clients could see and approve changes. The editor proofread these on-the-fly changes.

Comment: Editing at BMC Software and SAIC

At BMC Software and SAIC, the editorial procedures are comparable though different products, settings, and personnel change some details.

- *Both editors edit comprehensively.* They edit not just for grammar, punctuation, and consistency but also for content, organization, and design. They see the document whole, as users will use it, not just at the sentence level.
- *Both editors are part of the product teams from the start.* When editors begin work on a project at the development stage, they can suggest good ways of presenting information and can prevent problems. If they join the project later, when the document is almost ready for publication, there is time only to repair superficial problems. Requests for changes after the document is written may discourage writers and may build stress between writers and editors. By entering the project early, editors can contribute to the vision, not just to the *re*vision, of the document.
- *Both editors see themselves as collaborators with the writers and with other team members.* They have good working relationships with other team members because they share the same goal of developing a document that will work for the users.

The differences in their jobs reflect different documents, settings, and personal strengths. Kathy edits on paper, while Charlene edits on the computer. The difference in part reflects the fact that Kathy's document is printed on paper, while Charlene's is published on a computer screen. The editors need to see the documents as the readers will see them in order to edit well. But both editors are concerned with content (completeness, accuracy of details), organization (relation of parts to each other and to the whole), style, page or screen design, usability (ease of use and effectiveness in enabling comprehension and performance), and correctness of grammar and punctuation. All of these qualities determine whether the documents will enable readers to use the product they have purchased or to learn about computer security well enough to pass the test and protect their files.

The comprehensiveness of their tasks, integration into the product teams, and collaboration describe optimal situations for editing. Both BMC Software and SAIC demonstrate good management practices based on understanding of the editorial function.

Editorial Functions and Responsibilities

Why are there editors? What does an editor do? For technical editors, these questions can be answered by summarizing their two primary functions: text editing and preparing documents for publication.

Text Editing

Editing the text means making it complete, accurate, correct, comprehensible, usable, and appropriate for the readers. An editor, whose specialty is language and document design, can suggest ways to make the document easier for readers to understand and use. The editor knows how to use style, organization, and visual design to achieve specific goals. Even when a writer is sophisticated in the use of language, an editor can bring objectivity to the reading that the writer may lose by knowing the subject too well. The editor works with the text from the perspective of the reader. The editor serves as a readers' advocate. As advocate, the editor is concerned not just with comprehension and access but also with fairness and safety. The editor has an ethical responsibility to readers.

Not all documents receive the comprehensive editing that has been described so far. The amount of editing depends on the importance of the document and on time constraints. A company newsletter that won't be distributed publicly may be photocopied without much editing. The editor may simply check that every page is present and correctly numbered. For another document, the editor may check spelling, grammar, and consistency but not completeness and accuracy of information, organization, or visual design. The editor's supervisor or client establishes the expectations and limits of the job.

Text editing responsibilities may be classified as *comprehensive editing*, when the editor works with the content, organization, and design of the text as well as with grammar and punctuation, and *basic copyediting*, when the editor works with grammar, punctuation, spelling, mechanics, and labeling of illustrations.

Comprehensive Editing

When the editor shares responsibility with the writer for document content and usability, the editor is editing comprehensively. Other terms for this type of editing are *developmental editing, macro editing, analysis-based editing,* and *substantive* (sub'•stan•tive) *editing.* The term *developmental* emphasizes the process in which the editor works from the start with the writer in developing the content and organization. *Macro* distinguishes comprehensive text features from "micro" features such as punctuation. *Analysis* foregrounds the process. The term *substantive* emphasizes document content or "substance." *Comprehensive* is used in this textbook because it suggests both the process and the focus of editing.

In comprehensive editing, the editor analyzes the document purposes and makes decisions about the best ways to meet these purposes. The editor may add or delete material and evaluate the reasoning and evidence. The editor also reviews organization, visual design, style, and use of illustrations in order to help readers find and comprehend the information they need.

Some tasks of comprehensive editing may take place before the first draft is written, so that the writer and editor can develop a shared concept of the document and its readers and purpose. The editor may create outlines, templates, and a glossary of key terms. The editor may review early drafts to enable necessary reshaping of the document before a writer has invested too much time in its development. The editor may also edit for content after the document is almost complete. The editor may advise the writer about revisions or even rewrite some sections. Because comprehensive editing addresses the content, the editor must know something about the subject matter.

Basic Copyediting

Copyeditors check for correct spelling, punctuation, and grammar; for consistency in mechanics, such as capitalization, from one part of the document to the next; and for document accuracy and completeness. The copyeditor may mark the document to indicate typeface and type size, column width, and page length. Basic copyediting assumes that content, organization, visual design, and style are already established.

A good copyeditor has an eye for detail as well as a command of language. The editor refers to handbooks, style guides, and other printed or online sources and queries the writer or a technical expert to resolve inconsistencies or other text questions.

Preparing Documents for Publication

An editor is the link between the writing and the publishing of a document. Writers may lack the means for publishing and distributing their materials. They may also have little interest in preparing the document for publication. Their purpose is to develop the content.

Companies may complete a number of publishing functions in house, or they may contract for services. They may employ desktop publishing or web design specialists with the expertise to convert drafts of the document to the way they will look on the page or screen. (*Desktop publishing* means that some of the type and layout tasks previously completed by commercial printers are now completed at the desktop of the writer and editor or production specialists in the company developing the content.) Production specialists may prepare templates and style sheets to define the type and page size of the document. For example, all level-one headings will use the same font size and spacing. Alternatively, the company may contract for professional typesetting and printing or web design.

Whether the work is done in house or by contract, the editor communicates with the production specialists. (The writer is probably not involved.) The editor reviews the final draft of the document to verify that the templates have been applied correctly or to mark the document with instructions for type and layout and placement of illustrations.

The production editor contracts with vendors. The editor obtains bids for production based on length, illustrations, color, and turnaround time. For print documents, these estimates also include number of copies, paper, and binding. If the cost estimate exceeds the budget, the editor will either modify some decisions about the publication, such as choosing cheaper paper for a printed document, or negotiate for a larger budget if the choices made seem necessary.

Document Development and Production: Summary of the Process

Technical editing is part of the process of developing documents that solve problems or enable readers to use products. Editing requires knowledge of language and procedures of marking documents, but good editorial decisions also require knowledge of how those decisions affect the rest of the process and the effectiveness of the document as readers will use it. Editing is one type of quality control. The ultimate goal is an effective document as measured not just by language standards but also by ethical and usability standards.

In spite of good editorial judgment, the editor may miss some needs for improvement in the document. Technical reviews, inspections, and usability tests provide additional information. A technical review is a review for content by a subject matter expert, such as a computer programmer or medical researcher. The

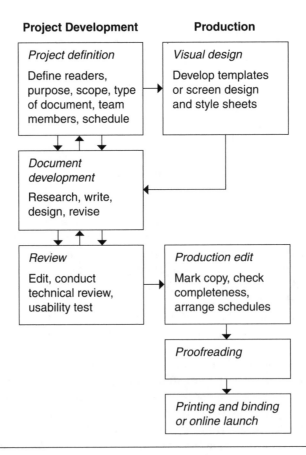

Figure 1.1 Project Development and In-House Production

reviewer looks for accuracy and completeness of information. An inspection is a kind of technical review, but after their independent reviews, various experts meet together as a group to discuss the results of their review and to make recommendations to the editor. They resolve together any conflicts from their own reviews.

In a usability test, representative users are observed performing the tasks that the document teaches under the intended conditions of use. The users may try to use software by following instructions in a manual. If they can't figure out how to do something, the testers know that the manual (or the software) requires revision. Informal usability tests for this textbook took place in classrooms, with college students as the evaluators. The students had some ideas for the book and found some confusing places in it that the writer, editor, and faculty reviewers had missed.

The flowchart in Figure 1.1 summarizes the document development and production process. The process is not entirely linear, as the two-way arrows between the boxes on the left imply. Writers draw on responses from editors, suggestions from subject matter experts, and results of usability tests in their revisions. As the document develops, the project definition may change. Some documents develop more simply, without technical reviews and usability tests.

The production process varies depending on whether the document will be printed or distributed in some other medium. In addition, the process depends on whether a company produces its own documents or contracts with commercial printers or web development specialists.

The Technical Part of Technical Editing

The text editing and production responsibilities are common to all kinds of editors: magazine and newspaper editors, academic journal editors, and the editors who work in commercial publishing houses on novels, trade books, and textbooks as well as technical editors. All editors share some responsibilities for helping to make writing effective and for arranging for the publication and distribution of documents. The adjective *technical* does, however, distinguish some defining characteristics of the specific type of editing that this textbook teaches.

Technical Subject Matter and Method

Technical editors work on documents with technical subjects. *Technical* connotes technology, and typical subjects are computer science and engineering. However, technical editors also edit in medicine, science, government, agriculture, education, and business. A technical editor may be employed in any field for which the documents aim to help readers solve problems or gain information. Because of the specialized subject matter, editors ideally have technical (subject matter) knowledge as well as language expertise. A technical editor working for a software firm would know some programming, while a medical editor would know some biology, chemistry, anatomy, and physiology.

Technical suggests not only the subject matter but also the method of working with the subject matter—to analyze, explain, interpret, inform, or instruct. The word derives from the Greek word *techne,* meaning "art and skill," which is also the source of *technique.* The art and skill of editing require specialized knowledge of the use of language and the methods by which we make sense of information.

Technical Genres

Technical editors typically work with the document genres (or types) that permit the transfer of information or that enable readers to act by making a decision or by following instructions. Examples of such genres include instruction manuals and online help, proposals, feasibility studies, research reports, and websites. The documents may be produced on paper, but technical editors increasingly edit online documentation for a computer program, a slide show that is part of an oral proposal for a grant, or a website.

In-House or Contract Setting

Only a small percentage of technical editors work for large commercial or academic publishing houses—that is, places whose primary function is the publication of documents. Rather, technical editors are likely to work in a company in which the primary function is software applications, engineering, scientific research, or business. In these settings, publishing is a secondary business function. Alternatively, editors may work for companies that accept assignments on contract.

Qualifications for Technical Editing

In small companies, editors complete the full range of editing and management tasks. In larger companies, several people share the responsibilities. One editor may be responsible for development, a second for copyediting, and a third for production, while a publications manager coordinates the process. A beginning editor will probably be assigned copyediting and manuscript coordination tasks. The editor who demonstrates competence at the beginning level can advance to greater responsibility for the text and for production. As in all professions, expertise and responsibility grow with experience.

The primary qualification for basic copyediting is to understand language and know its rules. Editors must also be able to read carefully and focus on details. They need some knowledge of the visual characteristics of text, such as spacing and type.

Editors with responsibility for document content and development also need to analyze and evaluate the subject matter. They can imagine readers using the documents, and they know something about how readers comprehend information and use documents. They know options for visual design and media and reasons for making choices among the options. They are visually, as well as verbally, sophisticated. They have some understanding of the subject matter of the documents they edit. They accept responsibility for the quality and ethical integrity of the document.

Editors with management responsibilities must be organized and well disciplined. They encourage top performance from the people who work for them.

People who enjoy editing collaborate well with people and respect the contributions of people in different jobs. They set high standards for themselves, but when there isn't time to be perfect at everything, they set priorities and remain flexible.

Using Your Knowledge

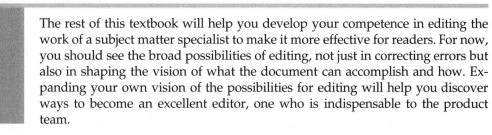

The rest of this textbook will help you develop your competence in editing the work of a subject matter specialist to make it more effective for readers. For now, you should see the broad possibilities of editing, not just in correcting errors but also in shaping the vision of what the document can accomplish and how. Expanding your own vision of the possibilities for editing will help you discover ways to become an excellent editor, one who is indispensable to the product team.

Acknowledgment

Paula Green, a technical writer at BMC Software, helped me update sections of this chapter. I am grateful for her knowledge and help.

Further Reading

Arakelian, Caroline. 1998. Developmental edits—a quick reference. Society for Technical Communication: *Proceedings, 45th Annual Conference.* 441–444. The author describes the concept and process of developmental editing and offers several checklists.

Bay Area Editor's Forum. *Editorial Services Guide.* www.editorsforum.org/ what_do.html. This site defines various types of editing, including copyediting, developmental editing, and production editing. It distinguishes light, medium, and heavy copyediting.

Corbin, Michelle, Pat Moell, and Mike Boyd. 2002. Technical editing as quality assurance: Adding value to content. *Technical Communication* 49.3: 286–300. The authors, all editors, discuss content editing as quality assurance.

Tarutz, Judith A. 1992. *Technical Editing: The Practical Guide for Editors and Writers.* Addison-Wesley. A technical editor offers tips and tricks as well as practical guidelines.

Weber, Jean Hollis. *Working with a Technical Editor.* www.techwr-l.com/tech-whirl/ magazine/writing/technicaleditor.html. The author identifies editing tasks and distinguishes "rule-based editing" from "analysis-based editing."

Discussion and Application

1. Using the flowchart in Figure 1.1, relate the development and production of the SAIC online tutorial to the different steps. What, for example, constituted research in the document development step? What steps, if any, were omitted, and why? How might production of a website differ from production of a print document?

2. Write three questions or comments you have as a result of reading this chapter. Be prepared to share them in class.

3. Compare the definitions of editing at the Bay Area Editor's Forum site and the definition by Jean Hollis Weber at the *Techwr-L* site (see Further Reading) with the definitions in this chapter. How might the distinctions between the different levels of copyediting—light, medium, heavy—help to clarify the differences between basic copyediting and comprehensive editing? How do rule-based and analysis-based editing relate to the chapter and Editor's Forum definitions?

4. Editors may receive the document to edit after it is written (at the end of the project), or they may participate in the planning and edit while the document is being developed.

   a. In your own words, explain how these different procedures might affect the editor's responsibilities. Which of the procedures will invite more superficial editing, and why? Could superficial editing ever be an advantage over substantive editing?

   b. Speculate on how the different procedures might affect the editor's relationship with the writer.

   c. Could the practice of involving the editor only at the end of the project provide an advantage in terms of what the editor is able to perceive about the work? Think about what happens when a person becomes familiar with the text.

5. Some terms in this chapter may be new to you. Make a list of terms that you need to understand better. Use the glossary at the back of this book to find definitions and mark them as terms to review. These terms may be unfamiliar: *comprehensive editing, contract, desktop publishing, in-house, legacy document, template, usability, version control.*

2

Readers, Users, Browsers, Problem Solvers . . .

The title of this chapter purposely suggests multiple roles for the people who will use the documents that you edit. These people will read, but reading is not their main purpose. Rather, they will *use* the document to find specific information so that they can complete a task. They may browse as much as they read. Reading, using, and browsing are means to the end of problem solving or decision making or operating equipment or getting information.

Just as writers begin their writing by considering who will read or use the document, in what setting, and for what purpose, editors make the best editing decisions if they consider why someone needs the document and how it will be used. Editorial decisions follow from this knowledge as much as from knowledge of handbooks and style manuals. Users and uses influence choices about the type of document (manual, poster, memo, website), medium (print, video, hypertext), organization and visual design, and even basic copyediting. In turn, all editorial decisions affect the reader's ability to use the document.

This chapter appears early in the book to establish the context for editorial decisions. As an editor, you must be able to imagine and interpret the situation outside the text as well as to make appropriate textual choices.

This chapter suggests links between textual choices and the ways people use documents. It places texts in their contexts to explain reasons for textual choices. It also reviews research on reading and comprehension and the implications for designing documents. It offers a theoretical framework for making decisions at all levels of editing by reviewing reading patterns, comprehension strategies, access devices in documents, and the implications of all these for designing documents for use.

Key Concepts

Documents enable readers to solve problems, complete tasks, or add to their knowledge. Reading is a means to an end. Readers interact with the document by using their prior knowledge and established reading patterns and by seeking specific information. Editors imagine the document in use in order to make choices about content, organization, visual design, style, and medium. The editor is readers' advocate and designer as well as language expert.

Texts and Contexts

The text of a document is its words as they are arranged in sentences, paragraphs, and sections. Editors are textual experts: they know and apply principles of grammar, style, organization, and design. Yet a document can be perfect in a textual sense and still fail to achieve its purpose of enabling a person to make a decision or use equipment or find information. The text may not meet the needs of users in the situation of use. This situation is the document's context. As the prefix of this term suggests, the *context* is *with* or *around* the text. The text never exists in isolation from people, places, values, and needs. Figure 2.1 illustrates the relationship between text and context.

The context includes origins and impact of the document, readers and conditions of use, the culture in which it is used, and constraints on development and production. Table 2.1 identifies some attributes of the text and the context. The rest of this section elaborates on these attributes.

Origins and Impact: The Problem and Solution

Context includes the reason for which the text is created. In technical and professional writing, this reason is probably a problem to solve. A problem does not necessarily mean something bad, but it does imply some gap between what exists and a goal. The document is the means of achieving the goal. For example, the purchaser of a new laptop computer may not know how to install the printer. The problem is the gap between current knowledge and the goal of successful installation; the user's manual or online help is the means of solving the problem because it tells the purchaser how to install the printer. Another problem could be a company's wish to inform employees of company initiatives, policies, and benefits and to develop loyalty to the company. An employee newsletter may enable the company to achieve those goals. Figure 2.2 illustrates the text bridging problem and solution.

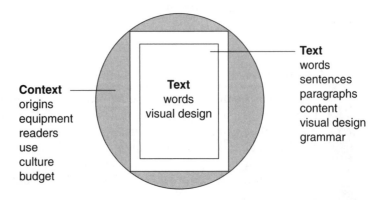

Figure 2.1 The Relationship between Text and Context

TABLE 2.1	**Attributes of Text and Context**

Text

content	headings	grammar	paper quality	menu bars
organization	index	mechanics	paper size	hypertext links
visual design	appendixes	punctuation	binding	screen layout
illustrations	page numbers	spelling	color	screen resolution
style	table of contents	citation style	typography	HTML or XML code

Context	**Examples**
origins and impact	problem or need for information; solution
readers and use	prior knowledge, interests, abilities
	work tasks
	conditions of use (low light, outdoor setting)
	equipment and software available
	storage and disposal
	reading pattern (for reference, tutorial)
culture and expectations	social values (attitudes toward work, time, gender, space)
	discourse conventions (formality, patterns of organization, directness)
	language
	metaphors and cultural values
	national symbols and historical events
accessibility	accommodation for user's visual or other impairments
constraints on development and production	staff: abilities, time available (competing projects)
	standards, policies, laws to which the document must comply
	budgets
	equipment available
	related documents
	copyrights

The impact refers to what happens as a result of the document. A proposal for funds to support research succeeds if the writer gets the grant; a letter of application succeeds if a candidate wins an interview; instructions for installing the printer succeed if the user installs the printer correctly without phoning technical support. Although textual features influence the impact, the measure of success is outside the text.

Editors aim to make a document correct but also to make it functional and efficient in helping users solve problems. Because documents are not ends in

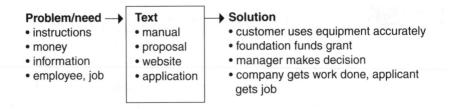

Figure 2.2 Text Facilitates Finding Solutions

themselves but means to other ends, measures of document success are described in terms of usability and persuasiveness more than textual features such as grammar. A document may be correct grammatically but still miss its purpose.

Readers and Use of the Document

A document's users focus on the outcome of reading, but reading is a necessary first step. Readers want to solve the problem—install the printer, make the decision, or learn about the new health insurance policy so they can sign up for it. Although they might read novels and magazines for fun, they read technical documents in order to learn or to act. The reading is closely tied to work, whether it is work for a salary or work in the sense of completing a household task.

Because reading of technical documents is linked to work, editors can assume that readers are busy, that they have other responsibilities competing for their time, and that they are more interested in the task than in the document. They may be impatient with delays and distractions caused by reading. Unnecessary information, difficult words, clumsy sentence patterns, unusual structures or style, or difficulty in finding information divert readers from the content and task. On the other hand, lack of information can also make the text hard to understand. Thus, editors and writers assess what readers probably know and what expectations they have for the document.

Anticipating how the reader will interact with the document will help an editor determine what physical features will work best. For example, low light requires large type, and outdoor reading requires weatherproofing. Media choices depend on whether users have equipment (such as a slide projector) and software. Choices of binding and paper size depend on whether users will use a document flat on a desk, store it on a shelf or in a file cabinet, carry it in a shirt pocket, or recycle it. Websites should comply with the Web Accessibility Initiative so that users with disabilities will have access to them (see www.w3.org/WAI). That means, for example, offering text alternates for images (speaking browsers can read text aloud to blind users) and using relative rather than absolute font sizes so that users can make the size bigger or smaller.

Because of the variety of contexts and uses of a document, it is impossible to create universal templates and formulas to define what a document should be, though conventions for document structure and design will help editors determine what textual choices are likely to work. A good editor at least imagines the document as readers are interacting with it to anticipate the likely uses and needs of readers. Even better, the writer or editor consults with the user to learn about needs and hopes for the document. Onsite research, at the site where the reader will use a document, provides the most complete information about probable uses. This method of getting information is called contextual inquiry.

Culture and Expectations

All readers take some of their cues for reading from their cultures. *Culture* may refer to national culture, such as the culture of Japan; to a professional culture, such as medicine; or to a corporate culture, such as the work environment at National Instruments or Microsoft. Culture can even be defined by products, such as computers that use the Windows operating system. Readers may be parts of different cultures simultaneously—for example, programmers (profession) working at an American office (nation) of National Instruments (corporation).

Just as different cultures and organizations have dress codes (explicit or understood), they have codes for communication. Readers develop expectations for the structure, visual design, and style of documents. Because of implicit and explicit standards and conventions for science writing, the reader of a scientific research report anticipates that the article will define the research problem, the methods of investigation, and the results of the experiment, and will discuss the significance. Furthermore, that reader anticipates references to other related research reports, a formal writing style, and the author-date documentation style. A writer who fails to meet these expectations probably will not be published, and even if the publication reaches print, it will have to be extremely significant to overcome the liability of unconventional patterns of argument and presentation. Readers become aware of these conventions both deliberately (as in writing classes) and by experience.

The Asian cultures expect business correspondence to begin with courtesies regarding weather and the correspondent's health, whereas correspondence from Western cultures conventionally gets straight to business. The various international cultures interpret colors in different ways, and words may have nuances of meaning that do not translate easily. Documents for international cultures may need to be more visual than verbal to minimize problems with translation. Translated documents may be shorter or longer than the originals in English. For example, German words and sentences are longer in translation than their English originals. Writers and editors make style and page design choices to accommodate these differences.

Jerrod Larson (see Further Reading) learned from experience how different time zones, customs regarding rounding prices, and the use of U.S.-centric terms could jeopardize customer relations for companies doing business with international customers. Even the simple courtesy of offering the country prefix in telephone contact information is a way to accommodate users. (See Chapter 20 for more details on editing in global contexts.)

Variations from expectations may distract readers from the meaning or diminish trust in the writer and information. Therefore, editors try to make the document appropriate for the values and expectations of the culture.

Accessibility

Some users have sight or motor impairments or reading disabilities. Accessibility means that users with a wide range of abilities can still use the material. Accessibility of reading material is a concept that has developed with the Internet. For example, text alternatives for illustrations will enable users with visual impairments to hear the content in audio translations of web content. Organized content rather than long lists of unsorted data will accommodate users with reading disabilities who have trouble skimming. Some guidelines apply to print as well. For example, clear and simple language and consistent terms and navigation devices will make the document easier to use. These strategies will also help users who are reading in a second language. As Chapter 21 will discuss in more detail, web developers and editors can make other choices to increase accessibility. An editor tries to include readers and not set up textual barriers.

Constraints on Development and Production

Context also includes any constraints on development and production of a document. Budgets will determine whether color printing is an option, and product liability laws require warnings if safety is involved. Website download time will influence how many color images can be included. Related documents may establish an optimum page size and layout so that all the documents will look like parts of the same set, which will help readers identify them. These external constraints may make the document better for readers (as in the case of safety warnings), or an editor may have to negotiate on behalf of readers to override constraints that cripple quality.

Using Context to Improve the Text

Good editors envision the context of use and the reader using the document. This vision gives editors reasons for making textual choices and for understanding why these choices matter.

The rest of the chapter focuses on ways in which readers use the text to get the information they need or to solve the problem that brought them to the text to begin with.

How Readers Use Documents

Designing usable texts requires knowledge of ways in which readers will respond to texts, whether print or online. Research in reading and cognitive psychology reveals that reading is much more complex than passive reception of information.

Instead, readers are busy making their own meanings. They depend on their prior knowledge, interests, and attitudes as well as on the text. A reader's understanding will not necessarily match the writer's intended meaning, even if the sentences and organization are clear.

Although writers and editors cannot know exactly how readers will respond, they can be certain of two things: readers create their own meanings, and they read selectively. Good design, including selection of content, organization, and visual cues, accommodates these predictable habits of reading.

Creating Meaning

The meaning comes not just from the words and other symbols on the page or on the screen but also from the knowledge that readers bring to the text and the way readers relate this knowledge to the information in the document. For example, people learning to use a cell phone probably use the telephone as a reference, creating their own understanding of procedures written in a manual. Two readers with different experiences and different memories of facts and concepts may create different meanings from the same text. Imagine, for example, how differently people who are anti-abortion and people who support abortion rights would read the same information on stem cell research.

Readers also differ in their attitudes toward the material and the task, their emotional states at the time of reading, and their reading environments. They may be distracted from their reading by other thoughts, ideas, or tasks. They may skip around, miss some information, and fill in the gaps with their own creations.

Nevertheless, the document and the writer's interpretation of data will influence the reader's interpretation. Precise terms, analogies, and background information can help readers make connections between familiar ideas and new ones. Placing key concepts in prominent places in the document (such as the first sentence in a paragraph or the independent clause of a sentence) will reinforce the importance of these concepts. The document structure will give the reader an idea about the information itself.

Reading Selectively

Because readers of a technical document are oriented to the task or problem that the document addresses, they look for shortcuts in reading. Or they may be interested in only a portion of the document. An experienced computer user may skip over introductory material in a manual and look up the directions for setting up a wireless home network. A manager may read only the summary and recommendations of a feasibility study. Users of websites may skip from site to site or through sections in the same site. Thus, they read selectively, skipping what they think they do not need. Writers and editors can help readers find what they need with menus, indexes, cross-references, headings, and other navigation cues. If readers will perform a task while they read, editors will make it easy to get from the text to the task and back—perhaps by using numbers to identify the task's steps or white space on the page to set off the steps.

Both print and online documents require navigational devices. A table of contents, index, chapter titles, and headings help readers find specific sections. Color and tabs on pages are more expensive options in print. Numbering and use of space or rules between sections are visual signals about meaning, but they also aid in access. A menu, site map, navigation bars, and color cues of an online document provide information about content and its organization. Each screen should mark the location in the site and provide tools for moving forward or back. The menu may appear on each screen, for example.

Placing important material early in a section is a way to make the most useful information the most accessible. Readers who read only the beginning of the section will be sure to read the important material. Information of secondary importance may be placed in an appendix. Such placement decreases the odds that readers will use the material, but it makes the information of primary importance more accessible by removing the clutter of secondary information.

With both print and online documents, imagining the document in use is a good editorial strategy because it will help you make good decisions about the structure and display of information.

Reading to Comprehend: Content, Signals, Noise

Whatever the practical outcome of reading, readers will try to get information. In order to act, they will have to know something. The information may be conceptual; that is, readers may need to understand and remember concepts and relationships between ideas. These readers read in order to learn. Or the information may be factual; readers may merely need to find a switch on a machine. These readers read in order to do.[1] Knowing that one broad purpose of reading is to get information, editors aim to make documents comprehensible. They make sure that terms are accurate, that conceptual information precedes detailed information, and that the level of detail is right for the readers. All levels of editing, from developmental planning to checking for correctness, work toward the goal of comprehension.

Documents give two types of information to readers: the content and the signals that help readers interpret that content. The content may be a review of research related to an experiment, a description of procedures or of a mechanism, a recommendation to make a purchase, or a parts list. Signals can relate either to the content or to the document itself. Verbal signals, such as the phrase *in conclusion*, show one's place in the document as well as the relationship of ideas. Structural signals, such as the table of contents and placement of important information at the beginnings of sentences and sections, orient readers to the document and help them recognize what is important. Visual signals, such as boldface or italic type or numbering of steps, indicate emphasis as well as structure. All of these signals help readers use the document and understand its content.

---

[1] T. Sticht. 1985. Understanding readers and their uses of text. *Designing Usable Texts,* ed. Thomas M. Duffy and R. Waller. Academic Press. 315–340.

Content

The document's content enables readers to solve the problem or take action—assuming readers understand and interpret it accurately. Content must be complete and at the appropriate technical level for the readers. Careful organization facilitates accurate interpretation. Writers and editors cannot know for certain how much information to include and at what level, but they will make better decisions if they imagine how readers will respond to the document as a whole and at each point of reading.

Anticipating Readers' Questions

The content must be complete enough to make sense in light of the readers' previous experience and learning, as well as complete enough to enable the readers to do the task that motivates the reading. Editors have to anticipate what readers know and want. Imagining readers' questions is a good strategy for evaluating completeness. Readers ask predictable questions such as "What is it?" "How does it work?" "How does part A relate to part B?" "Where do I go to sign up?"

Linking New with Familiar Information

Readers learn when they can associate new information with remembered information. Readers will struggle with comprehension if the context is unfamiliar, too much prior knowledge is assumed, and terms are unfamiliar. Analogy, background description, and reviews of literature are explicit ways of making the new information relate to something familiar. New information becomes familiar or "old" once readers have absorbed and interpreted it. Then additional new information can be linked to information previously presented.[2]

Organizing Information

Readers do not memorize separate facts but arrange them into patterns, or *schemata.* They remember, for example, a sequence of steps in a process, not random steps, and the position of parts on a piece of equipment. These structures distinguish concepts (the process, the equipment) from details (the step, the part) and show the relationship of parts. Content will be easier to comprehend if readers can sense its structure. The information in an organized document is easier to learn than that in a disorganized one because the structure of the document suggests the structure of the content.

Good organization reveals a hierarchy of information—the most important points and the supporting details. It also indicates how pieces of information relate, as in a cause–effect relationship, a temporal (chronological) one, or a spatial one. (See Chapter 17 for a more detailed discussion of organization.)

---

[2]Herbert H. Clark and Susan E. Haviland. 1977. Comprehension and the given-new contract. *Discourse Production and Comprehension,* ed. R. O. Freedle. Ablex.

Signals

Signals, as well as the content itself, communicate information about the content and the document. These signals can be verbal, structural, or visual. Readers use the information communicated by the signals to help interpret the content. Editors use signals to point readers to the correct meaning.

Verbal Signals

Verbal signals are words or sentences that suggest the structure of the whole or relationship of parts of information to other parts or to the whole. Introductions and overviews establish a framework for the new information that will follow. A forecasting statement or information map in an introduction indicates the way the document will develop. This forecasting statement may be just a list of topics that will be developed in the document. It helps readers form a mental outline of the key issues.

The writer can signal the hierarchy of ideas with phrases such as *the most important fact is* or *the significance is.* Verbal signals also include transitional words to indicate the relationship of ideas. Examples of such words are *however*, which signals contrasting information; *thus,* which signals a conclusion; and *then,* which signals a time relationship. Such signal words help readers understand the content by revealing the relationship of facts and ideas.

Structural Signals

Structure refers to the arrangement of words into sentences, sentences into paragraphs, paragraphs into sections, and sections into whole documents. Accurate structural signals help readers interpret the content by showing the hierarchy of ideas and distinguishing main ideas from supporting details. Some explicit structural signals are the table of contents, which shows the main divisions of the document, and headings, which reveal the divisions.

The arrangement of words and sentences implicitly cues readers about the relative importance of the words and sentences. The first item in a list or first words in a paragraph will get attention because of their positions, and readers look for key information at the beginnings of sections.

The sequence of content items shows their relationships as well as hierarchy. Items may be ordered chronologically, spatially (such as top to bottom), in order of importance, or from general to specific. The sequence, along with the words themselves, helps readers interpret the material by revealing the way the parts are related. Even when the sequence the reader follows cannot be predicted, as in hypertext, the top level menu items suggest the hierarchy. This hierarchy plus categories of information revealed by menu items bind the parts into meaningful relationships.

Visual Signals

Visual signals can be seen on the page or on the screen. A common visual signal is indention to signify a new paragraph, which, in turn, leads readers to expect a new concept or piece of information. Numbers and headings marked by variations in typography can also identify sections of a document and mark them visually as well as structurally. Boldface or italics indicate words of particular importance.

Graphs are visual signals that guide interpretation of data. An incline in a line graph, for example, signals growth or increase. Even the paper and the type quality are visual signals about the content. High-quality paper signals importance and encourages attention and respect. Neat work encourages readers to take the same care in reading that the writer and editor have taken in production.

The signals in documents to be read on a computer screen are often more visual than verbal: icons direct procedures, and color identifies different types of information, such as instructions for navigating the document and hypertext links to other texts.

Undesirable Signals: Noise

A document sometimes includes verbal, structural, or visual signals that interfere with comprehension of the content or that create a negative response to it. These signals are noise in the document, in the way that static in a radio broadcast is noise that partly covers the music or talk.[3] Just as listeners are annoyed and distracted by static, readers are annoyed and distracted by document noise. If the noise becomes too great, listeners will turn off the radio, and readers will stop reading. Noise interferes with comprehension by distracting readers. It makes interpretation more difficult and increases the chance of errors in interpretation.

Verbal noise can be misspelled words and grammar errors, which distract readers from the content. Inconsistencies in the use of terms or in capitalization represent noise if readers have to interrupt their reading in order to interpret whether the inconsistency signals a distinction in meaning. This book, for example, tries to be consistent in using the term *readers* rather than mixing the term with *audience* lest some readers wonder whether a distinction is intended between the two terms. A medical columnist used the terms *heart beat*, *heart rate*, and *pulse* interchangeably, but a reader, presuming the terms referred to three different things, queried the author for clarification. The variation gave a misleading signal about meaning.

Noise can also consist of irrelevant information, such as a digression from the main theme, unnecessary background information, or too many definitions. Flashing ads on a website may annoy readers. In addition, an inappropriate writer voice, or persona, can cause distracting noise in a document. For example, if the writer offends readers by seeming to be prejudiced or uninformed, readers may react negatively to the content, or they may not hear it at all.

Structural noise can result from the arrangement of content in a text that inaccurately reflects its actual structure (such as placing steps in a task out of chronological order). Structural noise can be hypertext links that indiscriminately send readers to both significant and trivial sources. Visual noise could be smudges on the page or an exaggerated mixture of typefaces that calls so much attention to itself that readers see only the chaos, not the content. It could be a busy screen pattern that diminishes contrast of words and background.

---

[3]Claude Shannon and Warren Weaver. 1964. *The Mathematical Theory of Communication.* Illinois University Press.

Although verbal, structural, and visual signals can help readers interpret and find information, too many signals can create noise. Not every sentence needs a transition word, nor does every paragraph need a heading. If readers are constantly processing signals rather than content, they will soon come to focus on the signals themselves. Signals should be relatively unobtrusive—integrated logically into the document and recognizable only when readers consciously seek them. When noise becomes so great that it dominates the content, the noise *is* the content, and the document has failed.

Researching Readers, Usability Testing

Editors can improve documents by understanding readers, by imagining uses, and by applying their knowledge of how people learn through language. But sometimes this mental work is not enough. As good as you are as an editor, there will be reader needs and responses that you don't anticipate. You may need to go directly to representative readers, probably as part of a product development team. This research can occur when a project is defined, throughout its development, and after the first round of editing. A rich source of information about users is contextual inquiry, or research where the users work. Seeing users at work helps a team understand the need for a product and the materials that accompany it. The users can participate in explaining their tasks and goals. A needs assessment leads to a plan for documents or other deliverables that the project team promises. As the product is developed, the intended users can be consulted.

After the product is ready or the document has been drafted, an observation of users working with the products and documents can show where the information is confusing and how well the users can complete tasks using the materials. This usability test may suggest places where additional editing and revision would help. In addition, because you are concerned with the accuracy of information and the user's safety, you will probably seek a technical review of the content by subject matter experts.

Contextual inquiry, needs assessment, and usability testing work together with editing to improve quality of documents, but editing is especially significant because of the editor's orientation to readers. You, as editor, represent the reader and advocate document design choices that enhance the reader's experience.

Designing Documents for Use

The more writers and editors know about how people will use a document and about the readers' prior knowledge and expectations, the more likely they will be to design useful products. Just as architects design buildings and engineers design roads, tools, and machines, technical writers and editors design documents.

Design is more than a visual or aesthetic concept: a good designer considers the people who will use the building or the road or the document and plans the design accordingly. A conception of the functional whole leads to choices about specific features, such as materials and space. Document design parallels architectural and engineering design in its scope and purpose.

A document designer's tools are words (including organization, style, grammar, and mechanics); visuals; layout and typography; color, paper, size, and binding; and (in the case of online documents) sound and motion. Good design results in documents that readers can use efficiently and successfully.

Using Your Knowledge

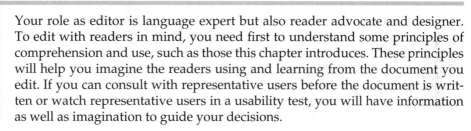

Your role as editor is language expert but also reader advocate and designer. To edit with readers in mind, you need first to understand some principles of comprehension and use, such as those this chapter introduces. These principles will help you imagine the readers using and learning from the document you edit. If you can consult with representative users before the document is written or watch representative users in a usability test, you will have information as well as imagination to guide your decisions.

Further Reading

Johnson, Robert R. 1998. *User-Centered Technology: A Rhetorical Theory for Computers and Other Mundane Artifacts.* SUNY Press.

Kneifel, Alix A., and Carol Guerrero. 2003. Using participative inquiry in usability analysis to align a development team's mental model with its users' needs. Society for Technical Communication: *50th Annual Conference Proceedings.* 403–408.

Larson, Jerrod A. May 2004. Lessons in internationalization. *Intercom.* 10–11.

Mirel, Barbara. 1993. Beyond the monkey house: Audience analyses in computerized workplaces. *Writing in the Workplace: New Research Perspectives*, ed. Rachel Spilka. Southern Illinois University Press. 21–40.

Nielsen, Jakob. 2000. *Designing Web Usability.* New Riders. See especially Ch. 1, "Introduction: Why web usability," and Ch. 6, "Accessibility for users with disabilities."

Redish, Janice C. 1988. Reading to learn to do. *The Technical Writing Teacher* XV: 223–233.

Yeats, Dave, and Paula Kozlowski. 2003. Audience analysis and information design: Creating a needs assessment documentation strategy. Society for Technical Communication: *50th Annual Conference Proceedings.* 262–267.

Discussion and Application

1. Figure 2.3 is the home page for the National Institutes of Health as of July 2004 (www.nih.gov). Discuss how it reflects one or more of the following concepts from Chapter 2.

   a. Readers of technical documents read in order to act. (What specifically in the example encourages or enables action?)

   b. Readers read selectively. (What are the aids to selective reading?)

   c. Structural signals help readers understand the relationship of ideas. (What structural signals give clues to the relationship of ideas?)

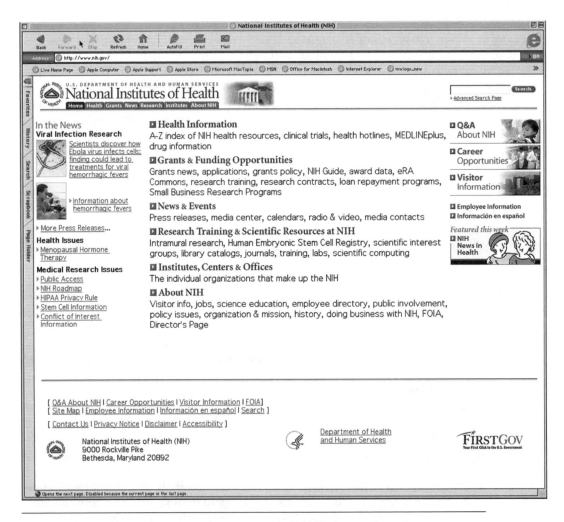

Figure 2.3   Home Page, the National Institutes of Health (www.nih.gov)

    d. Visual signals help readers understand information as well as locate it. (What are the visual signals, and how do they help readers locate and understand information?)

    e. Document noise may interfere with the message of the text. (Does this document contain noise? If so, how does it interfere with the message?)

2. Go to the current version of the website (www.nih.gov) to identify any changes that have been implemented since July 2004. What are they, and what evidence do you find that the designers were considering the needs of users?

3. Locate a document of a technical nature, such as a textbook, proposal, or user's manual.

    a. Find examples in it that illustrate the writer's and editor's awareness of one or more of the concepts discussed in Chapter 2. If your document fails to acknowledge one or more of these principles, show where and how it fails. Bring your examples to class for sharing.

    b. Using the document as an example, explain the difference between *text* and *context* in your own words. Imagine some details of the context of your document. Did the person or people who developed the document reveal a good sense of the context of use?

4. Compare the websites of three or four universities. What types of information do the sites include? What assumptions about users do the sites imply—for example, categories of users, purposes for seeking the site, prior knowledge, needs for information? Assess how easily each one enables users to complete these specific tasks (typical reasons for checking a university website): get information on degrees offered, apply for admission, find the phone number or email address of a student, move easily between links at the site. What textual features (words, arrangement, color, visuals and graphics, typography, symbols, menus) facilitate the completion of these tasks? What features, if any, interfere with these tasks? As an editor, could you recommend improvements? If your instructor directs you to, write up the results of your analysis in a brief report.

5. Check the introductory paragraphs of the chapters in a textbook for forecasting statements or information maps—statements that tell what the chapter will cover and in what order. How will students use these verbal signals?

 6. Verbal signals indicate relationships of pieces of information. Some categories of relationships are listed on the next page with an example for each category in italics. Extend the list of examples. Explain why an editor would not simply delete such words to make a document more concise.

Contrast	*on the other hand*
Time	*after*
Space	*above*
Continuation	*furthermore*
Frequency	*sometimes*
Example	*for example*
Conclusion	*in conclusion*
Cause–effect	*because*
Explanation	*that is*

7. How might a table and a bar graph visually invite readers to compare?

8. *Vocabulary.* Check definitions of these terms if you are unsure of their meanings as they relate to documents and their readers or users: *context, design, signals, noise, selective reading.*

CHAPTER 3

Collaborating with Writers

In Chapter 1, you met two editors and learned about their roles in document development and production. In Chapter 2, you expanded your sense of who uses documents, how, and why. In this chapter, you will consider the third important person in the editor-reader-writer triad: the writer. All three people are linked in the development of effective documents. The writer and editor work together to create documents that readers can use and trust, but they are influenced by the reader's needs and perhaps even the reader's specific suggestions. Writers and editors bring different strengths to document development. Writers usually have more subject matter knowledge, and editors are more expert in language and publication procedures.

This chapter defines writers in the context of the workplace and factors that influence editor–writer relationships. It identifies strategies of editing, management, and interpersonal skills to make the relationships work. It also offers suggestions for how and when to phrase queries to writers and how to compose letters of transmittal to writers. Chapter 25 offers suggestions on conducting conferences with writers and clients.

Key Concepts

Collaborative editor–writer relationships follow from good editing and good management as well as from good interpersonal skills. Good editing makes a document better, not just different and certainly not wrong. Good management means establishing expectations early, communicating with all people involved with the project, and completing work promptly. Interpersonal skills and tactful communication encourage cooperation. The focus should be on the document, the task, and the reader rather than on the personalities of the writer or editor. If editor and writer collaborate, they create more effective documents than either could do alone.

Who Are the Writers of Technical Documents?

A workplace writer is primarily a subject matter expert (SME). That person may or may not have the job description of writer. Paula at BMC (Chapter 1) is a writer by education and job description who has developed subject matter knowledge through the experience of working with the subject over time. In other situations, the writer may be an engineer or a research scientist or a manager who also has responsibility for writing. Paula gets information for the manuals she writes from engineers, who are also writers as they provide information in words. But they may not be writers by education and job responsibility. They may have no special training in language or document design. Editors provide that expertise.

Writers may have different degrees of ownership of the text, and that connection of the document with an individual affects the editor's privileges and responsibilities. Research reports are identified with the names of the researchers, but a user manual is perceived to be the product of the organization, not of the individuals who developed it. That manual is likely to be developed collaboratively, with several SMEs contributing information, several writers organizing and developing it, and an editor to guide its development and to review the progress. On Charlene's project (Chapter 1), there were several SMEs: some with knowledge of safety training and others with knowledge of instructional design. Each brings a particular expertise to the project, as does Charlene. Together they create a team.

Writers benefit from the editor's language and publication expertise. Editors benefit from the writer's subject matter expertise. The collaborative ideal results from different specializations.

The Editor–Writer Relationship

Editors and writers depend on each other and need to develop productive relationships. Relationships succeed for three reasons: good writing and editing, good management, and collaborative partnerships. The subject matter experts provide content throughout the product development process that the editor uses to refine the information products. SMEs and editors communicate with each other frequently to clarify expectations and product goals. SMEs and editors respect each others' work and schedules. They regard each other as partners working together with the ultimate goals being effective products, satisfied users, competitive advantage, and sound decisions and policies.

Because editing involves evaluation and emendation of a writer's work, the relationship between editor and writer can become adversarial. This chapter aims to prevent that outcome by helping you think productively about your relationships with writers and by presenting some guidelines for developing cooperative relationships. Relationships fail for three reasons: poor writing and editing, poor management, and oversized egos. Writers justifiably resent unnecessary or incorrect intervention in a text. Unnecessary delays in work by either writer or editor or poor communication among all the members of the production team will create stress. Egos may interfere with the goal of making the document work for readers. A writer may regard editorial comments as personal criticism and become defensive. An editor may develop a contemptuous attitude about the writing that only encourages defensiveness.

Ernest Mazzatenta, former president of the Society for Technical Communication, has collected statements from writer-researchers about what they like and dislike about editors. They represent statements offered by hundreds of scientists and engineers who have participated in his technical writing courses. These statements are paraphrased in the following lists.[1]

What Writers Like Most about Editors

- restructures the report so that the train of thought is smooth and logical
- points out ideas and explanations in the report that are not clear to the reader and then rewrites them
- catches misspelled words
- improves readability
- usually returns the paper within five working days
- approaches the writer considerately concerning any changes; edits without malice
- shows patience

What Writers Dislike Most about Editors

- asks the writer to rewrite a section without giving any indication of what's wrong with it or any direction to take
- makes changes only to incorporate the editor's style of writing
- suggests qualifiers and disclaimers to analyses that, in the writer's professional judgment, are excessive
- uses words that are not acceptable to the writer and won't change them
- replaces words with synonyms
- requires too many iterations (for example, ten)
- makes comments that are inconsistent with the department head's comments

Relationships between editors and writers are potentially good—and potentially stressful. They can be characterized by collaboration or by competition. Editors and writers have choices about what type of relationship they will have, and editors can do much to create a productive working relationship.

Strategies for Working with Writers

If you analyze the statements in the likes/dislikes list, you will see that most of the likes and dislikes concern editing itself (the editor's work with organization, spelling, and style). Others concern good management (promptness, coordination with a department head or other reviewer). A few imply personal qualities, such as consideration or stubbornness. The strategies discussed here parallel this classification of likes and dislikes.

---

[1]Ernest Mazzatenta. 1975. GM research improves chemistry between science writers, editors. Society for Technical Communication: *Proceedings, 22nd International Technical Communication Conference.*

Edit Effectively

Editing effectively is the most important thing you can do to make your relationship with a writer productive and cooperative. As this textbook emphasizes throughout, effective editing requires you to be correct, informed, and sensitive to the needs of readers and context of use. You must read well and make emendations only when you know a reason for doing so. You must restrict your changes to those that will make the document more comprehensible and usable. You must resist making changes simply to satisfy your own style.

Writers are justifiably impatient, and even angry, when editors change their meaning. They are frustrated when editors simply change things to no apparent benefit. One editor who edited instructions on Basic Life Support wanted to make the phrasing more instructional in the ABCs of cardiopulmonary resuscitation (CPR). The list read: "Airway clear of obstructions, Breathing or ventilation, and Circulation by cardiac compression." So she changed the phrasing to emphasize imperative verbs—but lost the ABCs. The problem is that ABC is a mnemonic device taught in CPR classes nationwide. The instructions had to be re-edited (or unedited), and the editor had to rebuild her relationship with that writer. An effective editor knows the subject matter well enough to avoid introducing errors and knows the resources to check when content questions arise. These resources include dictionaries and handbooks as well as queries to the writer. If you edit well, writers will appreciate you for rescuing them from clumsy, incoherent, or inaccurate writing.

Manage Efficiently and Communicate Well

Efficient management saves time. Efficient management also respects production schedules and other people's deadlines. As editor, you can take a number of steps to manage the job efficiently. Managing well requires good communication.

1. *Participate early.* An editor can offer helpful concepts at the early stages of project development. If you can participate in planning the project, you and the writer and team manager can agree on project goals from the start. This strategy helps to eliminate the conflict that arises when the editor enters the project at the end and has different ideas about the way it should have developed. The idea that editors are only fixers of errors at the end of development invites conflict.
2. *Clarify your expectations.* Guidelines that cover usage, punctuation, spelling of technical terms, documentation style, and visual design (headings, margins, spacing) should be available to writers before they write. You won't be able to anticipate all the possible variations in style until you edit, but the guidelines can cover general issues. Guidelines save editing time because writers prepare the documents correctly. Figure 3.1 shows the guidelines for authors published in every issue of *Technical Communication*. This page establishes standards for punctuation, style, manuscript preparation, and documentation. If your organization has a house style manual (see Chapter 8), alert writers about how to use it, and make it readily available.

CONTENT

Technical communication publishes the following types of original, peer-reviewed articles:

- results of original research
- original contributions to technical communication theory
- case studies of solutions to technical communication problems
- tutorials on processes or procedures that respond to new laws, standards, requirements, or technologies
- reviews of research, bibliographies, and bibliographic essays

Because the primary audience of *Technical communication* is informed practitioners, manuscripts reporting the results of research or proposing theories about topics in our field should include descriptions of or suggestions for practical application of the research or theory. In some instances it may be more appropriate to discuss practical applications of theory or research in one or more additional articles (for example, manuscripts intended for a special section or special issue on a topic) by the same author(s) or one or more collaborators. In such cases, the author(s) should discuss this approach with the editor while planning the manuscript.

STYLE

The purpose of *Technical communication* is to inform, not impress. Write in a clear, informal style, avoiding jargon and acronyms. Use the first person and active voice. Avoid language that might be construed as sexist, and write with the journal's international audience in mind.

Our authority on spelling and usage is *The American heritage dictionary*, 4th edition; on punctuation, format, and citation style, *The Chicago manual of style*, 15th edition.

MANUSCRIPT PREPARATION AND SUBMISSION

1. Submit an electronic file containing the text of your manuscript. Microsoft Word format is preferred; if you use another word processor, a Rich Text Format (RTF) file is also acceptable.

2. Provide the manuscript's title and the name(s) of the author(s) at the beginning of the text file.

3. Next include an informative abstract labeled "Summary" (150 words maximum).

4. Use up to three levels of headings, and indicate them clearly.

FIRST-LEVEL HEADING
(all caps, on a line by itself)

Second-level heading
(first word only capitalized, bold, on a line by itself)

Third-level heading   (first word only capitalized, bold, followed by two spaces, as part of the first line of the paragraph)

5. Within the file, include each table, along with a descriptive caption. Electronically paste a copy of each figure into the file, along with a descriptive caption.

6. Do not use footnotes. Instead, use author-date citations within the text, and provide a complete list of works cited (labeled "References").

7. Check all author-date citations and all entries in the reference list for both accuracy and conformance to *The Chicago manual of style* format (see pp. 593–754).

8. At the end of the text file, include a biosketch for each author labeled "Biosketch" (100 words maximum for each author) describing your professional experience, education, institutional affiliation, professional organizations to which you belong, and any other helpful information. Include your e-mail address (if applicable) and a telephone number where you can be reached during business hours.

9. If possible, transmit each file to the editor as an attachment to an e-mail message, one attachment per e-mail message. (Alternatively, files may be copied to a 3.5-inch diskette and mailed to the editor.) Include in the e-mail message (or in a letter enclosed with the diskette) the article title, authors' names, the exact name of each file transmitted, and the platform and software used to prepare it (for example, "Windows XP/Word 2003" or "MacOS X/Adobe Illustrator 10").

10. If your manuscript is accepted, you will be asked to submit one electronic file for each figure. These files should be in TIF format. Screenshots should be captured and output at 6 inches (width) by 4.5 inches (height) for full screens. Because illustrations will be reproduced in black and white, they are best captured in grayscale rather than color.

COPYRIGHT

The Society for Technical Communication holds the copyright on all material published in *Technical communication* but grants republication rights to authors on request. If your manuscript has been previously published or presented, please inform us when you submit it.

EDITOR'S CONTACT INFORMATION

George F. Hayhoe
Department of English
East Carolina University
Greenville, NC 27858
USA
george@ghayhoe.com
Voice: +1 (252) 321-7785
Fax: +1 (919) 882-9772

Figure 3.1   Guidelines for Authors Submitting Articles to *Technical Communication*
Source: Reprinted by permission of the Society for Technical Communication.

Editors and writers should agree in advance about readers and purpose, illustrations (size, number, whether rough or ready for printing), and length. Write down these assumptions as document specifications.

You can sometimes clarify expectations for content. If you know that a document will have to cover certain topics in a certain order, provide the outline. If some parts of the document will be boilerplate (that is, repetitions of existing text), include those sections with the outline. These aids can also prompt a writer to get started. The project may look manageable because something is already written.

A schedule, whether it is prepared by the editor or team leader or collaboratively, establishes due dates so that production can proceed on schedule. Estimate your time realistically, but be aware that people who are not editors underestimate how long it will take to edit. You have to educate them or they may abuse you innocently by completing their work late. (Scheduling and estimation of time are discussed in Chapter 24.)

If you are editing online, procedures for tracking editorial changes should be clear from the beginning. (See Chapter 6.)

3. *Work with the writer throughout development.* Ongoing collaboration follows from advance planning. The initial plan for document development should establish some points of review after sections are complete but before the whole is finished. If plans need to be adjusted for any reason, it's better to do so midway than to wait until the end. This plan keeps the writer on schedule, but it also allows rethinking of the plan before too much writing is complete.

4. *Don't surprise.* Share your plans for editing with the writer. If it is understood that you will edit only for spelling, consistency, grammar, and punctuation, don't surprise the writer with elaborate revisions of style and visual design. If you change your mind about your editing plans, have good reasons for doing so, and discuss your thinking with the writer before you show him or her a document smeared with red marks.

If you plan extensive revision, discuss the plans before you revise or ask the writer to revise. In a phone call you might say, "I would like to reorganize," and discuss plans for doing so. Expecting reorganization, the writer will be unlikely to resist it. Give the writer a chance to review the editing early enough in production to make changes. It's not fair or efficient to edit heavily and then show the writer the edited version as the document is about to be published.

5. *Be prompt.* Keep your expectations for yourself as high as for others. Get your work done on time. To avoid misunderstanding, let others know in advance when you will complete the work, and keep them informed about necessary schedule revisions.

Develop an Attitude of Professionalism

Even if the writer is not good at writing, you can respect the person and his or her expertise in other areas. (Chances are that you cannot do well what the writer does in his or her primary job.) You can focus on the task rather than on the person. Assume that your goals and the writer's are mutual: to make the document work for its readers and purpose. Assume that you are players on the same team, not that you are an expert correcting an incompetent person.

You can encourage professionalism in writers by setting an example. Complete your work on time and be prompt for meetings. Your language, dress, facial expressions, and posture can communicate your attitude toward your job and to your colleagues. Flexibility and good humor can help as well. If the writer's choices are as good as yours, or if they are relatively unimportant, you can yield. That doesn't mean you have to compromise on principles of good writing, but you can recognize that to risk a relationship or an entire project for a comma or variant spelling is to misorder your priorities.

Sometimes you will be wrong. You will misinterpret information or emphasis, and you will miss some errors. The review of edited copy by the writer or other technical experts acknowledges the possibility of editorial error. (In fact, the review should be seen in that light rather than as a time to show the writer all the failings in the document.) The way you respond to someone who discovers imperfect editing will model the way you want writers to respond to you. If you are defensive and try to blame someone else, you will encourage writers to respond in the same way to edited copy. You can be gracious. You can say, "Now I understand," rather than accusing the writer of being unclear to begin with.

If you are editing on paper, use clear, neat, and accurate marks to show respect for the writer as reader of your marks and indicate your professional commitment to your task. You may gain some psychological advantage by using a black or green pencil rather than a red one. Red reminds some writers of critical teachers and blood. Be cautious, however, about choosing blue, as blue does not photocopy well. If you are editing online, keep a copy of the original version so that you can revert to it if necessary.

Correspondence with Writers

Communication with writers prevents mistakes and builds good relationships. As you edit, you may find instances where you need to ask the writer for more information. When you return the edited copy to the writer for review, you may write a letter of transmittal. For planning and questions about substantial revisions, you are likely to meet the writer face-to-face, talk by phone, or correspond by email.

Queries

Sometimes you will be uncertain about the intended content and will not be able to edit until you do. You can look up some information, such as the meaning of an abbreviation, but when you cannot determine meaning with your own resources, you will contact the writer for the information. You don't have to interrupt the writer's work each time you have a question. Rather, you will write "queries," or questions about content, on the document. The writer will respond when you send the edited copy for review. If you are working on paper copy, you may write queries in the margin or attach sticky notes. If you are editing electronically, you can use "comments" in the document file.

In a query, you might ask questions like these:

What is CLHM?

The quotation in the previous paragraph refers to "dyadic" relationships, but you refer here to "dynamic" relationships. Is there a difference, or should the term here be "dyadic"?

You could ask for more information or for verification that an emendation is correct, or you might explain a substantive emendation, such as reorganization. You usually will not query or confirm capitalization and spelling and other language questions that are the editor's expertise. Thus, most queries will relate to content.

If you place the queries right where the questions arise rather than keeping a separate list, the writer won't have to match and hunt. If you keep a list, though, include page numbers. See Chapter 4 for a discussion of the mechanics of using query slips and Chapter 6 for instructions on using "comments" in electronic editing.

Evaluative comments, especially negative evaluations, are inappropriate because they do not direct the writers in revision. An occasional positive evaluation ("interesting" or "good") may be encouraging, but such comments—whether positive or negative—may imply that your goal has been evaluation of the writer's work rather than preparation of the document for readers and publication.

Letters of Transmittal

A letter or memo of transmittal with an edited draft, like any letter of transmittal, tells what the document is, what the writer should do with it, and when he or she should return it to the editor. A secondary purpose may be to motivate. Whether you are working with a writer at a distance or in house, you can introduce the editing and your goals. The letter or memo may be transmitted on paper or online. Use the conventional letter or memo format, as the situation requires, and be sure to edit your own writing. The tone of the letter should be professional. You gain credibility as an editor if your own writing meets the same standards you expect from writers.

Like all letters, a letter of transmittal, even an email version, has a three-part structure paralleling the three purposes. The introduction is a statement of transmittal. The body of the letter explains the goal of editing and the goal of revision. The conclusion asks for a response. Figure 3.2 illustrates how the body of a letter of transmittal might read.

The letter focuses in a businesslike way on the task at hand, not on evaluation. There is no purpose in telling the writer what is wrong. That will only make the writer feel inadequate. Rather, the letter should focus on goals that you and the writer share. The statement of transmittal orients the writer to this project, and the conclusion clarifies what the writer is to do. The letter does not substitute for queries at the point of question, but it can call attention to important needs.

statement of
transmittal

The edited version of the first draft of your proposal for the pulse welding
research project is enclosed.

explanation of
editing and
request for
revision

I have checked for accuracy of information and correctness of usage and
punctuation. I have also verified that the format conforms to the RFP.
Please check the editing, especially the rewording on pages 2 and 3.
The numbers in the table on page 3 don't add up to the total shown, so
please check to see whether the costs for each item are accurate or
whether the addition is incorrect.

request for
response

I will need the revised copy for final proofreading by July 2 if the
proposal is to be mailed to meet the deadline. Call at extension 292
if you have questions.

Figure 3.2   Letter of Transmittal to a Writer

Writers should have final review privileges. Do not send an edited copy directly to production without inviting the writer to check for content and accuracy.
By all means, alert the writer to substantial changes.

Email Correspondence and Shared Files

Email offers a convenient way to stay in touch with the writer or other members of
the product team. Because it is less formal than a letter, it may seem more spontaneous and less intimidating. An email message is less intrusive than an interruption
with a visit to someone's office. It is easy to correspond by email fairly frequently,
which may help to keep the project on schedule. You can attach documents to
email messages so that all team members can see exactly what the issue is.

If your group is networked, you may save updated versions of the document
on a server that various people can access. Be sure to use a file-naming convention, perhaps including the revision date, that enables everyone to tell which is
the most recent version.

Corresponding with International Writers

A global workplace makes it likely that you will collaborate with writers who live
in other countries or with resident writers who have an international heritage. Respecting different customs of communication and workplace schedules will help
working relationships. For example, the informality of American email messages

may make the person writing the messages seem abrupt and curt. Your correspondents in Asia and Europe will appreciate formal greetings ("Dear . . .") and closings. Wishes for a good day are appropriate in closing. On the other hand, the styles in some cultures are more blunt than the American style. If you receive such a message, recognize cultural influences and do not take the bluntness personally.

You may need to adjust your expectations for responses and for schedules of virtual meetings or conference calls according to different holidays and time zones. Keep an international clock accessible. Place on your calendar the holidays of countries where your collaborators live. You and your collaborator may exchange calendars at the beginning of your working relationship.

If you correspond by telephone or fax, include the country code in your phone number. In the United States, you would preface your area code with +011.

Because conventions for expressing dates, weights, and currency vary, as do spellings, establishing a style sheet early in the project will save time for imposing consistency later. (See Chapter 8.)

Some of these gestures are courtesies that enable people to work efficiently, and others reflect the respect that marks a good relationship.

Using Your Knowledge

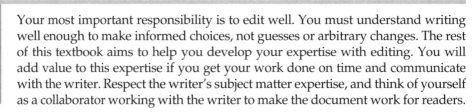

Your most important responsibility is to edit well. You must understand writing well enough to make informed choices, not guesses or arbitrary changes. The rest of this textbook aims to help you develop your expertise with editing. You will add value to this expertise if you get your work done on time and communicate with the writer. Respect the writer's subject matter expertise, and think of yourself as a collaborator working with the writer to make the document work for readers.

Further Reading

Mackiewicz, Jo, and Kathryn Riley. 2003. The technical editor as diplomat: Linguistic strategies for balancing clarity and politeness. *Technical Communication* 50.1: 83–94.

Discussion and Application

1. *Case studies.* Consider the following case studies as topics for class discussion.

   a. *Anthology.* You are editing an anthology of ten articles written by various subject matter experts. Part of your punctuation style is to use a comma before the final item in a series. When you return the edited copy to one writer who has omitted commas before the final item, he protests aggressively, challenges your knowledge, and threatens to withdraw the article from the collection unless you remove the commas you have inserted. What is your response? Before you decide what your final response will be, think about your overall goals for the anthology.

b. *Proposal and jargon.* You are a recent college graduate editing a research proposal for a senior staff member with a Ph.D. The proposal is full of academic jargon that you think may invite the proposal reviewers to smirk at the researcher rather than to respect her. You simplify some of the sentences and vocabulary. You return the edited proposal to the writer, who, in turn, delivers it to the typist. Later you learn from the typist that the writer has written, "Ignore the editorial comments. Type as originally written." You're furious at the put-down and the waste of your time; furthermore, you are convinced that the pretentious style will jeopardize funding. What do you do?

c. *Lack of writing skills.* The writer whose work you are editing is a very nice person, but his writing is terrible. You can't understand many of the sentences, not because the subject matter is unfamiliar but because the construction is so bad. Frankly, you're appalled at the lack of writing skill, and you know your attitude is getting in the way of your work. What can you do?

d. *Management.* For cases a, b, and c, identify specific ways in which efficient management might prevent the conflict.

2. *Strategies.* Describe a specific example of each of the three strategies discussed here for productive working relationships: good editing, good management, interpersonal skills.

3. *Experiences.* If you have had experience—good or bad—working as an editor with a writer or as a writer with an editor, analyze where the relationships succeeded or failed by considering editing skill and procedures, management, and interpersonal skills.

4. *Goal-oriented language.* Think of ways to express these critical statements in a way to encourage productive revision. Aim for a goal statement that is the counterpart of the critical statement or a description of the feature as it meets your expectations for effective writing. For example, instead of criticizing poor organization, suggest an organizational strategy. You will have to invent specific details.

a. The style seems too sophisticated for the intended readers. You've used too many abstractions, long sentences, and confusing words.

b. Third-person style and passive voice are inappropriate in instructions because they make readers think someone else is supposed to act.

c. These sentences are awkward.

d. The whole section of the report is unclear.

e. The brochure is poorly organized.

f. It's impossible to tell why the project you propose is important.

g. The instructions are confusing.

Methods and Tools

CHAPTER 4

Marking Paper Copy

Editors mark paper documents with instructions for revision of the text and for type and page design. These instructions tell the next person who works on the copy how to incorporate the editing. The person who uses the editor's marks may be the writer (in revision) or document production specialist, who prepares the pages as they will be printed or saved in PDF files. The editor's instructions, marked on the document itself, are written with a special set of symbols. Marking the document with these instructions in symbols is known as *markup*.

Marks for the writer may include suggestions for revision and queries to clarify meaning. Marks for the production specialist show where changes need to be made to make the document correct, consistent, accurate, and complete. In addition, markup on paper may include type specifications (font, size, and style) for headings, paragraphs, and other parts of the text; line length and page depth; and placement of illustrations. This information about the visual design of the text is necessary if the text is prepared as a typescript (not formatted as it will be printed). If the text to be edited already has the appearance of the final version, the editor checks that the visual design is correct.

When the editor transmits the text to a printer electronically, the editor may insert the text changes and tag the different types of text right in the electronic file. (See Chapters 5 and 6.) However, many editors prefer to work with paper copy at some point in production. Thus, all editors should know the accepted markup symbols.

This chapter gives examples of markup symbols for copyediting and for graphic design. This chapter covers only marking on paper; Chapter 5, "Marking Digital Copy," describes markup for online documents. Chapter 6 introduces procedures of electronic editing.

A note on terms: In this chapter, *production specialist* is used as a comprehensive term to refer to the person who implements the editor's directions and prepares the document for printing or online publication. In full-service typesetting, this person is a *typesetter* or *compositor*. (The compositor "composes" or sets the type and arranges the text and illustrations on the page.) Because much of the work of preparing the text for publication is now done in house, with templates and desktop publishing software, some employees fulfill the function of the compositor without holding that specialization.

Key Concepts

Editorial markup consists of directions for the development and production of the document. The reader of the marks may be the writer or the production staff. For the writer, the editor queries ambiguous content and verifies facts. For production, the editor marks for capitalization, spelling, and other variations and indicates type and spacing. The editor uses recognized symbols for these directions. Accurate markup should result in clean proof copy, the next stage of production.

The Symbols of Editorial Markup

Editors, production specialists, and graphic designers all understand a set of symbols to indicate emendations and design choices. These symbols are like a language: particular marks mean particular things, and the people who share the language understand the marks. Some of the symbols and methods of giving instructions may seem cryptic to you at first, and you may be tempted to write out fuller instructions. However, such variations in conventions will confuse rather than help designers and production specialists. Table 4.1 shows the symbols for indicating changes in letters, spacing, and type style of words. Table 4.2 shows symbols for marking punctuation. Table 4.3 shows how to mark for spacing.

Placing the Marks on the Page

When the typescript is double-spaced, the copyeditor inserts marks within the lines. Interlinear marks help the person making the changes in the text files because they appear right where the change must be made. If the typescript is single-spaced or already has the look of the final version, you will show in the text where insertions and replacements should be placed, but you write in the margins.

Instructions for visual design appear in the margin of a typescript. These directions cover line length, justification (whether the margins are to be aligned on the right or left or both), and typeface. They may also cover math symbols, extra space between lines, illustrations—anything not covered by standard markup symbols. Such directions often apply to whole blocks of text rather than to single words or phrases, which is one reason why they are marginal rather than interlinear.

Production specialists and writers appreciate marks that are neat, easy to read, and conventional. Small marks can be too small to see; big marks make the whole page look messy; unconventional marks (such as a circle rather than a dot over the letter *i*) look quirky. Marks should not obscure the correct type on the page. Clean copy will increase the chances of error-free copy at the next stage of production. You can help the production specialist locate specific changes by using a bright-color pencil, such as red or green. Faint pencil marks are difficult to read, and they suggest lack of self-confidence. Remove stray marks, such as a question mark you made for yourself, before you pass the typescript on for the next stage of revision or production.

TABLE 4.1	**Markup Symbols: Words, Letters**		
Symbol/Meaning	**Example**	**Result**	**Comment**
𝓎 delete	dele𝓎te	delete	Use the closeup mark, too, if the word could be spelled as two words.
𝓎 delete, close	proof⁀reading	proofreading	
— delete a word	in the ~~the~~ back	in the back	
∧ insert	in$_\wedge^s$rt	insert	Place the caret beneath the line. Write what is to be inserted above the line.
| insert space	insert|space	insert space	Usually the line alone will suffice; use the space symbol if there could be a question.
or |#	mark|#up a text	mark up a text	
𝒩 transpose	tra𝒩pose	transpose	If multiple transpositions in a word make the edited version difficult to read, delete the whole word and print the correction above it.
	Australia ~~Autsarlia~~	Australia	
⌢ close up	cl⌢ose	close	
≡ capital letters	ohio; ibm	Ohio; IBM	
= small caps	6 a.m.	6 A.M.	Since not all fonts include small caps, make sure they are available before you mark them.
/ lower case	Ƒederal	federal	
WORD lower case, whole word	FEDERAL	federal	
WORD initial cap	FEDERAL	Federal	
___ italics	Star Wars	*Star Wars*	Underline to change the type style from roman to italic or vice versa. Roman type is the opposite of italic, with straight rather than slanted vertical lines. Underline to convert from italic to roman, just as you do to convert roman to italic.
— roman type	*Star Wars* (rom)	Star Wars	
or (rom)	*Star Wars* (rom)	Star Wars	
∿ boldface	emphasis	**emphasis**	
V superscript	Masters degree	Master's degree	Use the superscript sign to identify apostrophes, quotation marks, or exponents.
	A2	A^2	
∧ subscript	H2O	H_2O	
//// delete an underline	revelry	revelry	

(continued)

TABLE 4.1	Markup Symbols: Words, Letters *(continued)*

Symbol/Meaning	Example	Result	Comment
⬭ spell out an abbreviation or number	② Ⓐssn ⓗp	two Association horsepower	Circle an abbreviation or number you want spelled out. Spell the word as well as circling if the spelling may be in question.
ⓢtet "let it stand"; or ···· ignore the editing	precede (stet) precede	precede	If you have edited in error or changed your mind, direct the compositor to set the copy in its original unedited form.

There may be more than one way to mark a change. For example, to mark the misspelling of "electornic," an editor could transpose "or" or delete the "o" where it appears and insert it after the "r." The choice depends on which marking is more clear and on the way the production specialist will keyboard the change. In this case, the transposition is the simpler mark, and the production specialist is likely to execute the change as one step, not the two steps of deletion and insertion.

Figure 4.1 shows a marked typescript, and Figure 4.2 shows how the same typescript looks after being typeset.

Marking Consistently

Generally you will mark each occurrence of change rather than depending on the production specialist to remember what you have done on previous pages. If, for example, you are deleting the hyphen in "on-line" throughout, mark each instance where the word occurs. Production specialists are taught to type what they see rather than to do global edits. The editing is your job. Furthermore, more than one person may work on the job. Prepare the typescript as though the next person to read it will begin reading only at the point of the mark you are making, rather than at the beginning.

Mark each heading to identify whether it is a level one, two, or three heading, and mark other types of text as well. In Figure 4.1, all the specific directions for type are written out, including the name of the typeface and its size, line length, and justification directions (FL = flush left, RR = ragged right). In Figure 4.3 on page 57, you will see a more efficient way to mark: each type of text is identified by an abbreviation (such as H1 for a level-one heading). The second type of marking requires that you provide the specifications for the type to the production specialist. If paragraph indentions are not clear in the typescript, mark each change.

TABLE 4.2	**Markup Symbols: Punctuation**		
Symbol/Meaning	**Example**	**Result**	**Comment**
⊙ period	...forever⊙	...forever.	Circle the period to call the compositor's attention to this small mark. Do not circle other punctuation.
comma	copper,iron,and,silver	copper, iron, and silver	Place an inverted caret over the comma. Do not place it over other punctuation.
: colon	following:	following:	
; semicolon	following; following; following;	following; following; following;	To create a semicolon from a comma or colon, draw in the dot or tail. Otherwise, simply insert the semicolon.
parentheses	(2002)	(2002)	The lines in the parentheses won't be typeset, but they do reinforce your intent to include parentheses rather than other lines.
[ ] brackets	[word]	[word]	Be sure to square the lines if the writer has used parentheses.
= or ✓ Mark (eq)	light=emitting diode computer= assisted	light-emitting diode computer-assisted	The underline or checking of the hyphen reinforces your intent to include a hyphen at that point.
			end-of-line hyphens for clarity.
equal sign	a b	a = b	Since the equal sign can look like the underlined hyphen, write *eq* by the mark and circle it to show that the information is an instruction.
M em dash or M	a pejorative disparaging word	a pejorative— disparaging—word	An em dash is as wide as the base of the capital letter *M* in the typesize and typeface used. It is used to set off parenthetical material or a break in thought.
N en dash or N	2000﹘01	2000–01	An en dash is as wide as the base of the capital letter *N* in the typeface and typesize used. Its primary use is in numbers expressed as a range.

Marking is especially important with spacing and graphic design marks if the design itself includes variations. Some designers, for example, indent all paragraphs except those that follow headings. You would check all paragraph indentions and mark any places where the pattern varies. You don't have to mark the ones that are typed correctly.

TABLE 4.3	**Markup Symbols: Spacing, Position**		
Mark	**Meaning**	**Example**	**Result**
¶	begin a new paragraph	...other design features. The editor's...	...other design features. The editor's...
⌐	begin a new line	numbers; abbreviations;	numbers; abbreviations;
⌒	run together (do not break the line or create a new paragraph)	...form your marks. It is not the time to express your...	...form your marks. It is not the time to express your...
⌐ or (FL)	flush left or justify left [Place the edge of the mark on the margin where text should move.]	⌐ The editor's choice...	The editor's choice...
¬	justify right	Book Title ¬	Book Title
][	center	]Book Title[	Book Title
(RR)	ragged right [Lines do not align on the right margin.]		
‖	align		
▭	indent one em		
▭▭ or ☑	indent two ems		
☑⌐	indent the whole block of text 2 ems		
⌐⌐	transpose a group of words	transpose of words a group	transpose a group of words
(	close up vertical space (as when an extra space has been skipped between paragraphs)	...too many lines skipped. Close up vertical space.	too many lines skipped. Close up vertical space.
#>	insert vertical space	**Heading** Insufficient leading follows.	**Heading** Insufficient leading follows.
⌇	set as a paragraph rather than as a list	numbers; abbreviations; and spelling	numbers; abbreviations; and spelling.

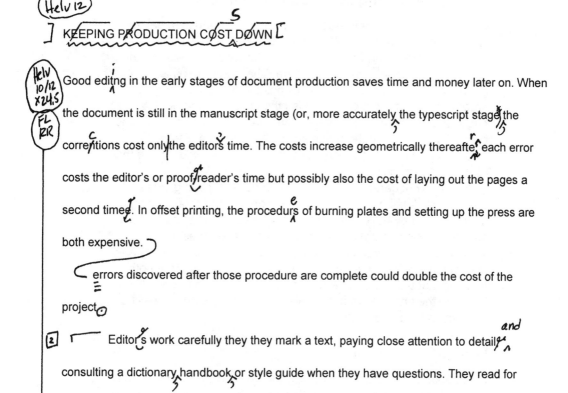

Figure 4.1   Marked Typescript

In Figure 4.1, the marginal line from the text specifications to the bottom of the example indicates that those type specifications apply to all the text identified by the line. If another heading appears, it will have to be marked again as will the body text. The need to mark each instance of text for graphic design explains why editors prefer to define type specifications for each type of text (heading 1, head-

Keeping Production Costs Down

Good editing in the early stages of document production saves time and money later on. When the document is still in the manuscript stage (or, more accurately, the typescript stage), the corrections cost only the editor's time. The costs increase geometrically thereafter; each error costs the editor's or proofreader's time but possibly also the cost of laying out the pages a second time. In offset printing, the procedures of burning plates and setting up the press are both expensive. Errors discovered after those procedures are complete could double the cost of the project.

Editors work carefully when they mark a text, paying close attention to detail and consulting a dictionary, handbook, or style guide when they have questions. They read for meaning to make sure the writer has not made careless errors, such as inadvertently substituting *in* for *on* or leaving out words. They are also careful to mark the text clearly and accurately so that both text and instructions can be read correctly. Thus, they can increase the chances of getting clean page proofs and of saving production time and costs.

Figure 4.2   Copy Set as Marked in Figure 4.1

ing 2, body text, bulleted list, and so forth) and then mark each section with a label (such as H1) rather than with all the directions for typography and spacing. (See Chapter 5 for more information on "structural" markup.)

Distinguishing Marginal Notes from Text Emendations

Marginal notes may be necessary to clarify your marks on the text. For example, if you want an equal sign but the marked text looks just like a hyphen with an underscore, you could write "equal sign" in the margin. But you don't want the production specialist to type "equal sign" into the text. So you distinguish instructions from text insertions by circling the instructions. If you need to include marginal messages to both writer and production specialist, you may distinguish these by using different colors of pencil for each category of message. Or you may preface the note with a label identifying the audience for the message—perhaps "au" for author and "comp" for compositor or production specialist.

Pierce's philosophy   *au: Correct?*

Influences on Darwin's Origin of Species   *comp: set rom*

Marginal notes may also give instructions for the placement of illustrations if those instructions are not clear in the text.

Special Problems of Markup

Though the markup symbols will be clear in most situations, marks or letters that could be interpreted in different ways require clarifying information. You may need to insert additional instructions when marking punctuation, distinguishing between hyphens and dashes, clarifying ambiguous letters and symbols or unusual spellings, and when marking headings, reference lists, and illustrations.

Punctuation

Because punctuation marks are so tiny, copyeditors add additional information to clarify which mark is intended. The conventions are to circle a period and to place an inverted caret (a "roof," a "house," or a "hat") over a comma. You can convert a comma to a period by circling it and rounding the shape of the comma, and you can convert a period to a comma by placing an inverted caret over it and drawing its tail. The circle and inverted caret are part of the information that tells what the punctuation is. These marks are no more interchangeable than are the comma and period themselves.

Conventions for colons and semicolons are less well fixed. *The Chicago Manual of Style*, 15th edition, recommends no extra signals around these marks (but, if necessary, clarify by writing the name—colon—in the margin and circling it); other guides put brackets at the top and bottom ($\wedge \atop \vee$) or draw an oval around them. You should find out the preference of the production department you work with and follow that pattern consistently. The main goal is that the production people understand your directions.

Hyphens and Dashes

Hyphens differ in use and size from dashes, but the distinction between hyphens and dashes is not always clear on a typescript and thus should be marked. Hyphens are the mark for combination words, such as *son-in-law*. They also appear in words that are broken at the end of one line and continued on the next. A line inserted under a hyphen during markup or a check placed over it indicates that the hyphen should be set as marked. For copy clarification in a typescript, you may need to underline all the hyphens, even those typed correctly, especially if the word is inconsistently hyphenated in the typescript. Mark all end-of-line hyphens to clarify whether the hyphen should be retained in case the word breaks differently in the final copy. Underline an end-of-line hyphen that should be retained. Likewise, use the close-up mark with the hyphen at the end of the line on a typescript to show that the word should be set closed. If you delete a hyphen that is typed in a word, mark to show whether to close up the words to make one word or to insert a space to make two or more words.

Marking the document with instructions is called mark up. The copy editor marks with the assumption that the publication specialist will enter text exactly as it is marked, letter for letter and mark for mark. End-of-line hyphens are particularly confusing and should be marked.

You can minimize the confusion of hyphens by preparing files without end-of-line hyphens (words that are always hyphenated will retain their hyphen). Instruct writers to turn off the hyphenation on their word processors; then no hyphens will appear at the end of the line. In the preceding example, "markup" could have been programmed to be typed on the second line without a hyphen, saving an editor time.

Em dashes and en dashes are longer than hyphens and have different meanings. Em dashes separate words or phrases from the rest of the sentence; they function like parentheses in casual style or add emphasis to words or phrases following an em dash. They are about the length of a capital letter *M* in the typeface in which they are set. Some people type two hyphens to create em dashes (a remnant of typewriter days). En dashes, which are the length of a capital letter *N* in the relevant typeface, are used in numbers to show a range. If there can be ambiguity in interpreting which dash is intended, you should mark each occurrence.

Place two lead weights—each weighing 4–7 grams—on the model car body between the rear wheels.

Hyphens and dashes are usually set without space on either side. If a typist has typed spaces around them, mark the copy to close up the space.

Dashes—marks of punctuation used to set off parenthetical material—are longer marks than hyphens. If they are typed as hyphens, the copyeditor must mark the em dash.

Ambiguous Letters and Symbols; Unusual Spellings

Some letters and symbols, such as the numeral *1* and the letter *l*, look similar. The editor should clarify anything about which the production specialist may have to make a judgment. If context establishes the meaning, the editor does not need to clarify. For example, the production specialist will recognize the insertion of letter *l* in "folow" to spell "follow" and will not insert the numeral 1. But in the following example, the production specialist may think "numeral" after typing the 2 and not recognize the letter *l* as an abbreviation for "liter." The note clarifies the editor's intent. The circle indicates information for a production specialist rather than text to insert.

Evaporate a 2-l sample.

The letter *O* and the numeral *0* may also be confused. You may need to write "zero" by the number and circle the word to show that you mean for a zero to be typed in that place. In equations, the *x* (indicating a variable or an unknown) must be shown to differ from the multiplication sign x. (See Chapter 12 on editing mathematical material for more information on marking equations.)

If spellings are unusual, you may write and circle "stet" next to the unusual spelling to indicate that the unusual spelling is intentional. *Stet* is the Latin term for "let it stand."

Headings, Tables, References, and Lists

Beginning copyeditors may intuitively read the paragraphs of a document more carefully than other types of text, such as the headings, tables, and list of references. Yet errors occur in these parts of the text. It is easier to make content and typing errors in a reference list or in a table than in paragraphs. Thus, you must check for the accuracy and completeness of the information in these parts of the text. Be sure also to check details such as type style (italics, roman, bold), accuracy of the numbers, and spacing. With lists, watch for incorrect end-of-line punctuation, capitalization, and indention.

Illustrations

The symbols used to mark tables and figures are the same as those for marking text. You may need to correct spelling by deleting or inserting letters, to adjust spacing, or to request alignment of numerals on their decimal points. Headings, labels, and titles need to be marked for correctness and consistency in capitalization and type style. Queries and marginal notes may be necessary for complex changes that the symbols do not address.

If the illustrations are attached to the end of the typescript for insertion at the time of page layout, mark the place where they are to be inserted. You can do this in the margin if the text does not already indicate the location. Simply write on a separate line, enclosing the words in square or angle brackets. You can specify only an approximate location because the page may not have enough room for the illustration at the exact point you have marked.

Marks for Graphic Design

The editor or graphic designer or the publication specialist may mark the document for its graphic design—that is, the face, style, and size of type, the spacing, and the line length. If you have some training in graphic design, you may make the decisions about design and mark them too. Or you may place marks on the document according to a graphic designer's instructions. You can mark boldface, italics, and capitalization using the marks displayed in Table 4.1. If you are editing copy prepared with templates in a word processing or page layout program, the design decisions are already incorporated in the template. You would mark only those instances that vary from the intended style, such as a level-two head inadvertently styled as a level-one head. There is no need to mark what is already correct and incorporated into the files.

The marks for typeface, type size, and line length will make more sense to you once you are familiar with typeface names and with the printer's measures of points and picas (see Chapter 23). The following example illustrates how you will mark such information that the graphic designer provides. The instructions direct the production specialist to set type of a particular size and face on a line of the specified length:

Here is what this note means:

set 10/12 x 30 Times

set	=	set type
10/12	=	10-point type on a line 12 points deep (there will be some extra space between the lines of type)
x 30	=	the line length—30 picas
Times	=	the typeface—Times Roman

These instructions would produce type just like what you are reading here. The instructions are circled to clarify that they are not part of the text. The note appears in the left margin so the production specialist will see them before typing the letters.

Whenever there is a change in the text, as from a heading to a paragraph, mark the change. You can do this either by marking the part, such as a level-one heading, or by providing the type specifications, such as size and style, each time. If you mark the part, include a type specification sheet with the marked copy.

Figure 4.3 illustrates a copyedited typescript page with marginal notes for the production specialist, alignment marks on the tabular material, and marks to identify type style. The flag at the top of the page is a query from the copyeditor to the writer. It asks for the writer's approval of a possible addition in content. The note by the abbreviation for inches directs the production specialist to use the symbol for inches, not for quotation marks. The note in the right margin tells the production specialist to use italics for the left column and offers an alternative if the text does not fit as marked. The straight vertical lines and the circled letters at the left indicate specific design elements. H3 means level-three heading, LT means list text, and MCL means multicolumn list. All of these types of text have design specifications: directions about typeface, spacing, type style, indention, and so forth. The abbreviations request that those particular specifications be applied. They save the editor the time of writing out all the specifications on each use, and they are easier for the production specialist to apply than individual directions if "styles" are defined in the word processing software for each element of text. (The next chapter will define styles and their use.) Some other abbreviations used frequently include the following:

CT	chapter title	BL	bulleted list
CN	chapter number	NL	numbered list
BT	body text	FN	footnote

The editor or graphic designer would create a list of text elements and their type specifications so that the abbreviations would be meaningful to others working on production.

Turnovers are phrases in the first column spilling onto a second line. The irregular vertical lines within the columns specify alignment for the three columns of examples. You can see the results of these marks by looking at the section on consistency of mechanics in Chapter 8.

All this careful marking is necessary to establish for the production people how to treat the text, both its language and its appearance.

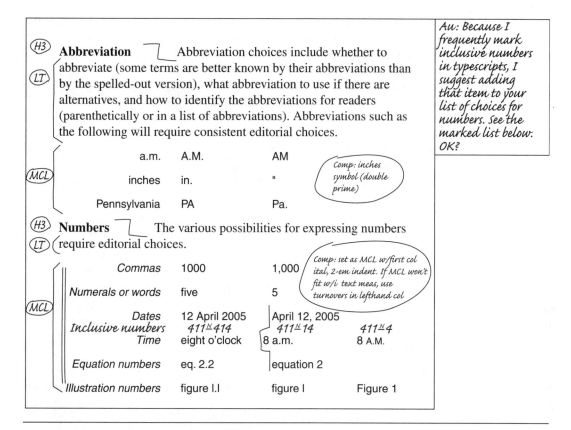

Figure 4.3   Edited Typescript Showing Notes to a Production Specialist and Author and Marks for Graphic Design

H3 means level-three heading; *LT* means list text; *MCL* means multicolumn list.

Queries to Writers

You may need to contact the writer for further information: to fill in a gap in content, to explain an editorial change in more detail, to advise on the display of an illustration, or to confirm that the changes are correct. A question to a writer is called a *query*. The term is also used to refer to all comments from the copyeditor to the writer.

Queries let you acquire information that you need to edit or mark correctly. They are likely to concern content. Questions that an editor should answer, such as how to capitalize or whether a term is hyphenated, are not usually appropriate for queries (unless they concern specialized content information). The editor is the language specialist and has the resources to look up what he or she doesn't know; the writer is the content specialist. However, queries let you explain marks that may puzzle a writer or to call the writer's attention to editing that may have unintentionally changed meaning. It's a good idea to query changes in terms or substantial rearrangement of sentences that might change meaning. Queries over obvious information or changes will be annoying. Not every mark you place on the page requires a question or comment.

In phrasing your queries, write directly and courteously, and avoid evaluative statements, especially when the evaluations are negative. For example, instead of writing "unclear" or giving the vague directions to "rewrite" or "clarify," tell the writer exactly what you need to know. Some examples follow:

Requests for Information

to check a discrepancy in an in-text citation and a reference list	Page 16 cites the date of 2003 while the reference list cites 2004. Please check the date and indicate the correct one.
to verify an unclear use of quotation marks and capitalization	May I assume that the quote marks signal a quotation and that ABC should be capitalized?
to clarify the reference for a pronoun	I'm not sure whether "it" refers to the program or to the previous step. Please clarify.
to verify design choices	Do you have any special instructions for this figure—e.g., single or double spacing, paragraph indentions or flush left? Please advise.

Explanations

to explain why a numbered list has been converted to text with headings	Other numbered lists in this book present very short discussions for each item. The importance and development of each of these topics warrant the use of headings. The headings will emphasize each topic more than the numbered list does. OK?
to explain changes in headings	I expanded the main heading and deleted the subheadings to parallel the pattern in other chapters. OK?

Marginal notes can work for simple queries, but some questions and explanations are too elaborate to be phrased on the text pages. For these—or for all your queries, even simple ones—you can attach query slips to the typescript. Sticky notes work well as query slips. You can attach them to the edge of the typescript, with the sticky part on the back of the page, at the place where the question arises. Write the query on the slip as it is open, and then fold the note over the edge of the page (the writing will be inside). When the writer opens the slip, he or she reads the query and responds by revising the text or confirming that the editing is correct. Ask the writer to indicate that he or she has considered the query by initialing or checking the slip. If the writer wants to think about the response, the slip can stay open, to flag the writer to come back to the question later. The slip remains in place until the copyeditor checks the response and makes the necessary changes. Then the slip can be removed before the typescript is forwarded for the remaining steps in production. The query slips prevent the clutter of notes on the typescript itself that could distract the production specialist. Figure 4.3 shows a marked page with a query slip attached.

If you note the typescript page number on the query slip, you will know where it belongs if it is accidentally detached. If you are writing queries to a designer or production editor as well as to the writer, also note on each slip who should read it. The note "au/24" identifies a query to the author on page 24 of the typescript. Some copyeditors attach the slips for the writer on the right side of the page and slips for the designer on the left or use different color slips for different readers. Then the designer and writer know which to read and which to ignore.

Using Your Knowledge

To develop fluency in any language, including the language of markup, you practice. While you are developing fluency, you refer to style manuals and monitor whether the writer and production specialist understand your directions. Eventually, you will seldom need to refer to the reference books. As with any communication task, you are effective when you think of your audience and purpose. You are moving the document further toward the goal of production, not "correcting" a writer's errors. You are working in partnership with the writer and the production staff to produce a document that is correct, consistent, accurate, and complete.

Further Reading

Stoughton, Mary. 1996. *Substance & style: Instruction and practice in copyediting.* 2nd ed. rev. Editorial Experts.

Many dictionaries include a table of proofreading marks, either in the word list or in the back matter. Style manuals, including *The Chicago Manual of Style,* 15th edition, also illustrate the marks and their uses. The marks for proofreading and copyediting are mostly the same though they may be used differently, as Chapter 13 explains.

Discussion and Application

1. Mark the words in the first column so that they will be printed like the words in the second column.

developement	development
interogation	interrogation
emphasis	*emphasis*
italic	italic
1/2	one-half
three	3
teh	the
ambivalance	ambivalence
on going	ongoing
tabletennis	table tennis
2mm	2 mm

semi-colon	semicolon
CHAPTER TITLE	Chapter Title
Cpr	CPR
Research Laboratory	research laboratory
testing confidential	confidential testing
Tavist D	Tavist-D
referees decision	referee's decision
m2	m^2
M2	M_2
We finished quickly – we had more errands to complete.	We finished quickly—we had more errands to complete.
end of sentence	end of sentence.
end of clause,	end of clause;
introduction	introduction:
quote	"quote"

2. Mark the double-spaced typescript so that it will be printed like the text that follows. The title is Arial 12, bold. The body text is set in the typeface Arial, flush left, ragged right, in 10-point type, 12-point lead, 26 picas wide. Paragraphs are indented one em.

COMPUTER VIRUSES

A malicious use of the computer is to insert a computer viris into an e-mail message in an attchment. When the recipient opens the attached file, the virus begins to do it's damage. One type of virus finds the all the ddresses in the recipient's email address book and sends a copy of the message to them, thereby spreading itself widely. Other viruses attack the files and directories on the harddrive. These infections are as destructive as a viral infection in the human body. They can damage if not destroy the software and data on individaul computers and even on whole networks of computers.

Prevention of these attacks requires multiple efforts. Anti-viral software can catch many viruses but new viruses may bypass the protection features. Users need to keep this software up to-date. Users need to be educated to not open attachments whose names end in ".exe.

(The file extension identifies executable code.) In fact, it's risky to open

any attachment if you are uncertain of it's source and purpose because

some viruses could come from apparently safe sources

Computer Viruses

A malicious use of the computer is to insert a computer virus into an email message as an attachment. When the recipient opens the attached file, the virus begins to do its damage. One type of virus finds all the addresses in the recipient's email address book and sends a copy of the message to them, thereby spreading itself widely. Other viruses attack the files and directories on the hard drive. These infections are as destructive as a viral infection in the human body. They can damage if not destroy the software and data on individual computers and even on whole networks of computers.

Prevention of these attacks requires multiple efforts. Antiviral software can catch many viruses, but new viruses may bypass the protection features. Users need to keep this software up-to-date. Users need to be educated not to open attachments whose names end in ".exe." (The file extension identifies executable code.) In fact, it's risky to open any attachment if you are uncertain of its source and purpose because some viruses could come from apparently safe sources.

3. Your job is to copyedit—to make the document correct, consistent, accurate, and complete. You do not alter word choices or organization. Yet you know enough about style to wonder about the choice of "catch" in the second paragraph of the text in "Computer Viruses." Given the metaphor of "virus," you wonder if "intercept" might be better here because "catching" a virus means to get sick, whereas the antiviral software "catches" the virus before it can make the computer sick. But changing words and editing for style are beyond your responsibilities. What can you do?

4. Circling an abbreviation instructs the production specialist to spell it out. Assume you want the abbreviation *STC* spelled out. Why might circling be inadequate? What should you do instead?

5. You personally prefer to spell *proofread* as a hyphenated compound (proof-read) rather than solid. Do you have the choice of spelling it according to preference if it appears in a document you are editing? Why or why not?

6. *Vocabulary.* These terms should now have meaning for you in the context of editing—*production specialist, markup, query.* If you are uncertain of their meaning, check the glossary or review this chapter.

5 Marking Digital Copy

With Carlos Evia

Traditional markup, as described in Chapter 4, gives instructions to someone else to implement—perhaps the writer, or a production specialist, or a typesetter in commercial printing. The marks on paper provide instructions for emendations to the language plus choices of type and spacing that affect the appearance of the document. Markup of paper copy for graphic design occurs near the end of document development as the document moves from revision toward publication.

Computers have changed both the methods and responsibilities of editors. Editors can work directly in the computer files both to emend the text and to mark for graphic design. Marking digital copy may reduce or even eliminate the need to mark paper copy. The editor may correct errors, rearrange paragraphs, modify the style, and query the writer within the files. Documents that are distributed electronically (on the Web or on CDs) have markup information embedded into the files.

Digital markup offers advantages over paper markup. It saves one step in production because the editor implements the changes directly. It can automate conversion of one form of the document to another, as from print to a website. And it enables the sharing of information that might be useful in more than one document, which provides the foundation for time-saving digital publishing practices, such as single-sourcing and content management.

This chapter is concerned with methods of marking digital copy for graphic design and content management. The next chapter, Chapter 6, discusses methods and policies for editing content electronically.

Key Concepts

Digital markup includes tags that identify different document parts. The tags can be linked to style sheets that include instructions for the appearance and meaning of those parts. An editor can change the appearance simply by changing the specifications in the tag. Digital markup facilitates presentation in multiple media. It also enables content management and single-sourcing by facilitating the retrieval of document parts that can be reused in other documents.

Procedural Markup versus Structural Markup

Markup of paper copy, as described in Chapter 4, is mostly procedural. That is, each mark directs a procedure (making the type bold, centering the type, inserting a letter). The editor is providing instructions for emending the text and establishing its appearance. In working with digital files, the editor would probably just make some of these changes. There would be no need to direct someone else to apply boldfacing or to correct a typo. (As Chapter 6 will explain, the editor would query the writer before changing content and would track the substantive changes so that the writer could review them.)

Procedural marks assume one type of output, probably a printed page. But many documents are published in multiple forms, perhaps as a website as well as print. The presentation (and even some content) for each version may differ. For example, the font used for print may not be the easiest one to read online. Marking the document for each version would take a lot of time.

In addition to possible publication of the document in multiple forms, parts of a document may be used in different documents. For example, a company may use a product description in its sales literature and also in a user manual. Companies see the efficiency of reusing information, not rewriting it each time similar information is needed. But they need some way to store and locate the information. Identifying the different kinds of information in a document (such as copyright statements and product descriptions) may enable the parts to be stored in a database or content repository for retrieval and reuse. Procedural markup for these purposes is limited. Procedural markup supports graphic design and text emendations but not publication in different forms nor content management. Each instance of publication (print, website) would require separate markup to apply a different design. For the goals of publication in multiple forms and content management, structural markup within digital files is a better option.

Structural markup, sometimes called descriptive markup or generic coding, identifies the parts (structure) of the text, not the procedures used to change the content nor to apply graphic design. Structural markup identifies the different text elements, such as levels of headings, titles, figure titles, index items, paragraphs, and even details such as the copyright notice. Each element has a different function in the document. When the editor marking paper copy identifies elements of text with a tag, such as H1 for a level-one heading, the editor is using a type of structural markup. (See Figure 4.3 in the previous chapter.) Each tag identifies a text element. A tag is information, like marks for graphic design, that is not part of the content.

Structural markup separates document structure from the appearance of the document. The tags mark the structural parts. As you can see from Chapter 4, marking "H1" 50 times in a typescript with 50 headings is much easier than marking "Helv 12 bf ctr 12 pts above 3 pts below" 50 times. The editor identifies the parts; a graphic designer can later determine the font, size, and spacing of the part.

The advantages of structural markup multiply in markup of digital copy. The tag embeds the specifications or procedures for the text element. The tag for a level-one heading, for example, identifies the font and size, spacing before and after, type style (bold, italic, roman), and justification (centered, left, right). A writer or editor can apply all the specifications simply by applying the tag from a pull-down menu to selected text. Tagging also ensures consistency of presentation from one instance to the next. You don't have to remember from one heading to the next, for example, whether it is in 10-point or 12-point type. Also, by identifying specific sections or sentences as, for example, "Heading 1," the document might be converted to HTML or other digital formats and the Heading 1 will always be treated as Heading 1. This is the semantic nature of tags, which many authors see as the future of the Web. (See Berners-Lee and Miller in Further Reading.)

The specifications for the tag can be changed, but these changes do not require a remarking of the text elements. If the specifications should change during development, the writer or editor can apply changes throughout the document by changing the tag. Because the elements are tagged alike throughout the document, there is no need to change each instance. Likewise, a different set of specifications for each tag can be developed for different publication media. The specifications for H1 for print might be Times 12 bold but Verdana 10 bold for online presentation.

Styles and Templates

In paper-based document production, procedural markup for the document appearance takes place after the text is established. With desktop and online publishing, document templates, including choices about typography, can be created early in the process. Instead of the writer creating a double-spaced typescript that is later marked and converted to its published look, writers compose using the type and margins of the final version.

Instructions for typography and page design can be created in the word processing or page layout program and embedded in tags and templates. A template is a collection of tags for a document type. Tags embed the instructions for display of the various text elements, such as level-one headings, bulleted lists, and paragraphs. Microsoft Word and other word processing programs as well as Adobe InDesign (page composition software) call the tags *styles*. FrameMaker uses the concept of paragraph designer; each item in its paragraph design catalog is analogous to a Microsoft Word style.[1] A style can include specifications for font size and style, indention, margins, tabs, columns, space before and after, borders, related styles, and the style to follow after a return. FrameMaker allows up to 66 different choices for each item in its catalog of paragraph designs!

If you are unfamiliar with the concept of styles, your reading of this section will be more meaningful if you pause now to explore your software and its styles.

---

[1]Microsoft Word, InDesign, and FrameMaker are used for illustration in this chapter and the next because the programs are widely used by technical communicators. Other word processing and page layout programs offer similar options.

A good starting point for beginning familiar with styles is to use the ones that are supplied with the software.

If you look at the toolbars at the top of your word processing program, you should see the name of the style that is being applied to the text you are working on. In Microsoft Word, for example, the style name is on the Formatting toolbar. The default style for Microsoft Word is Normal. If you press the arrow by the text box at the top of the screen that says Normal, you will see the names of other built-in styles, such as Heading 1 and Heading 2.

To apply one of these styles to existing text, select the text and then select the style. To apply a style to text you are about to type, place your cursor in the text, select the style, and begin typing.

By tagging text, you can apply all the formatting choices simply by selecting the text (such as a heading) and choosing the style to apply from the style menu. You will save a lot of time in typing if you don't have to apply each procedure, such as selecting the font, size, and spacing, each time you type a related document part, such as a level-one heading. With styles you can also be sure that each related part will match visually.

You can modify default styles or create your own new styles with your own names. For each style, you would choose its attributes: font and size, style (bold, italic), and spacing. To see or change the attributes of these styles, or to create a new style in Word, go to the Format menu item and select Styles and Formatting. A list of styles will appear in a column on the right. Mouse over the name of the style to see an arrow that you can press to get options for modifying the style or creating a new one.

If you decide to change the look of a heading or other text element, perhaps by using a different font or spacing, you can simply change the style definition. All the instances of its use will change. For example, if you changed Heading 1 to become centered instead of left justified, all the headings marked with that style would automatically be centered. You wouldn't have to go through the entire document to find and change each instance.

Another advantage is that you can automatically generate a table of contents, outline, and index if you have tagged the text elements. If you have marked headings as heading 1, heading 2, and heading 3, you can generate up to three levels of headings in a table of contents. (In Microsoft Word, do this through the Insert > Reference > Tables menu.)

A collection of styles can form a *template*. The template includes the styles for all the text elements in that document. Table 5.1 shows some of the styles in a template used for the draft of this textbook. The template was attached to the digital file for each chapter so that the appropriate styles could be selected. The publisher of this textbook modified the styles, so the specifications you see in Table 5.1 do not describe the elements of the published textbook.

Multiple versions of the paragraph style in Table 5.1 reflect different situations for using paragraphs. The style without an indent follows an example when no new paragraph is intended. The styles with "cell" as an extension have different indention values for use in the cells of a table. The hanging indent means that the line after the first line is indented, as in a numbered list. The border is the rule

TABLE 5.1	Template for the Draft of *Technical Editing*
Style name	**Specifications**
chapter title	Normal + Font: Helvetica, 18 pt, Bold, Space after 36 pt, Tab stop: 0.25"
example	Paragraph + Font: Helvetica, 9 pt, Indent: Left 0.25", Space before 3 pt after 3 pt
example gloss	Example + Font: 9 pt, Indent: Left 1.88"
example cell	Example + Indent: Left 0.25"
heading level one	Normal + Font: Helvetica, 14 pt, Bold, Kern at 14 pt, Space before 12 pt after 3 pt
heading level two	Normal + Font: Helvetica, Bold, Line spacing 1.5 lines, Space before 12 pt after 3 pt
list level one	Normal + Indent: Left 1.5", Hanging 0.25", Space after 4 pt
list level two	List level one + Indent: Left 1.75"
normal	Font: Times New Roman, 10 pt, Widow/Orphan control
paragraph	Normal + Indent: Left 1.5", First 0.25"
paragraph cell	Paragraph + Space before 6 pt
table title	Normal + Font: Bold, Indent: Hanging 1", Space after 3 pt, Border: bottom
table column head	Normal + Font: Bold, Space before 6 pt

beneath the table title and column heads. "Pt" is an abbreviation for points, a measure of vertical space.

Figure 5.1 shows part of a draft page of this textbook with styles from the template applied.

Many companies have standard templates for different types of documents—a template for a manual, another for an in-house newsletter, and a third for reports. The writers use the templates in creating first drafts. Following a standard company structure as embedded in the template may save time in organizing the material. Seeing the text as it will look in print—as the reader will see it—may help writers use visual cues effectively.

Tagging structural parts offers production efficiencies for print publication. With these tags, it's quicker to move from the draft to the production-ready version. Furthermore, the tags enable quick changes in appearance of information in case it is distributed in different versions. Only the tags have to be changed to alter the appearance. Furthermore, the tags support more than appearance. Because they mark structural parts, they enable the automatic generation of tables, outlines, indexes, and more. These efficiencies for print publication are even more significant (and necessary) with digital documents. Markup languages have been developed specifically to accommodate electronic distribution of information in different versions for different circumstances.

Identifying Abbreviations

Abbreviations that are not familiar to readers must be identified the first time they are used. A parenthetical definition gives readers the information they need at the time they need it. After the initial parenthetical definition, the abbreviation alone may be used.

> Researchers have given tetrahydroaminoacridine (THA) to patients with Alzheimer's disease. THA inhibits the action of an enzyme that breaks down acetylcholine, a neurotransmitter that is deficient in Alzheimer's patients. The cognitive functioning of patients given oral THA improved while they were on the drug.

If the abbreviation is used in different chapters of the same document, it should be identified parenthetically in each chapter, because readers may not have read or may not remember previous identifications. Documents that use many abbreviations may also include a list of abbreviations in the front or back matter.

Periods and Spaces with Abbreviations

Some abbreviations include or conclude with periods (U.S., B.A.); others, especially scientific and technological terms and groups of capital letters, do not (USSR, CBS, cm). The trend is to drop the periods. For example, NAACP is more common now than the older form N.A.A.C.P.

When an abbreviation contains internal periods, no space is set between the period and the following letter except before another word. At the end of a sentence, one period identifies both an abbreviation and the end of a sentence.

> While working as a technical editor, she is also completing courses for a Ph.D.

Figure 5.1   Draft Section of *Technical Editing*, Formatted with the Template in Table 5.1

The example shows level-two headings, paragraphs, and examples.

Markup Languages for Online Documents

Much publication of technical information is digital: websites, CDs, online help. Digital markup enables text and graphics to display online. A company may offer its information in multiple forms: print as well as digital and in various and changing digital forms. The multiple forms may reflect different uses for the same information. For example, instructions may appear in a manual but also, in smaller chunks, in online help. Digital publication has required a way to code documents so they can be displayed online and so they can be used with different hardware, different web browsers, and in different forms. Even within one medium, such as the Web, different users may have different needs for the way the same information is displayed. A user with a visual impairment, for example, wants large type; a website ideally offers the information so that users can adjust the size of the type they receive.

The solution to this need for multiple uses has been to develop markup languages that separate structure from appearance. The markup languages are structural: the purpose of the tag (such as H1, for a level-one heading) is to mark the document part, not to specify the appearance. If the document parts are tagged, the appearance can be modified for various situations much the way that you can modify the look of a print document by redefining its styles.

This section introduces three variations of markup languages: SGML, HTML, and XML. It briefly explains their different uses. These languages all enable structural markup. SGML (Standard Generalized Markup Language) is the oldest. It is complex and powerful. HTML (Hypertext Markup Language) was developed specifically for hypertext. XML (Extensible Markup Language) offers more options for text management than HTML. These markup languages are not used for emending style and other aspects of language; rather, their uses are for production and information management.

SGML

SGML, or Standard Generalized Markup Language, codes documents with identifying tags so that the documents can be distributed on any hardware and in any medium (print, CD-ROM, web). The "generalized" part of its name refers to this cross-platform capability. The tags, embedded into the document files, identify structural parts. Angle brackets around the tags mark the words as instruction and not part of the text to be displayed, much as circling a marginal note for the compositor in paper-copy markup indicates a query or procedure rather than content.

Tags occur in pairs—beginning and ending tags. A title would be marked in this way:

<title>Marking Digital Copy</title>

The slash in the end tag identifies it as the end of the title. Abbreviations or words within the brackets identify the part name. The tags are part of the digital file and accompany the text in any form it takes.

Unlike the tags in word processing and page layout styles, SGML tags are not concerned with appearance. Marking for structure rather than for appearance enables multiple uses of the document. Because the tags do not specify type and spacing, a document with SGML tags can be formatted for any output. A FOSI (Formatting Output Specification Instance) specifies the formatting. Changing the format for another publication requires only a new FOSI. All the level-two headings might be 12-point bold italic type in print but become 14-point bold roman type on the computer screen where italics are hard to read. The tags remain consistent; they are simply interpreted differently in the FOSI for different applications.

Because the tags are generalized (not particular to one type of hardware or software or output), a tagged document can be distributed through any system, so long as the system has software known as an SGML parser or editor that recognizes and interprets the tags.

SGML is especially useful for companies that publish more than one version of a document or that expect documents to have a long lifespan. An SGML-coded print document does not have to be redesigned for publication as an online version. Instead the codes themselves can be adjusted to make the text suitable for the medium.

SGML is an international standard, accepted by the American National Standards Institute (ANSI) and the International Organization for Standardization (ISO). Because the coding is recognized internationally, SGML may facilitate international distribution of documents.

SGML Tags and Document Type Definition (DTD)

The SGML international standard specifies tags for about 200 elements of text and other identifiers, including details such as a contract/grant number, price, and acid-free paper indicator as well as author first name, author surname, foreword, list item, and paragraph. Users may define additional elements.

Tags can be typed in manually, but some word processing and page layout software programs use pull-down menus to let a person select the tags and keep them relatively invisible. Applying the tags from menus is somewhat like applying styles in word processing. A writer, editor, or production specialist can code the text. A page that reveals the code would look similar to the source page for a hypertext page. (See Figure 5.2 in the next section.)

The document type definition, or DTD, is a list of elements for a specific document with rules for using them. The DTD for a memo, for example, would include the sender, receiver, subject, and date lines as well as paragraphs for the body. Those paragraphs might be further defined depending on the type of memo as problem statement, sales results, and conclusions. Each DTD is a kind of outline for each type of document. The DTD can specify rules, such as that each level-one heading must be followed by a paragraph before a level-two heading or that there must be at least two level-two headings within a section. HTML, discussed later, is a DTD of SGML. The rules for the HTML DTD specify that all documents will begin and end with the HTML tag (<HTML>,</HTML>) and that they will include a head (for preliminary material, including the title) and body.

The parser (software program) compares the actual document with the DTD to determine whether all the parts are present. It provides a good check of completeness at least from a structural standpoint.

SGML Documents as Databases

SGML tags do more than facilitate publication in different media and across computer platforms. They also identify document parts that can be searched and retrieved. The tags allow documents to function as searchable databases. Thus, SGML is a tool for information management.

Selected parts of a document can be retrieved for placement in another document. For example, the descriptions in various sources of the company's products could be compiled into a catalog. Or certain parts of a manual could be selected for a customized manual for a customer who doesn't want the full manual. The warnings could be updated in the manuals for all the company's products if liability laws change. All the titles of the publications could be compiled for a bibliography. The value of SGML is proportional to the amount of material to manage and the variety of uses for the material. It facilitates reuse of content and therefore saves the time of generating new content.

SGML Coding and Editing

An SGML specialist or the editor or the writer may code the document. Even if the editor does not do the work directly, the fact of coding affects the editor's work. If each document is structured according to a DTD, and a FOSI specifies the format, the company's documents should achieve structural and visual consistency. If writers can easily import sections of existing documents into new documents, those already-edited sections represent less work for the editor. The editor should then be free to pay the greatest attention to style, paragraph structure, and other uses of language.

But SGML by itself will not help the editor with style, paragraph structure, and accuracy of the information. For those tasks, the editor relies on knowledge of language and communication, as discussed in the chapters in Part 3 of this textbook.

HTML

HTML (Hypertext Markup Language) enables display of text and graphics on the World Wide Web. Like SGML, HTML includes codes with text in documents to give instructions. (HTML is a DTD of SGML.) If you open a website and choose View/Source, you will probably see the code behind the display of the website.

Web browsers interpret such tags as H1 with default attributes of type and spacing. HTML code looks like SGML code, and like SGML it marks structure. However, web designers, frustrated by their wish for more control over appearance than HTML can provide, have adapted HTML to make it a formatting language or to work around its limitations for specifying appearance. Forced to do something it was not intended to do, the language has become complex and clumsy.

HTML tags do not necessarily prescribe the exact look of the coded document. The appearance can be influenced by the user's hardware, browser (the software program that interprets the information), and preferences that the user sets. Different browsers create different output from the same input. Users also have the option to set the fonts and sizes of the displays. Thus, a heading coded as H1 might appear on your browser in one font and size, but the user receiving the file may have set the preferences for a different font and size. The limited ability to control formatting with HTML alone is one reason why style sheet languages have been developed. (See the section on cascading style sheets.)

In addition to identifying elements to be formatted in certain ways, HTML identifies the links from one document to another or from one part of the document to another part. These links make hypertext functional; that is, they enable readers to select or ignore components of the text and conveniently use from it only what they need. The HTML tags that link different documents are

<A HREF> tags, where A refers to anchor and HREF refers to **H**ypertext **REF**erence. The rest of the tag gives the file name (location) of the reference. The linking tags are conventionally underlined in the display. By clicking on the underlined text, a reader can jump to the document or section of the document that is linked.

Figure 5.2 on page 76 shows the source code for FAQs (Frequently Asked Questions) on editing. Figure 5.3 on page 77 shows the page as displayed through a browser. (Different browsers might change the appearance of the page.) The source includes both the code and the text. The tags are enclosed in angle brackets. The following list identifies meanings of some of the tags:

H1 = heading level one	UL = unordered list	P = paragraph
H2 = heading level two	LI = list item	

Editors who work with hypertext documents should understand HTML well enough to recommend options for using the tags to increase the usability, correctness, and accuracy of documents. For example, marking headings as "H1" rather than with font, size, and style attributes ensures some consistency throughout the document, as styles do for print documents. This procedure also reduces the amount of markup in the file and therefore the size of the file and download time. Design choices can be controlled (and updated) with cascading style sheets, discussed later in the chapter. One editorial task may be to check links to be sure that they all lead to the intended places, just as an editor working with paper documents will check to be sure that all the parts are present. HTML documents also require editing for spelling, grammar, and consistency. Because the coding itself makes the page hard to read, editors most likely emend the text through page development software that shows the page as their browser displays it rather than in the coded source.

Although it is easy to learn enough HTML to display information in a simple website, HTML lacks the power of either SGML or XML to enable reuse of information. XML offers more options.

XML

In *XML by Example,* Benoît Marchal explains that HTML has become a complex language, with more than 100 tags. Some of the complexity results from the workarounds of designers to specify appearance. Its rules are not particularly strict, meaning that browsers can interpret (or ignore) sloppy code.

XML, for Extensible Markup Language, aims to correct some of the limitations of HTML and also to enhance content management functions. XML, like HTML, derives from SGML, but it is less complex than SGML. XML has been called "HTML done right" and "SGML light." It is extensible because users can define their own tags. Users are not limited to the tags that are predefined in SGML and HTML. With some limitations, it is possible to convert SGML documents into XML with just a few lines of code. To use Marchal's example, there is no tag for price in HTML, but an e-commerce company could use information on prices for multiple purposes: to display for customers at the website, to use for determining income and expenses, and to put in a printed catalog.

Code

```
<HTML>
<HEAD> <TITLE>Editing FAQs</TITLE>

<STYLE TYPE="TEXT/CSS">
<!--
H1 {
    font-family: Arial, Helvetica, sans-serif;
    font-size: medium; font-weight: bold;
    text-indent: 0.25em;
    padding-bottom: 0.25em; padding-top: 0.25em;
    background: navy; color: ivory;
}
H2 {
    font-family: Arial, Helvetica, sans-serif;
    font-size: small; font-weight: bold;
    color: navy;
}
BODY {
    background-color: ivory;
}
P {
    font-family: Arial, Helvetica, sans-serif;
    font-size: small; line-height: 120%;
    font-weight: normal;
    margin-left: 3em;
}
UL {
    font-family: Arial, Helvetica, sans-serif;
    font-size: small; line-height: 120%;
    font-weight: normal;
    margin-left: 4.5em;
}
-->
</STYLE>
</HEAD>
<BODY>
<H1>Editing FAQs</H1>
<H2>What are the "levels of edit"?</H2>
<p>The phrase <em>levels of edit</em> refers to a
taxonomy of editing procedures, ranging from checking
for completion of parts to evaluation of style and
content. Initially developed in the 1970s at the Jet
Propulsion Laboratory as a management tool, the
taxonomy identified five levels, which incorporated
nine types of edit. A level-one edit includes all nine
types, whereas a level-five edit includes only two
types.
<H2>How are the levels of edit used today?</H2>
<p>The taxonomy has been extended and modified, but
it is still a meaningful way of identifying editorial
tasks.
<UL>
<LI><em>Extension:</em> The levels have been
expanded to include tasks of editing websites and
other online documents.</LI>
<LI><em>Modification:</em> The five levels are
sometimes collapsed into three categories of editing:
light, medium, and heavy. </LI>
</UL>
</BODY>
</HTML>
```

Comment

The opening HTML tag + title + body are required in HTML.

"CSS" declares that the style language is Cascading Style Sheets.

The **comment tag** <!— encloses style declarations so that older browsers won't print all the code.

Three different **font names** establish preferences. If a user's browser does not have Arial or Helvetica, then it is directed to search for another sans serif font.

Padding in H1 provides background space so that the letters of the heading don't touch the edges of the background stripe.

Sizes are relative ("small," "medium") rather than absolute. The user can adjust the size, and all the elements will remain proportionally spaced.

Colors are named, but for fine variations of colors, designers may use codes. "Ivory" is "FFFFF0."

The left margin value indents the text under the headings.
The comment tag is closed.

Now the content appears. All the style choices for H1 are applied to every element tagged as H1.

"Em" has a different meaning in the style sheet than it does here in the body of the text. If you compare the code with the output in Figure 5.4, you will see what it means here.

Closing tags end the major divisions: BODY and HTML.

Figure 5.2 Source Code for Editing FAQs, Including Cascading Style Sheet

Editing FAQs

What are the "levels of edit"?

The phrase *levels of edit* refers to a taxonomy of editing procedures, ranging from checking for completion of parts to evaluation of style and content. Initially developed in the 1970s at the Jet Propulsion Laboratory as a management tool, the taxonomy identified five levels, which incorporated nine types of edit. A level-one edit includes all nine types, whereas a level-five edit includes only two types.

How are the levels of edit used today?

The taxonomy has been extended and modified, but it is still a meaningful way of identifying editorial tasks.

- *Extension:* The levels have been expanded to include tasks of editing websites and other online documents.
- *Modification:* The five levels are sometimes collapsed into three categories of editing: light, medium, and heavy.

Figure 5.3   Display of Editing FAQs

This figure shows the display on the Web of the document in Figure 5.2. You cannot see the ivory background and navy letters, but you can see the font, indention, and list display that the style sheet describes. The style sheet would be adequate for a much longer document, assuming it had the same features as are represented here.

XML documents, like SGML documents, require certain rules to define what tags are being used. In SGML, these rules are included in a DTD (document type definition). While XML documents would work with DTDs, they can be more effective if their rules are written in the XML schema language. XML schema is also based on the concept of tags, and its syntax is pretty similar to that of XML data files. The main advantage of using schemas over DTDs when working with XML is in the *datatype* declarations, which allow you to be specific about the format and content of entries in the XML document. A *datatype* declaration, for example, can specify that a certain element in the XML file has to include an integer, and even determine if the integer should be negative or positive.

XML and schema files can be typed into any text or word processing application. However, there are many editors and parsers available that will validate the code as you type and would even connect the XML documents to their schemas automatically in order to verify consistency. Furthermore, some word processors (like Microsoft Word 2003) have integrated XML functionality that makes the coding invisible to the user but maintains the tags and single-sourcing capabilities.

Like SGML, XML can be used to create searchable content repositories from documents because it marks structure of the content, not appearance. Because it is simpler than SGML and because it is readily adapted for Internet publication, XML is likely to be part of any single-source documentation system. Like HTML, XML works with cascading style sheets that specify appearance. However, the real single-sourcing power of XML is in XSL (Extensible Stylesheet Language), which will also be discussed in the following section.

Cascading Style Sheets

SGML, HTML, and XML emphasize document structure, not appearance. In SGML, the FOSI provides the instructions for interpreting the tags for a particular presentation. HTML includes some tags for appearance, but a designer has far less control over screen display using HTML than over paper display using word processing or page layout programs.

Cascading style sheets (CSS) were developed by the W3C (World Wide Web Consortium) to give designers power to influence the appearance of web documents. The style sheets also separate markup for appearance from markup for content. *Cascading* means that several style sheets may apply to the same document: the designer's, the user's, and the browser's, as well as a style sheet for a document set and one for the specific document. They "cascade" in that a browser searches them in order. The style sheet within the document overrides the imported style sheet, which overrides the user's and browser's. The browser looks for style instructions in the general style sheets only if the elements are not specified in the document and imported style sheets.

The code for cascading style sheets looks much like the code for HTML, and a designer who works with HTML should be able to learn it readily. CSS are compatible with both HTML and XML. Many of the code terms are natural language terms. For example, a designer can specify the color blue for type without using the six-character hexadecimal code for blue (#000080), assuming the designer does not want some slight variation of blue. Cascading style sheets can also be applied through page design software. If you view the source code of websites, you may see style directions embedded in a comment tag (<!--, -->). That's because older browsers may not recognize CSS, but they know not to print material within a comment tag.

Each CSS instruction includes two parts:

Selector (the element that is being styled)

Declaration (property + value, expressed within curly brackets)

In the following example, the selector is H1. Four declarations are being made for this selector for properties including the color of the type, the font, the font size, and weight. The values for each of these properties are specified after the colon. This code says that all the H1 headings are to be navy, in a sans serif type, preferably Arial or Helvetica, in a large size, and bold. The declarations don't have to be on separate lines, but they are easier to read if they are. The semicolons are separators.

```
H1 {
color: navy;
font-family: arial, helvetica, sans serif;
font-size: large;
font-weight: bold;
}
```

Just as the declarations may be grouped for one selector, multiple selectors may be grouped for one or more declarations. H1, H2, H3 {font-weight: bold} says that the first three levels of headings will all be bold.

Figure 5.2 shows some style information embedded into an HTML file. The style tag encloses the specific instructions and separates the instructions about appearance from the document content. Figure 5.3 shows the output from these instructions, minus the color. More information on the codes and applying them is available from references in Further Reading.

The ability of the designer to control output, including font, size, and color of type and indent and line spacing, takes some of the control for display from the user. There are balances to seek between designer control and aesthetics on the one hand and user control and usability on the other. The way you define the size and spacing values in the style sheet can allow both designer and user control. Cascading style sheets allow for relative rather than absolute specifications of size and spacing. For example, because the em as a measure is proportional to the size of type, specifying a 3-em indent means that the space will adjust for the size of type that the user may choose. Specifying that a level-two heading is "smaller" than a level-one heading, rather than defining it as 10-point type, will let users select type size and keep the different types of text in proportion to each other.

Style sheet languages continue to be developed. XSL, Extensible Stylesheet Language, builds on the formatting features of CSS and the single-sourcing capabilities of XML. XSLT (for Transformation) displays the contents of an XML document, but it can also change the order of the file's elements or ignore some of them. For example, an XML file including information on clients all over the United States and their contact information can be transformed through an XSLT asking to display only the names of customers in Florida in a new XML document or an HTML file. CSS would only affect the format of the display, but XSL changes its content. Of course, XSL and CSS can work together. XSL-FO (for Formatting Objects) can paginate certain elements of an XML document and produce outcomes in audio or printed formats (like PDF, for example). XSL files can be complex to design, and a few WYSIWYG (What You See Is What You Get) editors are now available to create XSLT and XSL-FO, eliminating the need to code lengthy files.

Editing and Information Management

The editor's responsibilities for helping to develop a document for publication may overlap with information management goals of the company. As you have seen, the markup of digital copy may include some codes to facilitate publication in various media and retrieval of parts of the document that may be reused.

When companies produce many documents and when they publish the information in various forms, they need a system for managing the information. Otherwise, they may have multiple files containing essentially the same information, and they may constantly re-create information that already exists. One writing group may update information that appears in multiple documents but forget to update all the documents, creating the confusion of two versions. Or they may laboriously update all the documents, one by one. They may have to develop two sets of documents for publication in different media rather than relying on the same source. The users of their online documents may work on different types of computers that interpret digital information in different ways, meaning that the documents have to be customized for different users. This problem is typical of large companies with a lot of information to manage.

Many companies that produce a lot of documentation are aiming for single-source documentation. Content is written once and stored in a repository where it can be reused in new publications. The source can be a repository from which writers copy and paste, but in the most advanced content management systems, the publications link dynamically to the source. If the source is updated, all the publications that use the information are also updated. Achieving the goal of single sourcing and reuse requires that the documents be marked and stored so that the data can be retrieved. The documents also must be marked not just for a specific output but in a generic way, so that they can be published in different forms without extensive reformatting and restructuring.

In *Managing Enterprise Content* (2003), Ann Rockley explains reuse like this:

> Most organizations already reuse content, though they copy and paste it. This works well until the content—and everywhere it appears—has to be updated. Then it can be time consuming to find every place the content has been copied and reused and change it. Not only is this time consuming, but some occurrences may be missed, resulting in inconsistencies and inaccuracies. In addition, over time, inconsistencies tend to layer themselves, until original inconsistencies become buried and you end up with two completely different content sources.
>
> Reuse . . . is the process of "linking" to an element of reusable content. The reusable content is displayed in the document in which you are working, but it does not actually reside in the document. (p. 24)

Editors are not necessarily in charge of information management, but the work they do, especially with digital copy, is one part of information management. The computer files for documents are likely to include not just the editorial emendations for content and appearance but also information about the document's structure. Thus, it is important to understand what is happening within the files and why.

The key to this information management is tagging structural parts of documents according to the function they perform. Multiple-use documents require generalized markup rather than the document-specific markup of styles in word processing and page composition. Single-sourcing and content management systems are used to support reuse of information.

Using Your Knowledge

One of the most important concepts you can take from this chapter is the distinction between procedural and structural markup and between marking for appearance and marking for structure.

You should know how to create and apply styles and templates in word processing and page layout programs. Styles will make your life easier even while you are a student and writing term papers. You will set the type specifications just once for each different type of text and then apply the styles in a single procedure to the relevant text. This will save you time in a long paper and create consistency. As editor you will almost certainly work with templates and styles. Thus, it would be prudent to develop comfort with them before you graduate.

Even if you are not sure how to create a content repository using XML or how to fetch and transform pieces of the repository into new documents, you should understand that tagging text elements enables searching and document reuse in different media and that XML is fundamental to single-source documentation. The more you know about technology, however, the more value you offer as an editor and the more you can participate in high-level decisions about a company's documents.

Templates, style sheets, and tags add efficiencies to production and options for publication. However, they do not address the quality of writing, and for that editing goal, editors continue to use conventional (procedural) markup on paper or to edit the text electronically.

Acknowledgments

Carlos Evia contributed substantially to the content of this chapter. The students in Nancy Allen's editing class at Eastern Michigan University also offered helpful suggestions. For this help, I am grateful.

Further Reading

Applen, J. D. August 2002. Technical communication, knowledge management, and XML. *Technical Communication* 49.3: 301–313.

Berners-Lee, Tim, and Eric Miller. October 2002. The semantic web lifts off. *ERCIM News* No. 51. www.ercim.org/publication/Ercim_News/enw51/berners-lee.html.

Cascading Style Sheets Home Page. www.w3.org/Style/CSS.

Lie, Håkon Wium, and Bert Bos. 2005. *Cascading Style Sheets: Designing for the Web.* 3rd ed. Addison-Wesley. The authors developed cascading style sheets. Ch. 2 is available at www.w3.org/Style/LieBos2e/enter.

Marchal, Benoît. 2001. *XML by Example.* 2nd ed. Que.

Meyer, Eric A. 2004. *Cascading Style Sheets: The Definitive Guide.* 2nd ed. O'Reilly.

Rockley, Ann. 2003. *Managing Enterprise Content.* New Riders.

XML Cover Pages: Online Resource for Markup Language Technologies. www.oasisopen.org/cover/sgml-xml.html. Sections on CSS and on XSL.

Discussion and Application

1. *Definitions.* Illustrate the difference in structural and procedural markup by marking the first two pages of this chapter (or other document your instructor provides) in both ways. For example, in procedural markup, you might mark the title as 14-point bold Helvetica with 18 points beneath. In structural markup, you would mark the title simply as a title.

2. *HTML/CSS code and output.* Compare the documents in Figures 5.2 and 5.3 to relate HTML and CSS code to output. Identify codes for the headings, paragraphs, and list in Figure 5.2. By looking at the output, try to identify the two meanings of "em."

3. *Tutorials on styles.* Follow one or more of these tutorials on styles:
   - www.ablongman.com/rude. Select "Software Tutorials," then "Microsoft Word: Introduction to Styles."
   - education.socialaw.com/wordguide. Select "Understanding Styles."
   - www.addbalance.com/usersguide/styles.htm. Select "Understanding Styles in Microsoft Word."

4. *Vocabulary.* You should be able to define these terms: *procedural markup, structural markup, tag, style, template, single-source documentation, HTML, SGML, XML, cascading style sheet, schema, XSL.*

Electronic Editing

By David Dayton

Chapter 1 highlights a number of interesting contrasts in the work lives of two editors, Kathy and Charlene. One difference that stands out is that Kathy marks up paper copy while Charlene makes changes directly to computer files. Apparently, Charlene does not mark her edits for later review and approval, which represents a fundamental change in the writer–editor relationship that usually operates in technical communication workplaces.

The traditional writer–editor relationship in technical communication defines the writer as the "owner" of the document's contents. Using paper-copy markup, as Kathy does at BMC, safeguards the writer's primary responsibility for the contents by making each of Kathy's edits, in effect, a *suggested* revision. The writer responsible for the document incorporates the edits she agrees with and ignores or questions others. In a common variation on this procedure, the writer reviews the markup, indicating on it those she approves and those she rejects, along with any additional changes. The editor then enters the approved changes. (If she edits the writer's revisions, these changes have to be approved by the writer as well.)

Kathy explains that she edits on paper to get a "reader's-eye-view" of the documentation, which is published as printed manuals or as PDF files that resemble the printed page. If we asked Kathy why she doesn't edit directly in FrameMaker, she might mention one or more advantages of editing on paper that technical communicators who prefer paper-copy editing often cite:

- A manuscript is more portable even than a laptop, and she can read and mark up anywhere.
- She finds reading a printout easier on the eyes than reading on-screen text.
- She feels more confident about her perceptions of the text when editing on paper, because she gets a better global sense of the whole, and she can navigate back and forth through a document more efficiently by flipping through pages than by scrolling through screens.
- She can spot more errors on paper than she can when reading on screen.

In addition, there is the question of control over the source file of the document in FrameMaker. Paula and the other writers Kathy works with at BMC may not want an editor to edit directly in FrameMaker for fear that errors will be introduced that they will not catch when reviewing the changes, or they may simply resist giving up the content owner's control over the source file of the document. BMC itself may have a policy against the editors changing the FrameMaker files because they have not found a way to manage the electronic editing process in a way that equals the security, reliability, and archival convenience of marked-up paper copy. In short, editors and the companies or clients they work for often choose to keep the editing process on paper for a variety of reasons.

In Charlene's scenario, key factors defining the workplace context are completely different from Kathy's. In entering edits to the digital text without first seeking the writer's approval and without clearly marking what changes she has made, Charlene functions like a newspaper editor who makes the final changes to a story without seeking the reporter's approval. Charlene and her team have good reasons for modifying the traditional role of the technical editor:

- Available tools for marking suggested edits in web pages without directly changing the digital files would require more time, training, and software costs than the "invisible" editing that Charlene now does.
- The collaborative nature of the process used to develop the tutorial blurs the boundaries between writer and editor and promotes a sense of team ownership of each page.
- The tutorial pages contain only brief chunks of information; the writers probably find it an easy task to compare a clean copy of an edited page to the previous version.
- If the tutorial is delivered in plain HTML, the pages can be quickly corrected if someone discovers an error introduced by the editor.

In sum, Charlene's method of editing electronically is well suited to the context created by her work team and the constraints presented by the tools and process used to accomplish their goals.

The scenarios about Kathy and Charlene illustrate contrasting endpoints of a continuum of editing practices. Although editing on paper is still the most common practice in technical communication, editing on screen has become widespread and will undoubtedly increase. Most technical communicators who edit others report that they edit others electronically at least occasionally. Untracked edits, however, are *not* the most common method of editing electronically in technical communication. Other widely used methods preserve the traditional boundaries between the roles and responsibilities of editor and writer by including some form of on-screen copyediting markup.

The purpose of this chapter is to provide a knowledgebase and framework for understanding electronic editing practices, along with an orientation to the various on-screen markup options available. I have organized the discussion into five sections:

- The first section continues the examination of different electronic editing practices encountered in technical communication workplaces.
- Sections two and three examine the advantages that electronic editing can provide and the tradeoffs that have led many technical communicators to limit their on-screen editing or to blend it with paper-copy markup and/or proofreading.
- The fourth section describes several methods of on-screen markup and describes how these methods are implemented in software applications fre-

quently used by technical communicators. Some newer on-screen editing tools are also described.

- The final section of the chapter offers a condensed guide to using Microsoft Word's Track Changes, the most widely available tool for on-screen editing. The textbook website contains additional resources for hands-on learning of e-editing options in Word and other software tools commonly used by technical communicators.

Key Concepts

For many editing jobs, editing on screen can be faster, more efficient, and involve less drudgery than editing paper copy. For other jobs, however, editing paper copy will make more sense because it fits better with the established policies, priorities, and relationships in a given workplace. The specific method an editor uses for on-screen editing is often dictated by options available in the software being used, the nature of the documents being developed, and the factors that combine in a variety of ways from one workplace to another and even from one editing job to another.

How Do Technical Communicators Edit Online?

In 1999, I conducted a survey of 992 randomly selected members of the Society for Technical Communication (STC) who had put themselves in the "writer–editor" membership category.[1] I received completed surveys from 580 STC members, 444 of whom reported that editing others was an important job function. Most of these respondents were either writer-editors (43%) or peer-editing writers (34%). Only 4% of the 580 respondents reported that their sole function was to serve in the traditional role of technical editor. Compared to results from a similar survey in 1990 (Rude and Smith, 1992), my data suggest that editing others has become more important to a wider range of technical communication jobs, while jobs in which you "only edit" have become rarer.

One item on my survey asked respondents to indicate which method of showing or suggesting edits to others they most often used. Table 6.1 presents the results.

To simplify this statistical snapshot of editing practices, I divide the respondents into two categories: those who most often mark only paper copy and everyone else. This puts about 54% in the paper-copy only group and 46% in the group that uses a variety of methods in making changes directly to digital copy. At this sample size, a 54–46 split constitutes a statistical dead heat.

Though the editing respondents were about evenly divided between those whose primary editing mode was paper-copy markup and those whose main editing mode involved electronic editing, I read the data to mean that editing paper copy must still be considered the standard practice among STC members. For

---

[1]David Dayton. 2003. Electronic editing in technical communication: Practices, attitudes, and impacts. *Technical Communication* 50.2: 192–205.

TABLE 6.1	Method Most Often Used to Show/Suggest Edits	
Method	Count (N = 444)	%
paper-copy markup	238	53.6
changes to digital copy, summary of changes	35	7.9
changes to digital copy and paper-copy markup	44	9.9
change bars in margin beside edited lines	8	1.8
automated typographic markup: change tracking	50	11.3
manual typographic markup: various methods	12	2.7
automated typographic markup: doc comparison	1	0.2
do not track edits: silent editing	34	7.7
other: both or combined digital/paper-copy methods	6	1.4
other: changes to digital copy	16	3.6

one thing, while paper-copy markup is a standard method, electronic editing methods are quite varied, as Table 6.1 shows. In addition, about one in five of those in the e-editing group include a paper-copy markup to record on-screen changes. On a related survey item about e-editing frequency, 30% of the respondents who edited electronically at least occasionally reported that their usual e-editing procedure included a paper-copy markup to track changes.

Here are some other pertinent findings from an analysis of the survey data:

- Editors and writer-editors were more likely to use electronic editing than other job-role groups. Peer-editing writers were much more loyal to paper-copy editing than were editors and writer-editors.
- Age, years of experience, and gender were not significantly associated with primary edit mode or frequency of electronic editing.
- Greater frequency of editing online documents was significantly, but weakly, associated with more use of electronic editing.

What's It to You?

These survey findings have a number of implications for technical communicators.

- *Editing others is likely to be an important part of what you do.* Over three out of four of the respondents to my survey indicated that editing others was an important job function.
- *You need to develop the skills to edit efficiently on paper as well as on screen.* Editing paper copy must still be considered the standard practice in technical communication, a conclusion also reached by Lois Johnson Rew.[2] However, in my survey, 70% of the editing respondents edited others electronically at least occasionally, and over a third edited electronically often or very frequently.

---

[2]Lois Johnson Rew. 1999. *Editing for Writers.* Prentice Hall.

- *You can expect that you will sometimes have to edit on paper when you would rather work on screen, and vice versa.* You need to look at the larger picture of how editing fits into the information development process in a given workplace and be prepared to use the method that works best for everyone involved.

People tend to frame discussions about electronic editing as an either/or issue: editing on screen versus editing on paper. For most technical communicators, however, it is a both/and proposition: both editing modes are part of their regular work routine. The following two sections will help clarify why editing on a computer has thus far supplemented, rather than supplanted, paper-copy markup for most technical communicators who edit others.

Benefits of Electronic Editing

Electronic editing can provide important benefits for technical communicators by

- providing an efficient way to work with writers at a distant location
- speeding up the editing process
- semi-automating tedious tasks
- improving job satisfaction

In this section, I briefly discuss these benefits; in the section after this one, I look at the tradeoffs of electronic editing that compel many technical communicators to combine on-screen editing with paper-copy markup and/or proofreading, or that keep them from making the computer their primary editing tool.

Working Efficiently at a Distance

When sending paper copy back and forth between editors and writers involves unacceptable time and expense, electronic editing is clearly a more efficient form of collaboration. Common sense suggests that technical communicators working at a distance from those they edit will be more likely to edit electronically than counterparts whose writers are working in the next cubicle. Results from my survey confirm this assumption. Only one out of ten in the paper-copy editing group worked at a distance from those they most often edited; the corresponding proportion for the e-editing group was one out of three. To that data, I would add that the most evangelistic proponents of electronic editing I have encountered are freelance editors who always work with colleagues or clients at a distance. Though half a world away from their writers, these editors can communicate and exchange files by e-mail over the Internet nearly as fast as they could if they were on different floors of the same building and connected to the same Local Area Network.

Speeding Up the Process

Technical communicators who prefer to edit on screen maintain that this is superior to paper-copy markup because it is faster. That claim is obviously true if the changes are made to computer files (1) without on-screen markup or (2) with an on-screen markup system that allows the writer to review the edits and quickly convert approved edits into the new text while changing rejected edits back to the original. In some situations, editing paper copy might be faster for the editor than editing on screen; however, incorporating the edits into the digital file—by the writer, the editor, or someone else—adds to the total time to complete the process. To achieve meaningful gains in efficiency with on-screen editing, the writer or other person responsible for reviewing and incorporating the edits also has to learn to use the e-editing tool proficiently.

Semi-Automating Tedious Tasks

While editing on screen can be faster than editing paper copy, the advantage cited equally often is that working at the computer makes editing less tedious. The basic find-and-replace command in document production software can make quick work of global changes concerning details of style, mechanics, formatting, or word usage—changes that would cost an editor of paper copy considerable time and monotonous labor to scan for and pencil in by hand. In addition, software that allows for the quick creation of editing macros or scripts can really speed up copyediting compared to doing the same work on paper.

Improving Job Satisfaction

Working efficiently at a distance, speeding up the editing process, and semi-automating tedious tasks are the most commonly noted benefits of electronic editing. There is another important benefit that often gets overlooked, perhaps because it is so obvious. In an interview-based study of writer-editors at a government laboratory, I found that all of the writer-editors much preferred editing in Microsoft Word over editing paper copy. In explaining their preference, they usually mentioned the three benefits I have already reviewed, but they also made statements indicating that editing on the computer made them feel better about their jobs. They associated their computer skills with professionalism and a sense of enhanced status within their organization, particularly in relation to subject matter experts (SMEs).

Tradeoffs of Electronic Editing

The benefits of electronic editing reviewed above are compelling. They are the main reasons that electronic editing has become so widely practiced in technical communication. But the benefits of any new technology usually come with tradeoffs: while important advantages are gained, some valued aspects of the old way of working are lost. In the case of electronic editing, the tradeoffs are associated with what many technical communicators perceive as the advantages of editing on paper. In this section, I discuss four of the most commonly mentioned tradeoffs associated with electronic editing:

- the lack of an on-screen markup method that most editors find to be as efficient and easy to use as paper-copy markup
- the difficulties of reading on-screen text and the attendant concerns about overlooking errors
- constraints related to the non-portability of computer workstations used to edit and the potential for hardware and software compatibility problems
- a possible increase in certain health problems related to heavy computer use

The Problem of On-Screen Markup

Many people assume that electronic editing always means the use of on-screen markup, but the results of the survey I conducted demonstrate that on-screen markup is not part of the usual e-editing procedure for many technical communicators. When editing others electronically, many technical communicators do not track the changes they make to source files, or they provide only a summary of the changes. Almost as many others make changes to digital copy and use the corresponding paper copy to record the changes—combining on-screen editing with paper-copy markup.

It is customary in technical communication to assume that edits will be tracked closely, but some widely used document production software applications do not provide an easy-to-use on-screen markup system. In FrameMaker, for example, editors can set the program to mark all changed lines with vertical bars in the margin, but the exact changes are not automatically tracked by the software. (Other methods exist for accomplishing this, as I explain later on.) Windows Help authoring programs and software tools used to edit HTML also lack an easy and efficient change-tracking option.

In their comprehensive appraisal of electronic editing, David Farkas and Steven Poltrock (1995) argue that on-screen markup will prove to be an obstacle to the widespread adoption of electronic editing in technical communication.[3] "It is hard to overestimate the importance and centrality of markup in any online editing tool," they observe. "It is how the editor works and how the document is changed. Markup is also a key means of collaborating with the author" (p. 114). To explain the loyalty of so many editors to paper copy, they note that the traditional markup system

- is easy to learn and use,
- enables speedy editing,
- clearly preserves the distinction between the original text and the editor's changes, and
- allows an editor to make extensive changes without making the copy difficult for the author to review. (p. 113)

---

[3]David K. Farkas and Steven E. Poltrock. 1995. Online editing, mark-up models, and the workplace lives of editors and writers. *IEEE Transactions on Professional Communication* 38.2: 110–117.

Microsoft Word's change-tracking mode represents the best on-screen markup tool readily available to technical communicators. It implements what Farkas and Poltrock call the "edit-trace model," which is commonly referred to as "redlining." When the change-tracking option is activated, Word marks all changes made to a text by highlighting them with formatting changes. The default setting for this option puts deleted text in red ~~strikethrough~~ letters and inserted text in red <u>underlined</u> letters. Users can select different formatting options that they feel better distinguish deletions and insertions, and multiple reviewers of a document can select different colors. (The final major section of this chapter provides some tips for using Word's change tracking efficiently.)

Farkas and Poltrock assert that on-screen markup systems like the one in Microsoft Word tend to encourage heavy-handed editing, especially if the editor uses the option to hide the change tracking while editing. In their view, a better on-screen markup system would be one that allows editors to use handwriting and voice commands to draw traditional copyediting marks on an electronic overlay of the document being edited. If the author could then review these marks and automatically reject or incorporate the changes with a click of the mouse or a voice command, such a system would achieve the advantages of using the computer to edit *and* the most important benefit of editing paper copy: a fast, easy-to-learn, easy-to-use markup system that protects the ownership rights of the technical author against editorial trespass.

On-screen markup systems like the one envisioned by Farkas and Poltrock are under development; researchers in the field of computer-supported collaborative work (CSCW) have developed and studied collaborative writing and editing tools, and Microsoft includes new handwriting options for the Tablet PC that seem a step on the way toward the editing software envisioned by Farkas and Poltrock. A tool that allows writers to automatically incorporate electronically overlaid editing marks into digital documents will someday be available; whether such a tool becomes widely used by technical communicators, however, will depend on complex sociotechnical factors in particular workplace contexts.[4]

In the meantime, Word's change tracking is the state-of-the-art in electronic editing. And yet, many technical communicators who use Word apparently resist using this markup system. From responses to open-ended survey questions and interviews with writer-editors, I learned that many technical communicators have decided not to use Word's change tracking on a regular basis. They usually cite one or both of these reasons:

- Many technical communicators do not like Word's change tracking because they find the marked-up copy difficult for writers, or even the editors themselves, to review. They often use adjectives such as "cluttered" and "messy" to describe text with typographically highlighted deletions and insertions.

---

[4]David Dayton. 2004a. Electronic editing in technical communication: The compelling logics of local contexts. *Technical Communication* 51.1: 1–16.

————. 2004b. Electronic editing in technical communication: A model of user-centered technology adoption. *Technical Communication* 51.2: 207–223.

- Writer-editors in some workplaces, such as scientific and engineering research and development organizations, often edit SMEs who are not adept users of word processing tools. Many of these SMEs do not want to learn to use the change-tracking function in their word processor efficiently. Their resistance to or misuse of the function sometimes leads the writer-editors working with them to seek alternatives, such as giving writers clean, edited digital copy along with a marked-up paper copy, or using manually inserted formatting changes to mark edits in digital copy.

Even when the markup problem is not an issue, many technical communicators resist electronic editing because of other tradeoffs they perceive in comparison to editing on paper: the difficulties of reading on-screen text, the lack of portability options for electronic editing, software and hardware compatibility issues, and health problems associated with heavy computer use. For each of these perceived downsides to electronic editing, paper-copy markup has a corresponding upside.

Reading Difficulties and Quality Concerns

Because of the relatively poor readability of on-screen text for close, careful reading, and the eye strain that many experience when reading computer screens for extended periods, many technical communicators avoid electronic editing. Many others, however, having experienced the benefits of electronic editing, mitigate this particular tradeoff by blending paper-copy proofreading with on-screen editing. Of the 310 respondents in my survey who reported editing electronically at least occasionally

- about four out of five regularly used paper copy to check elements they found difficult to edit efficiently on screen, such as complex graphics and tabular data;
- three-fourths used paper copy to perform final prepublication edits;
- two out of three specifically checked formatting details on printouts; and
- two-thirds also used paper copy to catch errors they knew they had missed while editing or proofing on-screen text.

Higher-resolution font technology is under development; in the not too distant future, the readability of our computer screens will more closely approximate the readability of computer printouts. In the meantime, for the sake of quality control, most technical communicators depend on paper-copy proofing to supplement their on-screen reading when they edit electronically. I should add, however, that it is also very common to hear technical communicators remark that for top-quality editing you need to read documents on screen *as well as* on paper, because you catch different types of errors in each medium. This is especially true for documents specifically designed for on-screen display, such as online help and web pages. Online documents must be checked thoroughly on-screen.

Portability and Compatibility Constraints

Using the computer to edit commonly imposes two other constraints, which are sometimes interrelated: a lack of portability and specific software and hardware requirements. If your writers are using FrameMaker on Unix machines, you will probably not have the option of taking your editing work home with you. Even an editor who owns a laptop suitably equipped to do an editing job will find portability constraints that would not hamper her if she were toting only a manuscript and a colored pencil. Nothing electronic that you can use for serious editing beats paper for portability. And the first "cross-platform" publishing medium was paper copy, not HTML.

The Hazards of Heavy Computer Use

Technical communicators loyal to editing on paper often argue its ergonomic advantages. When you read and mark up a piece of paper, you can assume any number of positions in relation to the text. You can easily and repeatedly change the position of your head and body and the position of the paper. This ergonomic flexibility is another important advantage of editing on paper that we lose when we edit on-screen text. And this tradeoff in versatile ergonomics can be hazardous to our health.

Computer work is not heavy labor, but it can be physically punishing nonetheless. Repetitive strain injuries such as carpal tunnel syndrome afflict all types of workers who spend many hours a week typing and clicking a pointing device. Technical communicators are no exception. Almost one in five of the 580 respondents to my survey reported that they had been diagnosed with a repetitive strain injury. More than two-thirds reported that they had experienced pain in the neck, back, or upper limbs to such a degree that they had taken off from work or consulted a health professional. One in four had experienced eye strain to this degree, and respondents in job roles requiring the most editing were the ones reporting the most eye strain.

These statistics underline the importance of enforcing good ergonomic standards in arranging our workstations and developing healthy work habits. For many technical communicators, this means switching frequently between on-screen work and work with paper.

Despite the tradeoffs that many perceive in editing at the computer, the potential benefits in speed and efficiency, especially when working at a distance, will make on-screen editing the preferred method for more and more technical communicators. In the next section, you will get an overview of the on-screen markup and annotation options available to technical communicators as of September 2004. After that, the final section provides a quick-start guide to using the on-screen markup and annotation tools most readily available to technical communication students: Microsoft Word's change tracking and comments. A more detailed and frequently updated supplement to this information is available on the textbook website.

An Overview of On-Screen Markup and Query Methods

In this section I describe these general methods for on-screen editing and identify how they are commonly implemented in software applications widely used by technical communicators:

- automated typographic markup
- manual typographic markup
- electronic overlay markup
- electronic queries

I also identify some tools for implementing on-screen markup that are not well known in technical communication but that appear to offer options worth considering.

Automated Typographic Markup

Automated typographic markup (ATM) is any on-screen markup method that automatically applies distinctive typographic highlighting (primarily under-lining, strikethrough, and changed font style or color) to show text insertions and deletions. There are two general methods of ATM: **change tracking** and **document comparison**. In change tracking, the software marks the deletions and insertions as they are made. Users have the option of seeing the typo-graphic highlighting applied as they make the changes, or they can turn off the highlighting on screen though the program continues to keep track of the changes. In document comparison, the editor makes changes directly to the file without marking the edits; afterward, the editor or writer uses a utility in the software or a standalone program to compare the original file to the edited file. The document comparison produces a third copy of the document in which deletions and insertions are highlighted typographically as in change tracking.

The exact nature of the highlighting in ATM can be adjusted to some extent to suit individual preferences and, in some programs, to distinguish edits made by different reviewers or editors. For example, if changes are going to be reviewed on screen, you could choose to show the insertions as blue underlined type and dele-tions as red strikethrough type. On the other hand, if the changes need to be easily readable in black and white, you could set the software to display insertions as un-derlined text and mark deletions with ballons in the margins, as shown in Figure 16.3. Regardless of the typographic highlighting you select to mark changes, it is wise to set the software to add thin vertical bars in the margins of each line that is changed. This makes it even easier for a reviewer to inspect the marked-up copy. (These vertical lines are called "change bars" or "revision bars.")

ATM in Word and FrameMaker

Two of the software tools most widely used by technical communicators are Microsoft Word and Adobe FrameMaker. Both offer the option of automated typographic markup; however, the current versions of these programs (Word 2003 and FrameMaker 7.0) offer options for ATM that are significantly different. Microsoft Word includes both change tracking and document comparison, and both these utilities let the user inspect editing changes one at a time; each change can be accepted or rejected with a click of the mouse. Optionally, all changes can be rejected or accepted at once.

FrameMaker, on the other hand, offers only document comparison as an automated markup option, and its document comparison utility does not allow semi-automated review and incorporation or rejection of changes. FrameMaker users can choose from two methods, both somewhat complex, for approving and rejecting changes and removing the condition tags from the accepted edits. (See Jean Hollis Weber's book *Electronic Editing* for step-by-step instructions on these methods and for an explanation of other methods for marking and reviewing edits in FrameMaker files.)[5]

The major drawback of automated typographic markup is that every single change is marked. For many editing jobs, it may be preferable that the writer review only those edits representing a potential change in meaning, style, or clarity. When that is the case and Word is the primary document-production tool, change tracking is the obvious method of choice because you can easily turn it on and off. (See the Tips and Techniques section for how to do this.) In FrameMaker, using a combination of "silent editing" for mandatory changes and on-screen markup for those changes needing approval is also possible with custom-made scripts or a third-party add-in tool such as the Track Changes plugin from Integrated Technologies (www.intech.com).

Document Comparison Programs

Technical communicators are not the only professionals to make extensive use of electronic editing. Lawyers today often use the computer to compose and edit documents collaboratively, and most legal firms use document comparison to mark and review changes in documents. Because it is critical that every change in a legal document be tracked with the utmost accuracy, many legal firms use specialized document comparison software in conjunction with their word processor. Workshare 4, a leading program in this category, works with Microsoft Word and offers an impressive set of on-screen document comparison, editing, and change-review tools. Another program used by smaller legal firms is Diff Doc from Softinterface, Inc. (For more information, see Websites for Products Mentioned at the end of the chapter.)

Document comparison programs are also available for inspecting changes made to web pages. Two such products are Araxis Merge 2000 and CS-HTMLdiff.

---

[5]Jean Hollis Weber. 1999. *Electronic Editing: Editing in the Computer Age.* Chatswood, NSW, Australia: Archer Press. Available as a PDF shareware document at www.jeanweber.com.

Document comparison of web pages can also be accomplished using Adobe Acrobat, as described under "Electronic Overlay Markup."

Manual Typographic Markup

Manual typographic markup (MTM) is any on-screen markup method using distinctive typographic characters or highlighting to set off insertions and deletions from the unchanged text. On the face of it, we might conclude that MTM would be inherently less efficient than ATM, but this is not necessarily the case. In some situations it might be more efficient to use manually applied typographic highlighting. For instance, let's imagine a writer who is a research scientist with limited patience for learning new word processing routines; she is happy to have you edit electronically but wants to review edits on paper and have you incorporate approved edits and any additional changes. With MTM, you would have more control over the typographic attributes used to highlight changes. You could adjust the highlighting method for optimal on-screen viewing while you edit, and then change the attributes of the styles applied to insertions and deletions so that the edits also show up clearly in a black and white printout. Keep in mind that though you might opt for a non-automated method of applying typographic markup for a particular job, you could still use software features like shortcut keys, menu icons, macros, and the find utility to accelerate markup and edit-review procedures.

The best way to accomplish MTM in Word is to create character formatting styles and assign them to menu icons and/or shortcut keys. For example, you could create a new character style (which is different from a paragraph style) called "Delete" that has the same settings as the default paragraph style but displays in red strikethrough letters. You could then assign the shortcut key **Ctrl + Alt + D** to this new user-defined character style. To mark a deletion, you would highlight the text you think the writer should delete and press **Ctrl + Alt + D**. The text you highlighted will change to red strikethrough characters. You could create a contrasting character style—perhaps bold blue type—called "Insert" and work with it in the same way.

To remove manually applied typographic highlighting on rejected deletions, you or the writer would select the appropriate text and press the shortcut key combination for removing all character formatting: **Ctrl + Spacebar**. Rejected insertions would be deleted one at a time as they were encountered. Finally, to automatically incorporate all accepted edits, you or the writer could use an advanced search routine to find all text with the special character formatting and return it to the default character format style. (This can be done by using the appropriate wildcards and Format options accessible in Word's Find and Replace dialog box.)

A similar MTM system can be set up in FrameMaker using either character tags or condition tags. See Weber's book[6] for a discussion of the advantages of

---

[6]Jean Hollis Weber. 1999. *Electronic Editing: Editing in the Computer Age.* Chatswood, NSW, Australia: Archer Press. Available as a PDF shareware document at www.jeanweber.com.

each method and step-by-step instructions for implementing them. (Weber also covers ATM using Microsoft Word and Lotus WordPro.)

Electronic Overlay Markup

In the final general category of on-screen markup methods, editors mark changes and write or type queries on a transparent layer superimposed on a copy of the document. A growing number and variety of software applications make this possible. They let you insert text boxes to type in queries or comments and use drawing tools to make shapes for highlighting. If you have a stylus and graphics pad, you can even mark up the on-screen text with traditional copymarking symbols and attach handwritten queries with electronic "sticky notes." The paragraphs below describe several of these tools. (See Websites for Products Mentioned at the end of the chapter for links to more information on these software applications.)

Adobe Acrobat

Available for both Windows and Macintosh computers, Adobe Acrobat lets you mark up and annotate PDF files. The reviewing/editing tool set includes a highlighter, a drawing tool, and a pen for underlining or striking through text. Users can also add pop-up sticky notes containing keyboarded text.

Because the formatting and pagination of a PDF file are fixed, the annotation layer created by Acrobat can be separated from the document as a standalone file. Multiple reviewers can email annotation files for the same PDF document to a writer, who can then import them into the document all at once or individually to inspect edits, comments, and queries. This method of electronic editing and review is popular with FrameMaker users who see unacceptable risks in letting reviewers or editors take control of the FrameMaker source files. (For a good introductory tutorial on the use of Adobe Acrobat for on-screen editing, see the *Intercom* article by Marjorie Joyce Radella listed in Further Reading.)

Overlay Markup for Web Pages

Specialty products for marking up and annotating web pages can be used to communicate suggested edits to a web designer or web development team. The tool set in iMarkup Client, for example, is very similar to the tool set described above for Adobe Acrobat. (Other such products can be found by using Google to search on the terms *annotate web page*.)

Electronic Queries

Methods for inserting queries into digital documents parallel the three markup methods: automated methods, manual methods, and electronic overlay annotation. I have already described the sticky-note tool that has become a standard feature of electronic overlay tools. In this section, I will describe three other methods for communicating queries linked directly to the digital text being edited. The first two are automated and manual methods for inserting queries into digital text, and the third is a new product that allows queries to be linked to specific points in a hypertext draft of a document created from Word or FrameMaker.

Utilities for automating the insertion of comments, which are then displayed apart from the text and optionally viewed and/or printed, have become a standard feature of most document production tools used by technical communicators. Word's Comment function and FrameMaker's Comment marker, for example, make it easy for editors to insert queries and even easier for writers to review them. Nevertheless, some editors prefer manual methods in which they type directly into the text, framing a query with a series of rarely used characters or applying special formatting attributes to make the comments stand out.

One obvious way to accomplish the manual query-insertion method in FrameMaker is to use the default Comment condition tag or design another condition tag for this purpose. To see how the document will look without the comments, the Comment tag can be set so that queries do not show. You can accomplish the same on-screen query method in Word by creating a character style called "Comment": apply the hidden text attribute to it, along with whatever other typographic attributes you want to make your queries stand out from the text. You or the writer can then use Show/Hide on the Standard toolbar to view the text with or without queries showing. (To show or hide only the queries, and not all nonprinting characters, click Tools > Options > View and check or uncheck the Hidden text box.)

For inserting a query in HTML code, you can use a comment tag: begin with <!-- and to mark the end use -->. The message will not be viewable in a browser and, thus, will not affect the layout of the page. You can span lines with your query, but be careful not to use the right-pointing angle bracket (>) until you want to end the query because some browsers could read one of these, even without the preceding hyphens, as the end of the comment container.

Policies aimed at ensuring quality and accountability in many technical communication workplaces deny editors access to source files for documentation created by SMEs and technical writers. The rationale for such policies is that editors making changes directly in source files might inadvertently introduce errors that the technical writers might not detect, or they might change the formatting of the document in unacceptable ways, causing additional formatting work. In such workplaces, documentation is usually edited on paper or using electronic overlay markup—most often with Adobe Acrobat. However, a new product called FinalDraft offers an alternative to the use of Acrobat for review and editing of drafts.

Using FinalDraft, technical writers working in Word or FrameMaker can convert a document into a hypertext version that preserves the page design, formatting, and visuals of the original. The hypertext file can then be emailed to editors and reviewers, or placed on a shared network directory, including a shared Internet directory. Reviewers and editors use a free browser plugin for Microsoft Internet Explorer to open the hypertext document and add comments linked to specific places in the text. Because reviewers do not have to have a copy of the FinalDraft application to be able to participate in the electronic review/edit process, the cost of using FinalDraft can be substantially less than the cost of using Adobe Acrobat to set up a similar electronic editing process.

Change Tracking in Word: Tips and Techniques

This final section of the chapter is a quick-start guide for using change tracking in Microsoft Word for Windows. The basic procedures can be duplicated in Word for the Mac, but in many cases the exact menu commands and corresponding shortcuts will be different. In this condensed presentation, I describe what I think is the quickest and most convenient way to execute a procedure in Word 2003 for Windows. In-depth guides to change tracking in Word explain more than one way to execute procedures. You can find alternate commands and a more complete presentation of Word's change-tracking options in Word's Help system, in countless books on using Word and Microsoft Office, and on numerous websites. In Further Reading, I have included URLs for tutorials on the Web that walk users through the use of change tracking, complete with screenshots for particular procedures. Finally, the textbook website includes links to online tutorials and reference material that will help you learn to use Word for editing documents so that the writer can easily review your suggested changes and accept or reject them.

Configuring and Activating Track Changes

Tracked changes and queries inserted using Insert > Comment are linked to the user information recorded in Word's setup. Thus, before you start Track Changes, you will want to make sure your user information is set the way you want it to appear. Go to Tools > Options > User Information. Before leaving that properties sheet, click on the tab labeled "Track Changes." On this sheet, you can change the typographic highlighting used to mark deletions, insertions, and text whose formatting you change. You can also turn on/off change bars and specify where they should be placed.

Though you can activate Track Changes and access configuration options by way of Tools > Track Changes, a shortcut is always present in the status bar at the bottom of the Word interface. Just to the right of the line (Ln) and column (Col) counters are four little boxes with grayed-out letters. The second one, TRK, is a shortcut to Track Changes:

- Double-click TRK to turn on change tracking; the TRK turns dark. To turn off change tracking, double-click TRK again.
- Right-clicking TRK displays a pop-up menu showing three choices:

  1. **Track Changes** simply turns Track Changes on or off.
  2. **Reviewing Pane** splits the screen horizontally to show comments and changes in a separate area at the bottom.
  3. **Options** brings up the same properties sheet accessed by way of Tools > Options.

Whenever you turn on Track Changes, the Reviewing toolbar should appear at the top of your screen—or wherever you last positioned it. To force the appearance of the Reviewing toolbar, click View > Toolbars > Reviewing. After a few minutes of exploring the configuration options available from the Show menu on the Reviewing toolbar, you will be ready to start tracking changes. Before you do, though, read through the tips below. If you need additional orientation to using Track Changes and Comments, use the F1 key to activate the context-sensitive help in Word, or use Google to find additional information on "Word track changes," or visit the textbook website and explore the area related specifically to this chapter.

Tips for Using Track Changes in Word

Do not use Track Changes for corrections and other changes that the writer and you have agreed do not need to be reviewed because they are mandatory. Such changes might include capitalizing or using numbers according to the style sheet.

- Having Word mark every little change in a document that needs a lot of copyediting creates a markup with so many insertions and deletions that reviewing the changes becomes a difficult and unpleasant chore. To quickly turn Track Changes on and off as you edit, you can double-click the TRK box in the status bar, as explained above.
- To avoid having to take your hands from the keyboard to turn on/off Track Changes, you can use the default shortcut key combination **Ctrl + Shift + E.** In my computer, I assign the more convenient **Alt + R** as another Word shortcut to the toggle command for Track Changes. To do this, click: Tools > Customize > Commands > Keyboard. Pick All Commands at the bottom of the list of menu categories and scroll through the list to select ToolsRevisionMarksToggle. Put the cursor in the Press new shortcut key text-entry box and press the new combination of keys you want to use as a shortcut to turn Track Changes on/off.
- If your workplace or client requires a record of every change, remember that you can always use the Tools > Compare and Merge Documents command at the very end of the editing process. You could compare the final edited document to the original, creating a copy that displays all the changes that were accepted.

Be wary of using Track Changes for editing passes that require substantial reorganization, rewriting, and/or reformatting.

- When you move a block of text with Track Changes activated, the text is marked as deleted in its original location and marked as inserted in its new location; there is no indication in the deleted text that it has been moved, and no indication in the inserted text that it comes from another place in the writer's original.
- You can set Track Changes to put typographic highlighting on formatting changes—like changing a word from bold to italics—but these could make review of the changes unnecessarily confusing for a writer already struggling to make sense of numerous insertions, deletions, and relocated text. It might be better to suggest formatting changes with comments.
- For a heavy edit, a digital-copy markup may be no easier for a writer to decipher and deal with than a paper-copy markup. When a document

needs many substantial changes, your best approach may be to sit down with the writer and discuss a strategy for a complete revision. If working remotely, you could use extensive notes by way of Insert > Comments to guide the writer through a substantive revision. Alternatively, if the writer is willing, you could edit silently and deliver a "clean" document without markup. A record of the changes, if needed, could be produced using Tools > Compare and Merge Documents.

To insert a query using Insert > Comment, turn off Track Changes.

- Using Insert > Comment with Track Changes activated makes the query a tracked change, which removes some flexibility and may cause confusion when your writer reviews the queries and then tries to remove them.
- The shortcut for inserting a query is **Ctrl + Alt + M**. You can change this shortcut using the same method described above for turning on/off Track Changes. The command you look for in the Customize Keyboard dialog box is listed under Insert category: It's called InsertAnnotation.
- To make sure your comment is clearly visible to your writer, use the cursor to select some or all of the text to which your query refers. Word 2003 will mark the selected text with highlighting. If your writer is likely to review the comments on a printout, the highlighting may not show up clearly. To ensure that clearly visible highlighting appears on commented text in a printout, you may need to manually select each passage of commented text and apply the highlight icon from the Formatting toolbar.

Consider how the tracked changes will be displayed to the writer. Make changes in a way that will make reading the markup easier.

- Change entire words or phrases rather than one or two characters at a time. Consider, for example, a sentence taken from the marked typescript sample shown in Figure 4.1. The sentence is shown below in three forms: (1) unedited, (2) edited by changing single characters, and (3) edited by changing words. Notice that in the case of "proof-reader's," the deletion of the hyphen is not visible when marked with strikethrough in Track Changes.

  1. The costs increase geometrically thereafte, each error costs not just the editor's or proof-reader's time but also the production specialist's time.
  2. The costs increase geometrically thereafter~,~; each error costs not just the editor's or proof-reader's time but also the production specialist's time.
  3. The costs increase geometrically **thereafte,**thereafter; each error costs not just the editor's or **proof-reader's**proofreader's time but also the production specialist's time.

Be careful about inadvertently creating extra word spaces; use a cleanup procedure in the final proofing stage to eliminate any that remain.

- Extra word spaces will mysteriously appear in a document once tracked changes have been reviewed and removed. You can avoid many of them by being careful and consistent in how you apply deletions and insertions.
- In your final editing pass, you can search for extra word spaces using the Find and Replace function. It's best to inspect each instance of double

word spaces before authorizing the replacement with a single word space. The writer may have used multiple word spaces for some legitimate purpose in the document.

If you work with writers who have not used Track Changes, prepare a set of step-by-step instructions for them to follow when reviewing your edits and queries. Or take a few minutes to show a writer who works in your location how to use Track Changes.

- You can develop the instructions in a separate Word document that you deliver with the first edited document you return to a writer new to Track Changes. You could adapt one or more of the tutorials on the Web listed in Further Reading, or simply refer your writers to one of the tutorials, perhaps emailing them the URL.
- Before saving an edited document to send to its writer, make sure Track Changes is turned on and the Reviewing toolbar is showing. The easiest way to activate any toolbar is by right-clicking on any blank area next to the icons at the top of the screen; just select the toolbar you want from the dropdown menu that appears.

After a writer has reviewed your suggested edits, use Word's document comparison utility to see what edits the writer did not accept and any further changes he made to the document.

- Save a copy of the edited file you delivered to the writer. Accept all changes using the Reviewing toolbar: click the down-arrow beside the checkmark icon and select "Accept All Changes in Document."
- Keep the file incorporating all your suggested edits as the active document and click Tools > Compare and Merge Documents. Select the document returned by the writer as the one to be merged. Click the arrow next to Merge, and then select the option to display the results of the comparison in the writer's most recent document—the active document you began with. Click Merge.
- You can now use the Reviewing toolbar to step through each change between the document you sent the writer and the one the writer returned to you.

Using Your Knowledge

One of the ways you can add value to your work in editing is to develop competence at electronic editing as well as paper-copy editing. To edit electronically, you will adapt available text production tools, such as those in Microsoft Word, FrameMaker, and Adobe Acrobat, for emending the text and for communicating with the writer. All of these programs, and others like them, allow an editor to maintain a record of substantive changes and to interact with writers through queries and comments. You can learn the tools through online tutorials or workshops, but you can also explore possibilities of the software on your own, looking for ways to improve the efficiency and accuracy of editing while still respecting the need to communicate with the writer and others on the production team.

Acknowledgments

I am grateful to the Society for Technical Communication for a research grant that funded the survey whose results I partially summarize in this chapter. For more information on the research, see the area of the textbook website dedicated to this chapter. Condensed reports on the results of the survey and qualitative portions of the study have been published in the STC journal *Technical Communication,* and these are listed in the bibliography below.

I am grateful to two professors and their students who tried out a draft of this chapter in undergraduate technical editing classes: Dr. Kelli Cargile-Cook and her students at Utah State University and Dr. Carolyn Rude and her students at Texas Tech University. The students read and discussed drafts of the chapter and provided helpful feedback. Finally, thanks to Matt Sullivan for generously sharing his time and expertise about on-screen editing options in FrameMaker.

Further Reading

Einsohn, Amy. 2000. *The Copyeditor's Handbook: A Guide for Book Publishing and Corporate Communications.* University of California Press.

Hart, Geoff J. 2004. Implementing onscreen editing. *Intercom: The Magazine of the Society for Technical Communication* 51.5: 36–37.

Kelly, Shauna. 2004. How does track changes in Microsoft Word work? www.shaunakelly.com/word/trackchanges/HowTrackChangesWorks.html.

Microsoft. 2000. Track changes. *Microsoft Word Legal User's Guide.* education.socialaw.com/wordguide. Select Track Changes from left-side menu.

O'Keefe, Sarah. 1999. *FrameMaker 5.5.6 for Dummies.* IDG Books Worldwide.

Petersen, Judy H. 2000. Online editing: Minimizing your turnaround time. *Intercom: The Magazine of the Society for Technical Communication* 47.3: 9–11.

Radella, Marjorie Joyce. 2000. Graphic electronic editing. *Intercom: The Magazine of the Society for Technical Communication* 47.9: 19–21.

Ruby, Jennie. 1999. *Electronic Editing with Microsoft Word.* IconLogic.

Websites for Products Mentioned

Adobe Acrobat and FrameMaker: www.adobe.com
Araxis Merge 2000: www.araxis.com
ComponentSoftware CS-HTMLdiff:
    www.componentsoftware.com/Products/HTMLDiff
Diff Doc: www.softinterface.com
FinalDraft: www.quadralay.com
iMarkup: www.imarkup.com
Microsoft Word: www.microsoft.com/office/word
Workshare 4: www.workshare.com

Discussion and Application

1. Use a text file that your instructor provides or one of your own papers to explore track changes, document comparison, and comments. If you are using a word processor other than Microsoft Word, look for the parallel functions. Save the original document, and then save the one that you edit in a separate file with a different name.

   Option 1: Follow all the tips in this chapter, beginning with the section "Change Tracking in Word: Tips and Techniques" (page 98).
   Option 2: Focus on the three techniques described below.

   a. Open the document. Double-click on the TRK box at the bottom of the screen to turn on Track Changes. Edit some sentences and observe what happens. Double-click on the TRK box to turn off Track Changes and observe what happens. Use the dropdown menu in the Reviewing toolbar to change the selection from "Final Showing Markup" to "Final." (If you are working on a Mac, find the corresponding menu and toolbar options to complete this exercise.)

   b. Turn off Track Changes. Use the comment function to query the writer about a puzzling point, ask for more information, or seek approval for an editorial change. Drag your cursor over the words in the file best suited to posing your query. Turn on the reviewing toolbar (View > Toolbars > Reviewing), and select the Insert Comment icon. Alternatively, select Comment from the Insert menu.

   c. Compare the edited version of your file with the original. First, on the Reviewing toolbar select the dropdown arrow beside the checkmark icon. Select Accept All Changes in Document. Save the document, but keep it open. Search the online help in Word to find how-to instructions for the Compare and Merge Documents function (found under the Tools menu). Experiment by merging your edited document into the original and vice versa.

2. Go to education.socialaw.com/wordguide, select Track Changes from the left-side menu, and work through the tutorial.

3. In your own words, summarize the issue of editing without a record of the changes. In what circumstances might it be appropriate to change the file directly? When should you mark the changes?

4. Read about one or more of the document comparison tools for web pages, using the URLs listed above in Websites for Products Mentioned. (Araxis and CS-HTMLdiff are two such tools.) Write a memo to your professor describing these tools and analyzing their use for editing a website compared to the electronic overlay method used by iMarkup Client. Identify the advantages and drawbacks of each method. State which method you would prefer to use and explain why.

Basic Copyediting

CHAPTER 7

Basic Copyediting: An Introduction

Copyediting bridges writing and publication. A writer focuses on content but may not pay attention to details, such as hyphenation and capitalization, and to page design. The copyeditor adds value to the content by making the text correct, consistent, accurate, and complete. The copyeditor also marks the text for page design. These interventions, reflected in publication, help readers understand and use the published document.

The copyeditor works to meet these standards for a quality publication:

- correct: spelling, grammar, punctuation
- consistent: spelling, terms, abbreviations, numbers (spelled out or figures), capitalization, labels on illustrations, matching of numbers and titles on the illustrations with the references in the text, visual design, documentation style, colors, icons
- accurate: dates, model numbers, bibliographic references, quotations (accurately restate the original), hypertext links
- complete: all the parts are present, all the illustrations that are referred to are in the document, sources of information are identified and acknowledged, all hypertext links lead somewhere

In addition, the copyeditor provides instructions about how to prepare the text for its final form. These instructions relate both to the words (whether to hyphenate, where to put punctuation) and to the form (whether to italicize, where to indent). If the document is transmitted by paper from the editor to the production department, the copyeditor marks the page using symbols for specific functions, such as insert, delete, or italicize. (See Chapter 4.) The process of marking the choices is called markup. If the document is transmitted electronically, the copyeditor inserts the choices into the digital copy. (See Chapters 5 and 6.)

This chapter provides an overview of copyediting and guidelines for copyediting books or manuals, illustrations, and online documents. Other chapters in Part 3 of this textbook review copyediting for consistency, spelling, grammar, and punctuation—attributes of all documents, whether verbal or visual, printed or online.

Key Concepts

Copyediting aims for a correct, consistent, accurate, and complete document. Copyediting requires attention to the details of the text, which may be neglected in writing and comprehensive editing. Good copyeditors know grammar, spelling, and conventions of style. They read closely and can tell when a change in punctuation or wording might change meaning. In addition, they are alert to possible ambiguities in language and are willing to check printed and online resources and to query the writer to verify their choices. The editor directs information and questions about content to the writer but directs information about page design and about details of capitalization, hyphenation, and other mechanics to the production staff.

Making a Document Correct and Consistent

A correct document conforms to the standards of grammar, spelling, and punctuation of the language in which it is published. These standards are described in dictionaries and style manuals. A consistent document avoids arbitrary and confusing shifts and variations in the use of terms, spelling, numbers, and abbreviations. It is also predictable in its visual features, such as placement of elements on the page or screen.

Copyediting to make the document correct and consistent is widely understood by writers and managers to be the essential function of editing. Errors in language use are more apparent to readers than are errors of argument or organization. Errors distract readers from the content and can cause misreading of the content.

Chapters 8, 9, 10, and 11 review consistency, spelling, grammar, and punctuation in detail, but the following list identifies some basic standards:

- Groups of words punctuated as sentences are complete sentences.
- Punctuation is complete: sentences end with punctuation; quote marks and parentheses are closed.
- Subjects and verbs agree in number.
- Pronouns agree with their referents.
- Modifiers attach logically to the word or phrase they modify (they do not "dangle").
- Words are spelled and capitalized correctly and consistently.
- Numbers are spelled out or in figures according to a plan.
- Identifying information, such as running headers or navigation menus in online documents, is in the same place on every page or screen.

Technical documents are conservative in their use of language and require the highest standards of copyediting for correctness.

Making a Document Accurate

Accuracy refers to content, while *correctness* refers to the use of language. Errors in content, such as using the wrong term, date, or model number, can make a document partially or totally worthless to readers, legally incriminating, and even dangerous. Numbers and illustrations as well as words require editing for accuracy:

- quantitative data: model numbers, formulas, calculations (such as addition of figures in a column), measures, dates
- words: names, titles, terms, abbreviations, quotations
- illustrations: labels, cross-references, callouts
- organizational information: table of contents, index, menus and links in online documents

Accuracy errors can occur even after a comprehensive edit or technical review because often the editor and reviewer have focused on the larger text structures—the argument, the meaning, the organization—and have only scanned details.

Paragraphs are easier to copyedit than quantitative data and names and titles, yet errors are as likely to occur in tables, reference lists, and display text such as headings. A good editor checks simple calculations, such as of the total of a column of numbers (just as you would look up the spelling of an unfamiliar word in a dictionary). If you discover an error, you should query the writer to determine whether the addition or one of the figures is wrong.

To check for accuracy, you must be familiar with content and alert to meanings. Even if you are not a subject matter expert, you can spot probable errors. Here's an example: in one long article on relationships between people, the writer referred to "dynamic" relationships. However, the quotations from other sources on the subject referred to "dyadic" relationships (relationships of dyads, or two people). The copyeditor wondered whether "dynamic" and "dyadic" relationships differed or whether the writer used one wrong term. Even without knowing advanced psychology, the copyeditor could spot an apparent discrepancy.

Not knowing for sure the meaning of the terms as used in the article, the copyeditor would have been premature in changing one term or in querying the writer immediately. A change could introduce an error. Frequent or unnecessary queries are irritating and raise questions about the copyeditor's competence. Instead, she looked for other text signals about meaning, such as parenthetical definitions of the terms or some other explanation of the difference. As she checked a quotation with the original source, she noted that the writer had substituted "dynamic" for "dyadic" in the quotation. With this evidence, her query to the writer could be specific. The copyeditor might also have consulted a psychology text index or glossary or a specialized dictionary of psychological terms to find definitions.

Breaks in patterns, facts that seem to contradict logic, and omissions are clues to potential errors. A reference to a photographic process in 1796 raises a question about the date—you suspect (even if you do not know) that photography came later in history. The inconsistency in the following model numbers should send the editor to the product catalog or even to the products themselves if they are accessible:

JC238–9Y

JN174–2Y

R2472–3

An error in one type of text alerts you to pay close attention to all instances of similar text. For example, a reference list entry that cites a date and an author's initial(s) incorrectly is a clue to look carefully at the other entries. The error in one entry suggests the possibility that the writer was not careful in compiling the list. Likewise, an error in addition in one column of figures in a table requires a check of the other columns.

You rely on careful, alert reading more than on rules you can learn in a handbook when you copyedit for accuracy. Do not dismiss your quiet hunches that something may be wrong; do not take for granted that simply because a reputable writer and editor have previously reviewed the text that it is free from errors—nor that they are incompetent for having missed errors. They were concentrating on other text features. Be especially careful in reviewing numbers and dates, reference citations, and headings. But be cautious about changing terms because you can easily change meaning. Check first with the writer.

Making a Document Complete

A complete printed document contains all its verbal and visual parts, including front matter and back matter. The copyeditor provides the quality control check so that a document doesn't get published lacking its index or a table cited in the text. The copyeditor also checks that all the pages are included and in order, including blank pages, before the typescript or camera ready copy goes to the printer. (*Camera ready* means that the pages have been printed on a laser printer and will not be modified before they are reproduced.) The document includes identifying and contact information, such as a mailing address and website address.

A complete website includes its home page, all the content pages that its menus promise, and any linked attachments, such as forms or supplementary documents. Each page includes identifying information and navigation aids, including menus to let users find other pages in the site.

Parts of a Book, Manual, or Long Report

Books, manuals, and other long printed documents have three major sections: the preliminary pages or front matter, the body, and the back matter. Within these major divisions are subparts.

The document plan establishes what parts a document will include. The copyeditor does not decide whether to include abstracts or a glossary but makes sure that the specified parts are included.

Only long documents will include all the parts identified in the discussion that follows here. Simple documents contain only some of these parts. A part should be included only if it is functional. A comprehensive style manual, such as *The Chicago Manual of Style,* 15th edition, offers guidance on how to prepare these parts and information on other, less common parts, such as dedications.

Preliminary Pages

The preliminary pages appear before the text begins. All books, manuals, and long reports should include title, contents, and copyright pages. Other preliminary pages depend on the nature of the publication. Page numbers of preliminary pages use roman numerals, with the first page after the cover being page i. Blank backs of pages in a document printed front and back are counted in the numbering even if the number does not appear on the page

Cover	Identifies the title and subtitle and may identify the author or editor and publisher. On reports, it may also include the date and other publication information.
Half-title page	Includes only a short title or the main title without a subtitle, immediately inside the cover. The type is generally smaller than it is on the title page, and the back of the page is usually blank.
Title page	Includes the title and subtitle; may include the name of the author or editor. Title pages of reports generally contain more information; for reports or other short documents, the cover and title page may be the same page.
Copyright page	Generally falls on the back of the title page; includes the copyright symbol, year of publication, and copyright holder. Here is an example:
	© 2006, Allyn & Bacon All rights reserved Printed in the United States of America
	No periods appear at the end of the lines. The copyright page also includes the address of the publisher and the ISBN (International Standard Book Number) for commercial publishers. This number identifies the country of publication and the publisher according to an international code.
Table of contents	Lists main divisions and the page numbers where the divisions begin; may be labeled simply "contents."
List of tables and figures	Lists the title, number, and page number of the illustrations; tables and figures may be listed separately and on separate pages as a "list of tables" and a "list of figures."
List of contributors	In a multiauthor book or periodical, lists the authors' names and other identifying material, such as position and place of employment. Alternatively, this information may appear for each contributor on the first or last page of the article or as a list in the back of the document.

Foreword	Written and signed by someone other than the author or editor; introduces the document by indicating its purpose and significance. It is more common in academic and literary texts than in technical documents.
Preface	Written by the document's author or editor; introduces the document by indicating its purpose, scope, significance, and possibly history. It often includes acknowledgments.
Acknowledgments	Usually incorporated into the preface; names people who have helped with the publication, especially people who have reviewed drafts or provided information.
Executive summary	Summarizes contents, including conclusions and recommendations, for readers who are decision makers.

Body

The first page of the body of the document is numbered page 1 (arabic numeral).

Chapters; parts	All should be present, in order; all pages for each subdivision should also be present and in order.
Titles, chapter numbers, author names	The contents and pattern of identifying information should be standard for all the subdivisions.
Abstract	Summarizes or describes the contents of the chapter or subdivision at the beginning; a standard part of academic publications, including research reports; usually paragraph length.
Illustrations	Tables and figures. Illustrations are generally integrated into the chapters, though supplementary illustrations may appear in the appendix.
List of references	Identifies the sources mentioned in the chapter or division. May be called "Works Cited" or "References," depending on the documentation style. May appear at the end of each chapter or in a reference section at the end of the book.
Half-titles	Sometimes placed between major parts of the document (groups of related chapters); identifies the part title.
Blank pages	May be inserted in the typescript that goes to a printer to signal blank pages in the printed document. These blanks will probably be *verso* (back or lefthand) pages, used when all chapters or divisions begin on a *recto* (righthand) page and no text spills onto the preceding verso page. When editing electronically, insert page breaks for the blank pages, to keep the automatic page numbering accurate.
Running heads	Headers in the top margin that name the chapter, author, or part title. They help people find their place in the book; they may differ on the recto and verso pages.

Back Matter

The back matter contains supplementary information. Page numbers continue sequentially from those in the body, in arabic numerals.

Appendix	Provides supplementary text material, such as research instruments (for example, a questionnaire), tables of data, or troubleshooting guides. Each appendix is labeled with a letter or number and title (for example, Appendix A: Survey of Users).
Glossary	Defines terms used in the document; a mini-dictionary that applies just to the document and its specific topic.
References	Includes the complete publication data for the materials cited in the document, unless this information appears after each chapter.
Index	Lists the key terms used in the text and the pages on which they are discussed or referred to; helps readers find specific information.

Parts of a Website

In "Writing for the Web" (see Further Reading), Ginny Redish identifies four types of website pages:

Home page	Identifies the site, tells users what the site is about, helps users move to the information they want.
Scan, select, and move on	Menu pages below the home page when the navigation possibilities are too numerous to list on the home page.
Content pages	Include the information that users seek.
Forms pages	Ask users to give or verify information.

Comprehensive editing and the document plan determine what content pages, menu pages, and forms pages need to be included, but the copyeditor checks that the specified pages are included. For example, does the site include a FAQs page? Should it? Do all the pages have navigation options that let readers know where they are in the site?

Copyediting Illustrations

Illustrations (figures and tables) require copyediting for the same reasons that text does: to make the illustration correct, consistent, accurate, and complete. If the illustrations are camera ready, the copyeditor also reviews reproduction quality and readability. The marks used in copyediting illustrations are the same as the marks for text. The symbols identified in Chapter 4 for deletion, insertion, transposition, alignment, and type style work the same for illustrations as for prose.

The following list identifies the goals and tasks in copyediting illustrations:

Correct and Consistent
- Check spelling, punctuation, capitalization, and hyphenation.
- Check the title of an illustration for a match with the contents. (If the title promises records from 2004 to 2008, the illustration cannot stop with 2007.)
- Compare references in the text with the illustration. The illustration must offer what the reference promises.

- Establish patterns for placement of callouts and titles. Align callouts if possible.

Accurate
- Add columns of numbers to check the totals.
- Visually measure segments of graphs to determine whether they are proportional to their numeric values. If they look wrong, measure with a ruler or protractor, or check that the right numbers have been put into the computer program that created them.
- Compare line drawings with the actual product if you suspect distortion.

Complete
- Check that all illustrations mentioned in the text are present and that the numbers go in sequence. (A gap in numbers suggests a missing illustration.)
- Make sure identifying information is present:
  parts on illustrations
  base measures (such as percent, liters) on tables and graphs
  meanings of shading, color, or line styles
  sources of information
  labels, numbers, titles
- In online documents, check that links from images are working. All pictures should have an alternate text equivalent so that users with low vision will know the content.

Readable
- Eliminate excess ink—lines that aren't needed to separate rows and columns, measures (such as %) after each item in a column rather than in a column head, superfluous callouts.
- Use space to distinguish rows and columns.
- Align columns of numbers on decimals or imagined decimals.

Parts of Illustrations

The words attached to illustrations allow for cross-reference, identify and explain content, and provide information about sources. See Figure 7.1 (p. 115) for examples. This information may enable the reader to understand the illustration without reading the text.

Labels, Numbers, and Titles

Illustrations are classified and labeled as either tables or figures. A table presents information in tabular form (in rows and columns). The information in the rows and columns may be numbers or words. All other illustrations are considered figures, whether the illustration is a photograph, line drawing, bar graph, flowchart, line graph, or other graph.

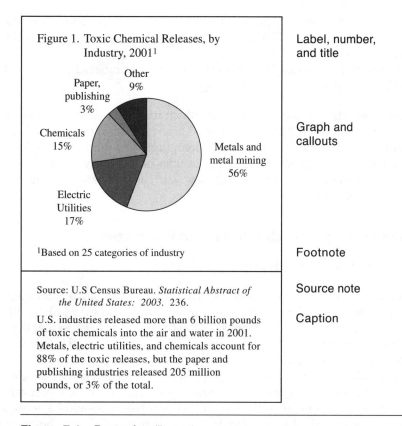

Figure 1. Toxic Chemical Releases, by Industry, 2001[1]   Label, number, and title

Other 9%
Paper, publishing 3%
Chemicals 15%
Metals and metal mining 56%   Graph and callouts
Electric Utilities 17%

[1]Based on 25 categories of industry   Footnote

Source: U.S Census Bureau. *Statistical Abstract of the United States: 2003.* 236.   Source note

U.S. industries released more than 6 billion pounds of toxic chemicals into the air and water in 2001. Metals, electric utilities, and chemicals account for 88% of the toxic releases, but the paper and publishing industries released 205 million pounds, or 3% of the total.   Caption

Figure 7.1   Parts of an Illustration

Labels. Labels ("table," "figure," or "step") and numbers enable easy cross-reference. Small and informal illustrations may not require labels and numbers if they appear right where they are mentioned and are not referred to again.

A consistent pattern of labeling illustrations should be established, and style guides may help. The American Psychological Association, for example, requires table titles and numbers at the top of the table and figure titles and numbers at the bottom. Not all documents in other disciplines will follow this pattern, but readers should be able to anticipate where to find identifying information in illustrations. Record your choices about placement of labels and titles on your style sheet to encourage consistency.

Labels on the illustrations themselves must be consistent with the references to them in the text. The reference to Figure 1 in the text must refer to the illustration actually labeled "Figure 1."

Numbers. Illustrations are numbered sequentially through the section or document, but tables and figures are numbered independently. Thus, a document may include a Table 1 and a Figure 1. Double numeration is common in documents with multiple sections. "Table 12.2" indicates that the table is the second numbered table to appear in Chapter 12.

Illustrations of steps in a process should be numbered if the text refers to steps by numbers.

Titles. Titles identify the content of the illustration in specific terms. For example, "Sales of Chip #A422, First Quarter 2005," gives more information than "Sales." Do not name the type of illustration (photograph, line graph) in the title (the type is evident from the illustration itself), but titles of line drawings and photographs may require information on point of view, such as top view, side view, cross section, or cutaway view.

Callouts, Legends, Captions, and Footnotes

Other identifying elements may be helpful.

Callouts. A verbal identifier, such as a part name, on an illustration is known as a *callout*. Callouts supplement the visual information, and they link the illustration with the text. Use callouts selectively. If readers do not need to identify a part or procedure, delete the clutter of unnecessary words. Query the writer, however, before you make such decisions about what is necessary.

The callouts on the illustration must match the terms used in the text. If a part is a "casing" in the text, readers will search the illustration for a part called a casing, not a cover. Generally all parts that the text names should have callouts, and all the parts identified with callouts on the illustration should be referred to in the text.

The callout should be next to the part or procedure it identifies. If necessary, a straight line can connect the part and the callout. If the object has major and minor parts, that structural relationship can be shown by callout placement or type style. For example, all major parts might be identified on the left of the object while minor parts are identified on the right, or major parts could be named in capital letters while minor parts are set in lowercase letters. If the object has many parts crowded together, the parts may be numbered, with a numbered list of parts following. This type of callout is common in an owner's manual that identifies all parts of the mechanism.

For aesthetic purposes and also for readability, use a standard typeface and size consistently for callouts on all illustrations in the document. The type must be large enough to be read but not so large as to overwhelm the illustration. Alignment of a vertical series of callouts, when practical, helps give the illustration a neat appearance.

Column and row headings. Headings in a table identify contents in the same way that callouts do for figures. When the cells of a table display quantities, the unit of measure (for example, dollars or inches) in a heading eliminates the clutter of repeated measures in each cell.

Legends. A legend explains shading, colors, line styles, or other visual ways to distinguish the elements on a graph from one another. The legend is a list of these variants with their meanings. The legend generally appears in a corner of the illustration. Because legends are separate from the elements they identify, they require an extra step in processing for readers. Label the components of the illustration, such as segments in a pie graph, when you can instead of depending on a legend.

Captions. Captions explain in a few lines the significant features of the illustration. The caption efficiently presents essential points of information; long and complex interpretations belong in the text itself.

Footnotes. Explanatory and source notes explain or give credit. A superscript number or letter identifies a footnote. The note itself appears at the bottom of the illustration, above the label, number, and title if these are also placed at the bottom.

If the information in the illustration or the illustration itself comes from another document, cite the source, including author, title, publisher, date, and pages, at the bottom of the illustration. If the title is at the bottom, place the source note above the title. If an illustration is reprinted from copyrighted material, get written permission to reprint. Add to the source statement the words "Reprinted with permission" or other words that the copyright holder cites. This statement also appears above the title but below the explanatory footnotes. This requirement applies to illustrations you obtain or use online as well as to printed illustrations.

Figure 7.1 illustrates the various parts of an illustration. There is no legend because the callouts identify the segments of the graph.

Placement of Illustrations in the Text

The text should refer to the illustration before the illustration appears so the readers will know what they are looking at and why. Illustrations should appear as soon after the first mention as is practical, given the limitations of page size and page makeup. If readers have to flip pages or scroll through screens to find the illustrations, they may not bother, and the illustrations will be wasted. The reference to the illustration may be incorporated into the sentence ("Table 3 shows that . . . ") or placed in parentheses. If the parenthetical reference is incorporated into the sentence, the period for the sentence follows the closing parenthesis. If the parenthetical reference stands alone as a sentence, the period is inside, and "See" begins with a capital letter.

> The paper and publishing industries account for 3% of the toxic waste released by all industries (see Figure 1). [cross-reference as part of the sentence; period outside parentheses]

> The paper and publishing industries account for 3% of the toxic waste released by all industries. (See Figure 1.) [cross-reference in a separate sentence; period inside parentheses]

Illustrations should face in the same direction the text faces; readers should not have to flip the document around to read the illustrations. If the illustration is too wide to fit on the page, it may be turned sideways (landscape orientation). Sideways illustrations on facing pages should face the same way so that readers have to turn the document only once. If an illustration appears on a verso (left-hand) page, place it with its bottom in the gutter (by the binding). On a recto (righthand) page, place the top of the illustration in the gutter. Readers will rotate the document clockwise a quarter turn to see any illustrations displayed at right angles to the text.

Instruction manuals often use a side-by-side arrangement of illustrations and text, with the illustration on the left and the explanation on the right.

Illustrations may be appended to the end of a typescript to allow an appropriate placement when the pages are made up in camera ready form. You mark the text to show where the illustrations are to be inserted by writing the placement instructions in the margin. For example, write "Insert Figure 4.2 about here." If you are preparing the pages in their final form, insert the illustrations where they should appear.

Quality of Reproduction

If a typescript includes camera ready illustrations, the copyeditor determines whether the illustrations are suitable for printing. Photographs need a high contrast of darks and lights. Line drawings and graphs need to be drawn professionally or printed from a laser printer. Printing will exaggerate poor quality in the original, such as feathery lines. The resolution of an illustration produced for a website is probably too low for printing, and the print will look coarse.

The same principles of reproduction quality apply to online documents. Check the display on several monitors using different browsers to see what users will see. Background colors or patterns should not distract from the illustration. Because graphics files are bigger than text files, they will slow the speed with which a document appears on a user's screen. File size can be reduced by minimizing the number of colors, lowering the resolution, and sometimes by changing the file type. Unless you are the computer specialist as well as editor, your job is not to make these changes but rather to notice and to suggest modifications when the visual quality seems poor or the document is slow to load.

Copyediting Online Documents

In online documents, the navigation aids and screen design as well as the content need to be correct, consistent, accurate, and complete.

Visual consistency in online documents helps readers maintain a sense of their place in a document. The same types of information, such as menus and returns to a home page, should appear in the same place on every screen and use the same cues of color and type. As with print documents, establishing styles and templates at the beginning of document development helps with consistency.

Also some standard screen layouts have become familiar, with functions and pull-down menus at the top of the screen and topics often in a column on the left or right. The bottom of the document may include some navigation aids—links to related documents or to points in the displayed document and an email address for the writer or website manager. Variation from these familiar displays may confuse readers.

Because screens are usually harder to read than print, text must contrast sharply with the background. Italics—hard to read on a computer screen—should be minimized or eliminated.

To check for completeness of an online document, use document specifications and the menus on the home page and the scan and move on pages. Try the links. Readers appreciate identifying information comparable to that on the copyright page of a printed book: the date of the last update and the name of the writer or website manager. Some online documents include a copyright statement. This identifying information may appear at the bottom of the document or on the home page. As with print documents, sources of information should be cited.

Steps in Copyediting

Because copyediting requires attention to so many details, it may require more than one editorial pass. In one reading, you may not be able to keep your attention focused on grammar and spelling and all the elements of consistency, accuracy, and completeness, as well as attending to instructions for printing, especially if the text is long and complex. So you may plan on going through the text two or three times, or making several passes, looking for different things each time: perhaps spelling, consistency, grammar, and punctuation the first time; illustrations and mathematical material the second time; and headings, type styles, fonts, and spacing the third time. Check for completeness (the inclusion of all document parts) separately.

1. **Gather information about the task.** Find out about the document's readers, purpose, conditions of use, relation to other documents, and any document specifications that are available. Specifications in the document plan may include styles for the various levels of headings, parts to be included in the completed document, and information about printing and binding or other reproduction. These specifications will help you draw up your checklist for the completeness check. If a style sheet has been started, work with it in copyediting. (A style sheet lists choices of capitalization, spelling, use of numbers, abbreviations, hyphenation, and other mechanics.) Your supervisor may provide you with this information, or you may need to inquire about it. Clarify the production schedule and your specific duties. For example, by what date will the writer have the chance to review the editing? If you are editing on the computer, should you make corrections silently—that is, with no record of

the changes—or should you track changes and insert comments? Are you expected to edit just for correctness and consistency, or do you also have responsibility for style?

2. **Survey the document.** A quick initial reading will identify features that need editorial attention, such as an extensive use of tables or an inconsistent documentation style. It will also reveal how extensively the document uses illustrations, lists, or other features. These preliminary impressions will help you make thoughtful line-by-line decisions.

3. **Run computer checks.** The spelling and grammar checks will find errors and let you clean up the document before you begin line-by-line reading. Include a check for extra spaces between sentences. Using the replace feature of the word processing program, replace two spaces after a period with one. If the document has been prepared with a publishing template and styles (see Chapter 5), you may check at this point that the right styles have been applied to the different types of text elements.

4. **Edit the paragraphs and headings for correctness, consistency, and accuracy.** Read for meaning; make emendations only when you are sure that they will clarify, not distort, the meaning. Before you guess at a change of terms or phrasing, query the writer. If the text contains a large number of errors, you may need to make more than one pass to complete the correctness, consistency, and accuracy checks.

5. **Edit the illustrations, equations, reference list, table of contents, and other front matter and back matter.** Edit for correctness, consistency, accuracy, completeness, and readability. Mark insertion points for illustrations. A graphic artist may provide instructions for treatment of illustrations—screening, doing color separations, cropping, or reducing illustrations. Attach these instructions to the paper copy.

   Check equations themselves for accuracy, enumeration, and references in the text. (See Chapter 12 for more information on editing quantitative and technical material.)

6. **Prepare the typescript for printing or other reproduction.** Confirm that the document is complete—all the parts are present and in order. If the text is to be transmitted by paper, mark headings to define levels, indentions, alignment, and vertical space. (Alternatively, a graphic designer may mark the copy for the graphic design.) For electronic transmission, confirm that styles for various types of text have been applied correctly. For online documents, check that links lead to where they promise.

At the point when the copy is delivered to the printer or other vendor for reproduction, the copyediting is complete. The editing from this point forward is production editing—coordinating the rest of the production process to keep it on schedule. The copyeditor may be responsible for production editing, but the emphasis in editing changes from preparing the text to managing reproduction for distribution.

Using Your Knowledge

 To be a good copyeditor, take pride in knowing language and reading well, but consult dictionaries, handbooks, and style manuals when you are unsure. Keep key resources at your desktop or bookmarked on your computer so that it will be easy to use them. When these resources cannot provide answers to questions, query the writer or subject matter expert. Edit with readers, writer, and content in mind.

Further Reading

The Chicago Manual of Style. 2003. 15th ed. University of Chicago Press.

Einsohn, Amy. 2000. *The Copyeditor's Handbook: A Guide for Book Publishing and Corporate Communications.* University of California Press.

Judd, Karen. 2001. *Copyediting: A Practical Guide.* 3rd ed. Crisp. See especially Chs. 1 and 3.

Redish, Janice C. June 2004. Writing for the Web: Letting go of the words. *Intercom*: 4–10.

Van Buren, Robert, and Mary Fran Buehler. 1991. *The Levels of Edit.* Society for Technical Communication. Provides good lists of things to check in copyediting.

Discussion and Application

1. *Parts of a book.* Find a book from a major commercial or academic press and identify the parts of it. Bring it to class and be prepared to point out its parts and their functions for that book. What parts have been omitted, and why? Are the parts labeled and ordered correctly?

2. *Group work in class.* In groups of about four, share the books you have collected for application 1. Determine:

   a. How many books and manuals have odd-numbered pages on the right side when the book is open and even-numbered pages on the left? How do you explain the patterns?

   b. What books use running heads? Do you see any patterns to them— that is, are left and right running heads the same? If the examples have different information in the running heads, is there any reason for the variations?

   c. What books have half-title pages? Are these pages at the front, or do they divide sections within the book? What is their function? How does a half-title differ from a title page?

   d. Does anything besides the half-title and title page precede the contents page?

e. What other parts besides a title page and contents page precede the body of the book?

f. What books have indexes? appendixes? glossaries? What are the different functions of these parts?

Summarize. Make an observation to share with the whole class about the construction of books from their various parts. For example, one group might compare different uses of running heads, another group might consider different types of appendix material, a third group might observe different uses of half-title pages, and a fourth might notice a part that isn't mentioned in this chapter.

3. *Technical manuals or reports.* Compare a book and a technical manual or report. Find the book, as directed above. Also find a technical document, such as a computer manual or technical report. How do the two types of documents compare in the variety and arrangement of parts?

4. *Websites.* Examine the website of your college or university. Identify some examples of "scan, select, and move on" pages. These pages might be pages for different academic programs or for different types of students (new, enrolled). Identify a content page linked to one of the scan, select, and move on pages. What differences do you see between the content pages and the "select and move on" pages?

5. *Vocabulary*

a. Distinguish between a preface and a foreword in the front matter by identifying their functions and writers. Note that both differ from an introduction, which is a substantive part of the book's body.

b. Distinguish between a glossary and an index in the back matter.

c. Define these terms (look up unfamiliar terms in the glossary)— *running head, half-title, style sheet, template.*

6. *Illustrations.* The illustrations in Figure 7.2 appeared in instructions for fastening a preprinted sign to a car door. Use the guidelines in this chapter and the symbols in Chapter 4 to edit the illustrations. Look for correctness (especially spelling) and consistency of placement of different types of information. For example, distinguish callouts from instructions. Determine whether the callouts and instructions should be inside or outside the edges of the sign that is represented.

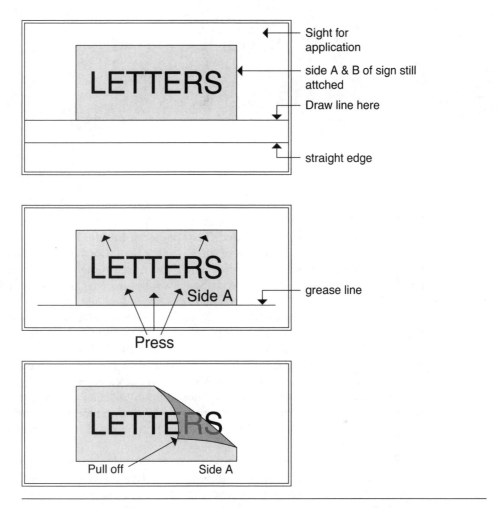

Figure 7.2   Three Illustrations from Instructions for Attaching an Adhesive Sign to a Vehicle

These illustrations require a copyeditor's attention for consistency and correctness.

8 Copyediting for Consistency

In 1999, the Mars Space Orbiter launched by NASA missed its destination, and a project valued at $125 million failed. The spacecraft builder had used the English system of measurement (pounds of force) in its calculations while the navigation team had used the metric system. When the measurements were mixed, the navigation calculations were flawed. (See www.space.com/news/mco_report-b_991110.html.)

Most instances of inconsistency do not result in failures of such magnitude. But every day, users depend on consistency in their texts to help them interpret meanings and find information. They are confused by shifting usages, such as different ways of referring to numbers. Consistency in copyediting means that the same text or the same type of text or illustration is treated uniformly throughout the document. For example, one term, not a synonym, refers to the same process throughout the text; in directions, pronouns referring to readers are either in second person ("you") or third ("they") but not both; and all headings of the same level use the same type and spacing.

Consistency gives useful information to readers. For example, the consistent type style and placement of headings help readers perceive the organizational pattern. Inconsistencies may suggest changes in meaning where no change was intended. Does a user "turn off" or "shut down" the computer? Do the two directions mean something different? Consistency enhances usability. For these reasons, one task of copyediting is to make consistent choices throughout a document.

Key Concepts

Consistency helps readers to locate and comprehend the information they need. Within all the categories of consistency—verbal, visual, mechanical, structural, and content—editors can make some choices from among acceptable alternatives. These choices are not simply personal preference; they reflect conventions recorded in dictionaries and style manuals. The document style sheet encourages consistent choices throughout a document. Planning for consistency before the document is developed saves time and stress during editing and production.

This chapter discusses five types of consistency in documents: verbal, visual, mechanical, structural, and content. It also discusses the use of style guides to achieve consistency.

Document Consistency

Table 8.1 lists and defines five types of document consistency. In addition to internal consistency (meaning that a document is consistent in itself), a document should be consistent with the document set to which it belongs, such as other manuals for the same type of equipment.

Verbal Consistency

Verbal consistency refers to the meanings and arrangements of words. Verbal consistency, whether semantic, syntactic, or stylistic, clarifies meaning.

Semantics

Semantics refers to meaning. Semantic consistency means that one term is used to represent a meaning. Referring to one switch as a "toggle switch" in one paragraph and an "on-off switch" in the next paragraph is confusing because the two terms indicate two switches. Your teachers may have told you to avoid repetition, but readers of practical documents prefer the consistency of repeated terms for clarity of meaning.

Syntax

Syntax refers to the structure of phrases and sentences. Syntactic consistency refers to parallel structure of related terms, phrases, and sentences. In the series "terms, phrases, and sentences," the choice of nouns for the items reinforces their relationship as three types of text that may have parallel structure. A nonparallel alternative might read: "terms, the phrasing, and whether sentences are parallel." The alternative seems sloppy, however, and the inconsistency of noun, gerund, and clause makes it more difficult to form a mental list of the levels at which structure matters. Likewise, series of steps in a procedure could be expressed either as commands or as statements but not as a mixture.

Style

Style refers to word choice, sentence patterns, and the writer's voice. Stylistic consistency means that the document does not mix formal language with casual language nor a sophisticated approach to a subject and reader with a simplistic one. A variation in style surprises a reader and calls attention to itself rather than to the content. It forces readers to adjust their sense of who they are in relation to the subject. For example, a shift from third-person description to second-person instruction moves the reader from observer (outside the action) to participant (in the action). A shift in person is disorienting unless it reflects an intentional shift from explanation to performance.

TABLE 8.1	Types of Document Consistency
Type	**Meaning**
Verbal	
semantics	meanings of words: one term represents one meaning
syntax	structures: related terms, phrases, and clauses use parallel structure
style	word choice, sentence patterns, and writer's voice: the document is consistent in its level of formality and sophistication and relationship to readers
Visual	
typography	typeface and type style are the same for parallel items, such as level-two headings
tables, figures	labels, titles, callouts, captions, and cross-references are treated the same for all similar illustrations; illustrations match content and purpose in quality, taste, and seriousness
print layout	repeated or sequential items, such as running heads and page numbers, appear in the same location on a series of pages
screen layout	elements, such as menus, search boxes, and contact information, appear in the same position on related screens; color and type style have specific meanings and are used consistently
color	colors consistently signal sections of a text or types of information; colors adhere to a company's color standards
Mechanical	
spelling	one spelling is used for the same term
capitalization	terms are capitalized the same in each related use
hyphenation	hyphenation is consistent for the same term and for like terms
abbreviation	use and identification of abbreviations follow a pattern
numbers	related numbers are consistent in punctuation, use of words or numerals, abbreviation, and system of measurement
punctuation	items in a series and lists follow a pattern of punctuation; possessives and accents are the same on repeated uses of the same term
cross-references	all cross-references follow the same patterns of capitalization, abbreviation, and punctuation
documentation	references conform in arrangement, abbreviation, capitalization, underlining, and spacing to one acceptable pattern
lists	related lists match in capitalization, punctuation, spacing, signals such as numbering and bullets, and phrases or complete sentences
type style	variations such as italics, capitalization, and boldface signal specific meanings such as emphasis or terms used as terms
Structural	related parts of documents, such as chapters, follow a pattern
Content	references to the same object or event do not contradict one another

Visual Consistency

Visual consistency refers to anything a reader can see on the page or screen, including typography, spacing, color, icons, and illustrations. A graphic designer may make the decisions about typography and layout, but the copyeditor marks the typescript to apply the design and achieve visual consistency. Choices about visual design create the "look and feel" of the document as well as useful signals for finding and interpreting information.

Typography

Typographic consistency means that parallel parts of a document use the same typeface and type style. Levels of headings are distinguished by their spacing, typography, and capitalization. Variations in type require a reason.

Layout: Print

Layout refers to the location of related items on the page and to the use of space. Page numbers appear in the same location on all pages, and the amount of space following a level-one heading is the same for each use. Figures are labeled consistently at the bottom or at the top, not sometimes at the top and sometimes at the bottom. The top, bottom, left, and right margins as well as the space between columns, if any, are uniform from page to page. Indention is the same not just for paragraphs but for other text elements, such as lists and block quotations. If the location of related items varies, it does so according to a pattern (for example, page numbers may appear in the upper left corner on a lefthand page but in the upper right corner on a righthand page).

Layout: Screen

Consistent layout of screens helps readers find information. Menus are consistently at the top or on the right or left, but the position does not change. Search boxes are probably at the top, and contact information is at the bottom. Screen design also establishes company identity. Colors, logos, and typography conform to company standards and create the look and feel of the site.

Figures and Tables

The use of numbers and titles and placement on the visuals should follow a pattern. Callouts (names of parts), captions, and cross-references should be set like others for similar illustrations in typeface and type style, capitalization, and placement. (See Chapter 7.)

The content and reproduction of figures and tables should also be consistent with the purpose and quality of the document. A poor reproduction or an amateurish drawing looks especially out of place printed on quality paper. The insertion of a cartoon is inconsistent with the goals of a serious financial report.

Consistency of Mechanics

Mechanics include spelling, capitalization, hyphenation, abbreviations, numbers, and type styles. The following examples illustrate issues in the consistency of mechanics.

Spelling

Dictionaries often provide alternative spellings for words. The first choice may indicate the preferred spelling, or the choices may be listed alphabetically. Editors choose one from alternatives such as these:

acknowledgment	acknowledgement
insure	ensure
copy editing	copyediting

Capitalization

An editor might look up capitalization of the following examples, and others like them, in a dictionary or style manual:

Crohn's Disease	Crohn's disease
French fries	french fries
Helvetica	helvetica
Web site	website

Hyphenation

Hyphenation distinguishes adjective phrases from noun phrases. Some terms always use a hyphen in their spelling, and in some terms, such as email (e-mail), the editor makes an informed choice.

long-term project (adjective)	in the long term (noun phrase)
two-meter length	length of two meters
cross section	cross-references
e-mail	email

Abbreviation

Editors decide whether to abbreviate (some terms are better known by their abbreviations than by the spelled-out version), what abbreviation to use if there are alternatives, and how to identify the abbreviations for readers (parenthetically or in a list of abbreviations).

a.m.	A.M.	AM
inches	in.	"
United States	U.S.	US

Numbers

The various possibilities for expressing numbers require editorial choices.

Commas	1000	1,000	
Numerals or words	five	5	five (5)
Dates	12 April 2005	April 12, 2005	
Inclusive numbers	411–414	411–14	411–4

Time	eight o'clock	8 a.m.	0800 hours
Equation numbers	eq. 2.2	equation 2	
Illustration numbers	figure 1.1	figure l	Figure 1
Measurement system	English (pounds, quarts, inches)	Metric (grams, liters, meters)	

Punctuation

All of the following text features require decisions about punctuation.

Items in a series

Should there be a comma before the final item?

Five types of consistency are verbal, visual, mechanical, structural, and content.

OR

Five types of consistency are verbal, visual, mechanical, structural and content.

Lists

Should list items be punctuated? Should they be numbered or bulleted? Capitalized?

1. Spelling;	• spelling
2. Capitalization;	• capitalization
3. Hyphenation; and	• hyphenation
4. Numbers.	• numbers

Possessives

Should an *s* be added following the apostrophe?

Paris' museums   Paris's museums

Accents

Should accents be used?

resume   résumé

Cross-references

Cross-references within the text require consistent use of abbreviations, capitalization, and parentheses.

See Figure 1   See Fig. 1
(see fig. 1).   (See fig. 1.)

Documentation

The styles for citing sources differ from discipline to discipline. The style establishes the arrangement of information, capitalization, abbreviation, punctuation, and the use of italics. Styles are usually variations of the author-date and reference note styles. The following two entries conform to the author-date styles of the American Psychological Association (APA) and the Modern Language Association (MLA).

APA: Brumberger, E. R. (2003). The rhetoric of typography: The persona of typeface and text. *Technical Communication, 50*, 206–223.

MLA: Brumberger, Eva R. "The Rhetoric of Typography: The Persona of Typeface and Text." *Technical Communication*, 50 (2003): 206–23.

Reference notes are often used in engineering publications. The notes are numbered in the order in which they appear in the text. The following example shows the style used by *IEEE Transactions on Professional Communication.*

[1] E. R. Brumberger. "The rhetoric of typography: The persona of typeface and text," *Technical Communication,* vol. 50, pp. 206–223, 2003.

Lists

Lists should be consistent in these respects:

punctuation

numbering

capitalization of the initial word in each list item

Type style

Type style (italic or roman, bold or regular) is a matter of visual consistency, but it relates to mechanics when it is used to identify terms used as terms or glossary words.

The word *proofreading* is no longer hyphenated.

The word "proofreading" is no longer hyphenated.

Inconsistencies distract readers and may lead to misinterpretations. Consistency contributes to the readers' impression of quality.

Structural Consistency

Structural consistency refers to organization. Related parts of documents, such as chapters, should follow a pattern. For example, the introduction of each chapter in this textbook ends with "key concepts." If a report reviews four topics and includes a table summarizing the topics, the order of topics in the report and in the table should be the same. Sometimes the pattern is external to the document. For example, a proposal that responds to a request for proposals includes the sections requested in the order that they are requested. If a company publishes user guides for a number of products, all the guides will include the same sections in the same order. As you will learn in Chapter 17, recognizing patterns is an important part of learning. Consistency in structures helps readers perceive the patterns.

Content Consistency

The check for content consistency eliminates contradictory information. For example, if a part of a mechanism is said to be made of steel at one point, it should not later be described as aluminum. If the initial date of a five-year project is 2005, readers will be confused if the ending date is identified as 2009. If readers are told to find a reference on page 18 of a document when that reference actually appears on page 19, they will be frustrated or may not find the reference at all. If a forecasting statement in an introduction establishes four major divisions in a document with five divisions, readers may wonder how the fifth division relates to the others.

Inconsistencies in a document often result from revision or from product changes. Inconsistencies can also result from typographical errors or carelessness.

The check for content and structural consistency somewhat overlaps comprehensive editing, but copyediting is concerned with the details of content, whereas comprehensive editing focuses more on the larger issues of organization, completeness of information, and style.

You can use the computer's search and replace functions to locate variations in spelling, capitalization, abbreviations, and numbers. You can also use the styles function to create the same typographical and spacing choices for related types of text. (See Chapter 5.)

A Foolish Consistency . . .

Consistency aids interpretation, reduces noise, and communicates quality. However, copyeditors can go too far in their pursuit of consistency and create errors and clumsiness. One editor, reading the following sentence, saw that two of the three items in the series ended in *-al*.

The reasons for the change are philosophical, technological, and economic.

Aiming for parallel structure, the editor added *-al* to "economic."

The reasons for the change are philosophical, technological, and economical.

The structure is parallel—but the use of the word is wrong. Changes might be economical, but reasons (which have no price tag) are economic.

Ralph Waldo Emerson observed: "A foolish consistency is the hobgoblin of little minds." A foolish consistency would be to insist that all the items in a series be plural, even when some are represented appropriately as singular, or that all graphs be bar graphs, regardless of their content and purpose. The goal of providing readers with the information they need in the most effective way supersedes any goal of making related parts of the document uniform.

Style Manuals

Copyediting for consistency often means choosing among acceptable alternatives. Copyeditors depend on dictionaries and style manuals to guide their choices. Style manuals establish conventions accepted in various disciplines or organizations. They save a copyeditor from guessing or from imposing personal preferences.

Style in the sense of conventions or mechanics differs from the meaning of the term as writing style and as type style. In fact, editors use the term in four different ways, as Table 8.2 illustrates. Style manuals, as discussed in this section, refer to style mostly as choices of spelling, number use, capitalization, citation, and other such variables in language, as well as to standards for visual presentation.

Style manuals range from comprehensive to selective. Selective manuals are for working in specific contexts: for international audiences, in disciplinary publications, and in organizations. As an editor, you will work first with specialized manuals and consult the comprehensive manual when the specialized manual does not answer a question.

TABLE 8.2	**Multiple Meanings of *Style***

Style is a term with multiple meanings for editors. You will use it differently in different contexts.

Writing style	Writing style refers to the choices of words and sentence structures and tone. These choices reveal the writer's personality and the formality of the document; they also affect whether the text is comprehensible.
Mechanical style	*Style* as it is used in "manual of style" or "style sheet" refers to choices about mechanics, such as capitalization and representation of numbers.
Type style	Type style is a variation of mechanical style. It refers to whether the type is bold or light, roman or italic, condensed or expanded.
Computer styles	In word processing and Internet publication, styles are tags for different text elements, such as headings and paragraphs. A style includes instructions for font, size, indention, spacing before and after, borders, and anything else from the format or paragraph design menu. This collection of choices is saved and given a style name. A user can apply all the choices by selecting the named style for a portion of the text.

Comprehensive Style Manuals

Comprehensive style manuals cover style choices that might apply in any publishing situation, no matter what the discipline or document. These manuals are longer and cover more topics with more examples than the more specific manuals in categories that follow. *The Chicago Manual of Style* (15th edition, 2003, University of Chicago Press) is widely used in book publishing. The list of chapter titles from *The Chicago Manual* (see Figure 8.1) shows how comprehensive and thorough is the manual's coverage. The guidelines for using numbers, for example, take 18 pages.

The Oxford Guide to Style (2nd edition, 2002, Oxford University Press), like *The Chicago Manual,* is oriented to book publishing. It is helpful for references to style preferences in Europe as well as in the United States.

International Style Manuals

The European Union has published a reference tool for written works, the *Interinstitutional Style Guide.* Versions are available in eleven different languages. It would be helpful if you are editing material for European audiences. (See eur-op.eu.int/code/en/en-000400.htm.)

The Canadian Style: A Guide to Writing and Editing (rev. ed. 1996, Dunham Press) was prepared by the Department of Public Works and Government Services of Canada.

Discipline Style Manuals

Some disciplines publish their own style manuals. Editors working in that discipline consult a discipline style manual for documentation style and for conventions

1. The Parts of a Published Work
2. Manuscript Preparation and Manuscript Editing
3. Proofs
4. Rights and Permissions
5. Grammar and Usage
6. Punctuation
7. Spelling, Distinctive Treatment of Words, and Compounds
8. Names and Terms
9. Numbers
10. Foreign Languages
11. Quotations and Dialogue
12. Illustrations and Captions
13. Tables
14. Mathematics in Type
15. Abbreviations
16. Documentation I: Basic Patterns
17. Documentation II: Specific Content
18. Indexes
     Appendix A: Design and Production—Basic Procedures and Key Terms
     Appendix B: The Publishing Process for Books and Journals
     Bibliography
     Index

Figure 8.1   Chapters from *The Chicago Manual of Style*, 15th ed.

With 956 pages covering many topics, *The Chicago Manual of Style* is a comprehensive style manual. It answers most questions about language and publication that an editor might have. For solutions to editorial problems in particular contexts, such as technical documentation, an editor would turn to a specialized manual.

Source: Used by permission of the University of Chicago Press.

for labeling illustrations, hyphenating terms, using numbers, and other mechanics preferred in the discipline. Some of the widely used manuals are as follows:

Biology, agriculture	*Scientific Style and Format: The CBE Manual for Authors, Editors, and Publishers,* 7th ed. In process 2005. Cambridge University Press. By the Council of Biology Editors. See www.councilscienceeditors.org.
Mathematics	*Mathematics into Type,* updated ed. 1999. American Mathematical Society. By Ellen Swanson, Arlene O'Sean, and Antoinette Schleyer.
Medicine	*American Medical Association Manual of Style: A Guide for Authors and Editors,* 9th ed. 1997. Williams & Wilkins. Ed. Cheryl Iverson.

Journalism	*The New York Times Manual of Style and Usage*, rev. ed. 2002. Three Rivers Press. By Allan M. Siegal and William G. Connolly. *The Associated Press Stylebook and Briefing on Media Law.* 2002. Perseus Press.
Psychology, social science	*Publication Manual of the American Psychological Association*, 5th ed. 2001. American Psychological Association.
Government	*U.S. Government Printing Office Style Manual*, 29th ed. Online at www.gpoaccess.gov/stylemanual/browse.html.

Four specialized style manuals are useful for editors who work with particular kinds of text:

Scientific and technical reports	*Scientific and Technical Reports—Elements, Organization and Design* (revision and redesignation of ANSI Z39.18–1987). ANSI Z39.18–1995. Oxon NISO Press. By the National Information Standards Organization. See www.niso.org/standards/resources/Z39-18-1995.pdf.
Computer industry publications	*The Microsoft Manual of Style for Technical Publications*, 3rd ed. 2003. Microsoft Press. See www.microsoft.com, downloads section, keyword "style." *Read Me First! A Style Guide for the Computer Industry*, 2nd ed. 2002. Sun Microsystems; published by Prentice Hall.
Illustrations	*Information Graphics: A Comprehensive Illustrated Reference*. 1999. Oxford University Press. By Robert L. Harris. Not a style manual but a great reference.
Websites	*Web Style Guide*, 2nd ed. 2002. Yale University Press. By Patrick J. Lynch and Sarah Horton.

Other specialized style manuals are available. The instructions for authors in a discipline's journals may name the preferred style manual. Reference librarians can also assist in locating style manuals.

Organization ("House") Style Manuals

An organization that publishes frequently, whether or not it is a commercial publishing house, will make similar choices for most of its documents. Choices that will be appropriate for all publications from that organization, such as technical terms and use of numbers and standards for the use of color and type, can be listed in a house style manual. The manual may incorporate the preferred choices in a comprehensive or discipline style manual, but because the house manual is more selective, it answers specific questions without the clutter of extra information. Instead of eighteen pages on the use of numbers, for example, there may be only two pages that list those instances in which the organization will use numbers. Digital Equipment Corporation, Bell Laboratories, and the publishers of *Oil and Gas Journal* are three organizations that have prepared their own style manuals, but such manuals are common wherever publication is frequent. They are prepared for in-house distribution, not for public distribution. Figure 8.2 shows a

Lists

Types of Lists

There are two types of lists:

- ordered lists
- unordered lists

In an ordered list, each item is numbered. Use an ordered list only when the order of the items is important. In general, procedural steps are the only type of list in which the order of the items is important.

At *all* other times, use an unordered list. If you list three features of a product, do not enumerate the features. Use bullets instead.

Use of Colon to Introduce a List

Use a colon to introduce a list if the introductory statement is a grammatically complete sentence and it is not followed by another introductory sentence.

Example: 3270 OPTIMIZER/CICS offers three major advantages:

- It operates transparently to CICS.
- It uses standard 3270 features.
- It uses minimal CPU resources.

Use a colon to introduce a list if the introductory statement contains *as follows* or *the following*, and the illustrating items (rather than another sentence) follow it directly.

Examples: The features of 3270 OPTIMIZER/CICS are as follows:

- transparent CICS operation
- standard 3270 compatibility
- minimal use of CPU resources

The following results occur after you press **F3**:

- The selected record is added to the file.
- The program writes a backup copy to disk.
- The primary menu appears.

Figure 8.2   Page from the *BMC Software Technical Publications Style Guide*
Source: Reprinted by permission of BMC Software, Inc.

page from the *BMC Software Technical Publications Style Guide*, prepared by BMC Software, Incorporated.

In establishing a style for a document, you will probably consult the house style manual first. It is most likely to address the choices that have to be made for that document, and it's probably short and easy to use. This manual will also help you achieve consistency with the document set (related documents). If writers within the organization use the house manual, they can incorporate the standards from the start and not wait for the editor to "correct" the document.

Document Style Sheet

To help you remember what you decided so that you can make the same choice later in the document, you can record the choice on a style sheet for the document. The style sheet does not list every editorial choice but only those that may recur in the document. It includes choices for which the editor had to consult a dictionary, handbook, style manual, or other source, as well as terms or spellings for which several options are available. For example, because you know that state names are always capitalized, you would not list your marking of a lowercase *c* in *California*. But if you were unsure of the spelling and capitalization in *Alzheimer's disease*, you would list the term on the style sheet. The style sheet is not a record of changes but rather a guide to editing the pages that follow in the document.

The style sheet includes choices about punctuation, abbreviations, spelling, capitalization, hyphenation, and other matters of consistency. The style sheet may be only one page long if the document is short. If you keep your document style sheet in a word processing file, you can always keep it alphabetized, complete, neat, and handy for ready access as you edit.

spelling, hyphenation, capitalization
backup copies
cross-references: Chapter 9, Figure 4.1
copyediting
double-space (vb.)
double-spaced (adj.)
email
half-title
hard copy
in-house (adj.)
in house (pred adj.)
Internet
lefthand (adj.)
markup
metric system
online
proofreading
righthand (adj.)
roman type (not capitalized)
typeface
type size
type style
World Wide Web, website, the Web

type style and punctuation
clause following a colon: no initial cap
introductory phrase or term: comma follows (Thus, editing . . .)
terms used as terms: italicize

illustrations
cross-references in text: initial cap (Figure 3.1, see Figure 3.1)
labels: double numeration (Figure 3.1)

terms
document = the entire publication, including hypertext
text = prose part of the document
font = typeface when the reference is to the computer
typeface = named alphabet of characters with recognizable and distinctive shapes

citations
Last name, First name. Date. Title of article. *Journal Name* Vol: pp-pp.
Last name, First name. Date. *Title of Book*. Edition. Publisher.
Website URL: omit http:// no underscore period at end of URL at end of sentence

Figure 8.3   Partial Style Sheet for *Technical Editing*

If your treatment of a term will differ depending on its part of speech, as from noun to adjective, include the part of speech on the style sheet. For example, the phrase *top-down* as an adjective is hyphenated, but as an adverb it is not. One might write, "Top-down editing helps an editor make decisions in light of the whole," and then be consistent in saying, "That person edits top down." The style sheet would look like this:

top-down (adj.)

top down (adv.)

In addition to listing terms in the alphabetical sections, you can include sections on the sheet for punctuation, numbers, and cross-references.

Part of the style sheet for this textbook appears in Figure 8.3.

If you can anticipate style sheet items when documents are being planned, you can share the style sheet with writers to help them incorporate the standards. The style sheet accompanies the document through subsequent stages of production, including proofreading. It gives other people working on the document a basis for answering questions and making decisions.

Using Your Knowledge

To become good at editing for consistency, first work on recognizing the types of consistency in a document. Become familiar with at least one style manual. Refer to it when you edit class projects and client projects. Practice using style sheets even when you are working with a short document. Even when you are familiar with the points of consistency, it may take you more than one editorial pass through a document to see the inconsistencies.

You can incorporate consistency into a document when the document is planned by offering a style sheet to writers. Then you will have less to do when the document is complete and less to negotiate if there are questions. Make the guidelines available online so that everyone will be able to find what is expected. With these procedures, copyediting for consistency will rightly become part of the design process, not just an end-stage correction process.

Further Reading

Farkas, David K. 1985. The concept of consistency in writing and editing. *Journal of Technical Writing and Communication* 14: 353–364. Even after two decades, the article cogently explains consistency and its value in technical communication.

Mackay, Peter D. 1997. Establishing a corporate style guide: A bibliographic essay. *Technical Communication* 44: 244–251. Identifies resources and discusses issues and procedures in developing a style guide.

Quesenberry, Whitney. 2001. *Building a Better Style Guide*. www.wqusability.com/articles/better-style-guide-paper.pdf. Claims that a style guide can promote a shared vision of a product design.

Discussion and Application

1. Locate two discipline style manuals from the list in this chapter (pp. 133–134) or by asking a reference librarian. Compare the directions for citing references and for spelling out numbers in the two manuals. What similarities and differences do you note?

2. Explain why a document style sheet should not list every editorial change made in a document. Explain why you might sometimes record choices that are already covered in a published style manual.

3. What features in the following two paragraphs will require a decision about consistency of mechanics? Why would the typo "continous" not appear on the style sheet?

   Assuming that the paragraphs are the introduction to a ten-page report that must be edited in its entirety, make a style sheet for these paragraphs so that subsequent decisions may be consistent. If you have questions about some of the choices, check your dictionary first. If questions remain, consider what other resources might be reliable.

   > Ignitron Tubes were first used in dc-arc welding power supplies. An ignitron tube is a vacuum tube that can function as a closing switch in pulsed power (using a pulse of current rather than a continous current) applications. A closing switch is a switch that is not *on* or will not pass current until it is triggered to be on. When the tube is triggered to be "on," it provides a path for the current. The ignitron tube is turned on by a device called an "ignitor." The ignitor sits in liquid mercury inside the vacuum tube. The ignitron tube creates a path for the current by vaporizing mercury. More mercury will be vaporized as the current crosses the tube. When there is no current the vaporized mercury goes to the bottom of the tube and turns back into liquid.

   > When an Ignitron Tube turns on and passes current, this process is referred to as a shot. The greatest amount of current that ignitrons can presently handle is nearly 1,000,000 amps. The tube can handle this amount of current for five shots before the tube fails. Industry wants a tube that will handle 1000 shots before it fails.

4. Type "corporate style guide" into an Internet search engine to identify an online style guide for an organization. Analyze the guide to determine its scope (topics covered) and purpose. For example, how detailed are the guidelines for consistency of mechanics? Which of the topics suggested in this chapter are covered, and which are excluded? How much emphasis is there on corporate identity (logos, colors, type, page layout)? How do you explain the emphases you have observed? In class, compare the results of your analyses. What have you observed as a class about the role of style guides?[1]

---

[1]Activity suggested by Professor Kelli Cargile-Cook, Utah State University.

CHAPTER 9

Spelling, Capitalization, and Abbreviations

If you have ever found a spelling error in a published document, you know the feeling most readers have when they discover errors. Errors in spelling and capitalization are like having a piece of spinach stuck between your front teeth when you ask your boss for a raise—the person responds by laughing, inwardly or outwardly. Such errors mean something is amiss; someone was not careful. Credibility suffers. Readers' comprehension may suffer, too, if only because they are distracted from the content.

Finding that lone error in a document that is printed or displayed on a screen is easier than finding all of the errors while you are copyediting. The error sticks out in the published document because most of the rest of the document is correct and probably attractive. In a typescript, however, you are distracted from one error by others and even by the mess of a heavily copyedited page. Or you may become so familiar with a typescript that your eye reads the words as correct even when they are not. The computer spelling checker, for all its help, may mislead you or miss words that are incorrect in a specific context.

This chapter identifies ways to recognize misspelled words and alerts you to times when you should check capitalization and abbreviations.

Key Concepts

> Experience with language and understanding of its rules will help editors work quickly and accurately on spelling, capitalization, and abbreviation, but good editors will also depend on dictionaries and style manuals. Spelling checkers are great resources, but they are not entirely reliable as they do not distinguish homonyms and understand context. Recording the correct version (or your choice when there are alternatives) on the document style sheet (see Chapter 8) will help you make the same choice consistently throughout the document and will save the time of rechecking when the word occurs again.

Spelling

Spelling is notoriously difficult in English because of the many irregularities and inconsistencies. Imagine trying to teach a non-native speaker how to spell and pronounce these words: *tough, though, through*. Or try to explain to someone why the *sh* sound in English can be spelled in at least ten different ways, as illustrated by the words *shower, pshaw, sugar, machine, special, tissue, vacation, suspicion, mansion,* and *ocean.* Editors depend on guidelines and tools to achieve correct spelling.

Guidelines and Tools

The following list outlines resources and practices to help you edit for spelling.

1. **Use a spelling checker on your computer.** By matching words in the document with its database of words, a spelling checker can recognize words that do not match and that may be misspelled. For each mismatched word, you must choose whether to change the spelling or to let it stand. Spelling checkers may suggest alternative spellings, but the user must key in changes that are not in the database.

    Using the electronic spelling checker is a good first step in editing any document in digital form. If spelling errors can be corrected at the beginning, they will not distract the editor from other editorial needs. Running the check as the last step in editing can identify errors that have been introduced in editing.

    An electronic spelling checker will not relieve you of the responsibility for checking spelling. Because it matches only characters and does not read for meaning, it will not know when words are correctly spelled but misused. It will approve *type script* even though you meant *typescript* because the two words match words in its database, and it will not catch *in* substituted for *on.* A computer would approve the spelling in this sentence:

    Their our know miss steaks in this sent ants.

    The memory may also not contain all the words in the document. A typical desk dictionary in English contains 200,000 words, including variations (such as different tenses) of the same word, whereas an online dictionary may store about 80,000 words.Thus, the computer may question words that are spelled properly, especially technical terms and names.

    Be cautious about accepting the spelling checker's recommendations for change. The newsletter *Simply Stated,* from the Document Design Center, reported in March 1989 the amusing results of a spelling change implemented by a computer. The Barclay Banks in England used a spelling checker on a document prepared for depositors. When the checker did not recognize "Barclay," it substituted "Broccoli." Depositors were confused—and amused—when they received correspondence from the Broccoli Bank!

    Don't depend entirely on the computer for your spelling check, but do use it to save time and to catch misspelled words that you might otherwise miss.

2. **Use a dictionary.** Look up the spelling of a term that is unfamiliar to you. Use a subject dictionary (such as a dictionary of medical terms) for specialized terms. When you look up a word, mark it in your dictionary to encourage review when you are using the dictionary another time. You can also add the word to your own list of commonly misspelled words and to the document style sheet to enable a quick check another time you are uncertain about the spelling.

TABLE 9.1	**Frequently Misspelled Words**	
accommodate (double *m*)	feasibility, feasible	occurred
achievement	forty (not *fourty*)	parallel
acknowledgment (no *e* before the suffix)	gauge	pastime
	grammar (the ending has an *a* in it)	personnel
acquire		plausible
a lot (two words, like *a little*)	grievous	precede (compare *proceed*)
all right	harass	questionnaire (double *n*)
analysis	inasmuch as	receive ("*i* before *e* except after *c* . . .")
approximately	indispensable	
argument	irrelevant	separate (*a* in the second syllable)
assistance	irresistible	
basically	judgment	severely (keep the final *e* before the suffix)
category (*e* in the second syllable)	liaison	
	license	sincerely
changeable	lightning	subpoena
complement	maneuver	resistance
convenient	miniature	subtly
defendant	mischievous	surveillance
embarrass (double *r* and double *s*)	misspell	undoubtedly

3. **Keep a list of frequently misspelled words.** Even good spellers have difficulty memorizing the spelling of certain words. Table 9.1 lists words that are frequently misspelled; you may have other words to add to your individual list. (In addition to these frequently *misspelled* words, see the list of frequently *misused* words on pages 143–145.) The list itself reveals problem areas:

endings (*-ar, -or,* or *-er; -ment* or *-mant; -able* or *-ible; -ance* or *-ence; -ceed* or *-cede; -ary* or *-ery*)
consonants—single or double
the *schwa* ("uh") sound—created by *a, e, i, u*
ie and *ei*

When you check spelling, look especially for the endings and consonants and the *schwa* sound. Spelling rules and mnemonic devices, like those in guidelines 4 and 6, will help you spell these words.

4. **Develop mnemonics.** A mnemonic is a memory device, usually by association. If you can remember the association, you can remember the spelling. Here are seven examples:

accommodate: When you accommodate someone, you do more than is expected; there are more *m*s than you expect.
all right: It's two words, like *all wrong*; Likewise, *a lot*: two words, like *a little*.
liaison: The "eyes" on either side of the *a* look both ways, the way a liaison looks to both sides.

principal (a school official): The principal is (or is not) your pal.

stationery (writing paper): It ends in *er* as does pap*er*.

dessert (sweet food): More of the letter *s* than in *desert* (a dry region); a dessert tastes so good that you want two, or more.

affect, effect: Effect is usually a noun, while *affect* is usually a verb. An auditory mnemonic: the **e**ffect (long *e* for both words); *the* precedes a noun, not a verb. A visual mnemonic: an *effect* is the result or the *end*.

5. **Learn root words, prefixes, and suffixes.** Words are frequently formed from basic words and from the addition of suffixes and prefixes. Recognizing the structure of the words can help you spell them. Here are some examples:

subpoena: Although you don't hear the *b* when you pronounce this word, you can recognize the prefix *sub-*, meaning "under." A *subpoena* requires a person to give testimony in court, *under penalty* if he or she refuses.

misspell: Instead of wondering whether the word has one *s* or two, consider that the prefix *mis-* is added to the root word *spell*. The structure of the word explains the double consonant. The words *misfortune, mistake,* and *misconduct* have only one *s* because their root words do not begin with *s*. The same principle applies to words with the *dis-* prefix: *dissatisfied* and *dissimilar*, but *disappear* and *disease*.

inanimate: One *n* or two? The root of this word is a Latin term, *animalis*, meaning "living." The prefix *in-* means "not." An inanimate object is not living. There is only one *n* in the prefix, and the root word does not begin with an *n*, thus the single *n* in the first syllable. There is only one *n* in the root word, as in *animal*.

millimeter: The root mille means "thousand." A millimeter is one thousandth of a meter.

Common roots in English are sometimes similar enough to be confused:

ante (before); *anti* (against): *ante*bellum (before the war); *anti*dote

hyper (above, over); *hypo* (less, under): *hyper*active; *hypo*glycemia (low blood sugar)

macro (large); *micro* (small): *macro*cosm; *micro*scope

poly (multiple); *poli* (city): *poly*gamy (multiple marriage); metro*poli*tan

6. **Apply spelling rules.** Spelling rules treat especially the addition of suffixes to words.

final *e:* A word ending in a silent *e* keeps the *e* before a suffix beginning with a consonant: *completely, excitement, sincerely*. Exceptions: *acknowledgment, judgment, argument, truly*. If the suffix begins with a vowel, the silent *e* disappears: *loving, admirable. Changeable*, however, keeps the *e* to establish the soft *g* sound.

final consonant: When a word ends with a consonant, the consonant is doubled before a suffix beginning with a vowel: *planned, transferred, occurrence, omitted*. If the final syllable is not stressed, the final consonant is not doubled: *inhabited, signaled. Busing/(bussing), focusing/(focussing),* and *canceled/(cancelled)* are correctly spelled with one consonant or with two.

able/ible: The suffix *-able* generally follows words that are complete in themselves, such as *remarkable, adaptable, moveable, changeable, readable, respectable,* and *comfortable*. The suffix *-ible* follows groups of letters that do not form words. There is no *feas, plaus,* nor *terr;* thus, we write *feasible, feasibility, plausible,* and *terrible.* Words that end in a soft *s* sound also take *-ible: responsible, irresistible, forcible.* A mnemonic: words that are *able* to stand alone take the suffix *-able.* EXCEPTIONS: *inhospitable, indefensible, tenable.*

ie/ei: The familiar rhyme works most of the time: *i* before *e* except after *c* and when sounded like *a* as in *neighbor* and *sleigh.* Exceptions should be memorized: *Neither foreigner seized the weird financier at leisure on the height.*

7. **Respect the limits of your knowledge.** Editors have made mistakes by substituting familiar words for similar but unfamiliar words. You can imagine the impatience of a writer in psychology whose editor changed *discrete behavior* to *discreet behavior,* not knowing the term *discrete* means "separate." Or the frustration of the writer in a social service organization whose editor changed *not-for-profit* to *nonprofit* when *not-for-profit* was the legal term used in the application for tax-exempt status. Or the writer in biochemistry whose editor changed *beta agonists* to *beta antagonists.* In all cases, the first term was correct in its context; the editor substituted a term that was more familiar but incorrect.

Wise editors respect the limits of their knowledge and check a reliable source, such as a dictionary or the writer, before changing the spelling of an unfamiliar term.

Frequently Misused Words

The following groups of words are so similar in appearance or meaning that they are frequently confused.

affect, effect: Both words can be nouns, and both can be verbs. Usually, however, *effect* is a noun, meaning "result," and *affect* is a verb, meaning "to influence." Because a noun can take an article but a verb does not, an auditory and visual mnemonic may help: th*e* *e*ffect. Because a verb is an action, the beginning letters of *act* and *affect* can serve as a visual mnemonic.

Your training in writing will have an effect on your performance in editing.
The Doppler effect explains the apparent change in the frequency of sound waves when a train approaches.
Class participation will affect your grade.

assure, insure, ensure: All three words mean "to make secure or certain." Only *assure* is used with reference to persons in the sense of setting the mind at rest. *Insure* is used in the business sense of guaranteeing against risk. *Ensure* is used in other senses.

The supervisor assured the editor that her salary would be raised.
The company has insured the staff for health and life.
The new management policies ensure greater participation in decision making by the staff.

complement, compliment: A complement completes a whole; a compliment expresses praise.

In geometry, a complement is an angle related to another so that their sum totals 90 degrees.
The subject complement follows a linking verb.
Your compliment encouraged me.

continually, continuously: Continually suggests interrupted action over a period of time; continuously indicates uninterrupted action.

We make backup copies continually.
The printer in the lab stays on continuously during working hours.

discreet, discrete: Discreet means "prudent"; *discrete* means "separate."

A good manager is discreet in reprimanding an employee.
We counted seven discrete examples of faulty reporting.

farther, further: Farther refers only to physical distance; *further* refers to degree, quantity, or time.

The satellite plant is farther from town than I expected.
The company cannot risk going further into debt.

fiscal, physical: **Fiscal** refers to finances; *physical* refers to bodily or material things.

The tax return is due five months after the fiscal year ends.

imply, infer: Imply means "to suggest"; *infer* means "to take a suggestion" or "to draw a conclusion."

These figures imply that bankruptcy is imminent.
We inferred from the figures that bankruptcy is imminent.

its, it's: Its is a possessive pronoun; *it's* is a contraction for *it is.*

The company is proud of its new health insurance policy.
It's a good idea to accept the major medical option on the health insurance policy.

lay, lie: Lay is a transitive verb that takes an object: you lay something in its place. *Lay* can also be the past participle of *lie. Lie* is always an intransitive verb. It may be followed by an adverb.

A minute ago, I laid the flowers on the table, and they lie there still. I will lay the silverware now. Then I will lie down.

personal, personnel: Personal is an adjective meaning "private" or "one's own"; *personnel* is a collective noun referring to the persons employed or active in an organization.

The voltage meter is my personal property.
Take the job description to the personnel manager.

principle, principal: A principle is a basic truth or law; a *principal* can be a school official (noun), or it can mean "first" or "primary" (adjective).

Our company is managed according to democratic principles.
Writing ability is the principal qualification for employment.

stationery, stationary: Stationery refers to writing paper; something *stationary* is fixed in place or not moveable.

Our business stationery is on classic laid paper.
The projection equipment in the seminar room is stationary.

their, there, they're: Their is a possessive pronoun; *there* is an adverb designating a place (note the root *here*) or a pronoun used to introduce a sentence; *they're* is a contraction for *they are.*

Their new book won a prize for design.
There are six award-winning books displayed over there on the table. They're all impressive.

whose, who's: Whose is a possessive pronoun; *who's* is a contraction for *who is.*

Whose house is this? Who's coming to dinner?

your, you're: Your is a possessive pronoun; *you're* is a contraction for *you are.*

I read your report last night. You're a good writer.

International Variations

Spellings differ in countries even where the language is the same. For example, British and American English spell words differently. A British person would write "whilst," "amongst," and "colour," whereas an American would write "while," "among," and "color." Choose the spelling for the countries where most readers reside. (See Chapter 20.)

Capitalization

The basic rules of capitalization are familiar: the first word of a sentence, titles of publications and people, days, months, holidays, the pronoun *I,* and proper nouns are capitalized. Proper nouns include names of people, organizations, historical periods and events, places, rivers, lakes, and mountains, names of nationalities, and most words derived from proper nouns.

Other capitalization choices require more specialized knowledge. Capitalization conventions vary from discipline to discipline. For example, if you work for the government, you might capitalize Federal, State, and Local as modifiers, but in other contexts, such capitalization would be incorrect.

Generally, capitalize nouns that refer to specific people and places but not general references:

- Capitalize names of specific regions (the *Midwest*) but not a general point on the compass (the plant is *west* of the city).
- Capitalize adjectives derived from geographic names when the term refers to a specific place (*Mexican* food) but not when the term has a specialized meaning not directly related to the place (*french* fries, *roman* numerals, *roman* type).

- Capitalize the specific name of an organization (*Smith Corporation*) but not a reference to it later in the document by type of organization rather than by specific name (the *corporation*).
- Capitalize a title when it precedes a name (*President* Morales) but not when it is used in general to refer to an officer (the *president*). And, capitalize names of disciplines that refer to nationalities (*English, Spanish*) but not to other disciplines (*history, chemistry*).

The trend over time is toward less capitalization. In the eighteenth century, many nouns, not just proper nouns, were capitalized. When words are introduced into the language, they are often capitalized to distinguish them as new words. For example, in 2005, *Internet* and *Web* are often capitalized. Gradually their meanings will be understood from the context alone.

The use of capitalization has declined in part because excessive capitalization distracts readers. Capitalizing for emphasis encourages word-by-word reading rather than reading for overall sense. The repeated capitals also look like hiccups on the page. Typing words in all capital letters destroys the shape of the word (all words become rectangles) and therefore obscures information that readers use to identify words. Other techniques of typographic emphasis, including boldface type, type size, white space, and boxing, are more effective.

Distracting Capitalization

Visitors Must Register All Cameras with the Attendant at the Entry Station.
VISITORS MUST REGISTER ALL CAMERAS WITH THE ATTENDANT AT THE ENTRY STATION

Preferred Styles

Visitors must register all cameras with the attendant at the entry station.

Visitors must register all cameras with the attendant at the entry station.

Abbreviations

An editor considers whether to use abbreviations or to spell out the terms and inserts identifying information for unfamiliar abbreviations. A related editorial task is to make consistent decisions about capitalization and periods within abbreviations. The best reason for using an abbreviation is that readers recognize it; perhaps it is more familiar than what it represents. Abbreviations may also be easier to learn than the full name, and they save space on the page. *NASA* is now more familiar to readers, and also was easier to learn when first coined, than is the spelled-out title, National Aeronautics and Space Administration. When an abbreviation becomes a word in its own right, it is an acronym. *NASA* has reached that status.

Abbreviations in the spelling of names should be respected. If a company is known as Jones & Jones (with the ampersand appearing on company signs and on the company letterhead), do not spell out the "and" to make it more formal.

Identifying Abbreviations

Abbreviations that are not familiar to readers must be identified the first time they are used. A parenthetical definition gives readers the information they need at the time they need it. After the initial parenthetical definition, the abbreviation alone may be used.

> Researchers have given tetrahydroaminoacridine (THA) to patients with Alzheimer's disease. THA inhibits the action of an enzyme that breaks down acetylcholine, a neurotransmitter that is deficient in Alzheimer's patients. The cognitive functioning of patients given oral THA improved while they were on the drug.

If the abbreviation is used in different chapters of the same document, it should be identified parenthetically in each chapter, because readers may not have read or may not remember previous identifications. Documents that use many abbreviations may also include a list of abbreviations in the front or back matter. Check your style manual for other guidelines about abbreviations.

Periods and Spaces with Abbreviations

Some abbreviations often include or conclude with periods (U.S.); others, especially scientific and technological terms and groups of capital letters, do not (CBS, cm). The trend is to drop the periods. For example, *CPA* (certified public accountant) is more common now than the older form *C.P.A.* The 15th edition of *The Chicago Manual* recommends using periods with abbreviations that appear in lowercase letters but not with abbreviations in all capital letters.

When an abbreviation contains internal periods, no space is set between the period and the following letter except before another word. At the end of a sentence, one period identifies both an abbreviation and the end of a sentence.

> She works for Super Software, Inc.

Because they represent complete words, initials substituting for names and abbreviations of single terms require spaces after the periods (for example, Mr. T. J. Corona). But single abbreviations consisting of multiple abbreviated words do not have spaces within them (see the next section, on Latin terms). When in doubt, check a dictionary. Most standard desk dictionaries include a list of abbreviations as part of the back matter.

Latin Terms

Abbreviations that represent Latin terms are used primarily in academic publications and may be unfamiliar to readers or seem pretentious outside a scholarly context. The Latin abbreviations in the list that follows are common in academic work. Editors who work in an academic environment should know them. Abbreviations of Latin terms use periods following the abbreviated term. Editors who work with user manuals should convert Latin abbreviations, except for the well-known A.M. and P.M., to their English equivalents.

> A.M. (*ante meridiem,* before midday); P.M. (*post meridiem,* after midday): these abbreviations may be set in lowercase letters or small caps. (Small caps have the shape of capital letters but are of a smaller size, as shown here.) The periods distinguish a.m. from the abbreviation for *amplitude modulation,* but small caps or context may do as well, and the periods may be omitted.
>
> *ca.* (*circa,* about): used in giving approximate dates
>
> *et al.* (*et alii,* and others): used primarily in bibliographic entries in humanities texts. Note that *et,* being a complete word and not an abbreviation, takes no period.
>
> *e.g.* (*exempli gratia,* for the sake of an example; for example): not interchangeable with *i.e.*
>
> *etc.* (*et cetera,* and other things, and so forth)
>
> *ff.* (*folio,* on the following page or pages)
>
> *i.e.* (*id est,* that is): used in adding an explanation
>
> *N.B.* (*note bene,* note well)
>
> *v.* or *vs.* (*versus,* against)

The overused *etc.* should be restricted to informal writing—or omitted altogether. It is often a meaningless appendage to a sentence, added when a writer is unsure whether the examples suffice. It may signal incomplete thinking rather than useful information for readers. In addition, it is redundant if a writer has also used *such as,* which indicates that the list following is representative rather than complete.

Measurement and Scientific Symbols

Three systems of measurement are widely used: the U.S. Customary System, the British Imperial System, and the International (metric) System. Specific abbreviations are recognized for the units of measure in all three systems. An editor checks that abbreviations used for measurement conform to the standards and that one standard is used consistently throughout the document. Conversions may be necessary if there are shifts.

The metric system, which provides a system of units for all physical measurements, is used for most scientific and technical work. It is also the common system for international use. Its units are called SI units (for Système International d'Unités, in French). The abbreviations for the metric system do not use periods. In the U.S. and British systems, the conventions about periods with abbreviations vary by style manual. For example, the abbreviation for inch may or may not include a period (*in.* or *in*). The abbreviation for pound (*lb.*) uses a period because it stands for the Latin word *libra.* The abbreviation is always singular even if the quantity it stands for is multiple. Thus, *3 lb.* is correct; *3 lbs.* is not.

Check a list of abbreviations in a dictionary or style guide to determine usage, and record your choice on the style sheet. Because the abbreviations for measurements are standard, they do not require parenthetical identification. Within sentences of technical and scientific texts, measures used with specific quantities may be abbreviated or spelled out. When measures are used alone, without specific quantities, they are spelled out. A sentence would read "The weight is measured in kilograms." In tables and formulas, however, abbreviations are preferred. As always, the document should be consistent.

Like the abbreviations for measurements, scientific symbols for terms in chemistry, medicine, astronomy, engineering, and statistics are standard. Consult a specialized style guide to determine the correct symbols. (Chapter 12 discusses quantitative and technical material in more detail.)

States

The United States Postal Service recognizes abbreviations consisting of two capital letters and no periods for the states and territories (for example, CA, NC). The old system of abbreviations for the states—usually consisting of capital and lowercase letters with a period—is obsolete. Use the postal service abbreviations on envelopes because both human and machine scanners recognize them quickly. In the headings of letters, either the abbreviation or the spelled-out state name is appropriate, but be consistent. Within the text of a document, spell out names of states. In references and bibliographies, the use of postal abbreviations, nonpostal abbreviations, or spelled-out state names depends on the style in use.

Using Your Knowledge

Even if you are a good speller, use a spelling checker, and know the rules of capitalization and abbreviation. Perfecting a typescript can be a challenge because you are also editing for other things. In a complex document with many editorial needs, use a separate editorial pass just for spelling, capitalization, and abbreviations. Know the principles of spelling to increase your efficiency. Check dictionaries and style manuals so that you won't change unfamiliar words to familiar but wrong words.

Online Resources

Acronym Finder: www.acronymfinder.com
The World Wide Web Acronym and Abbreviation Server:
    www.ucc.ie/info/net/acronyms/index.html
Merriam-Webster Collegiate Dictionary and Thesaurus: www.m-w.com
Multicultural Glossary: www.io.com/~tam/multicultural/glossary.html

Discussion and Application

1. Use markup symbols to make spelling corrections in these paragraphs.

> A feasability study is a way to help a person make a decision. It provides facts that help a researcher decide objectively weather a project is practical and desireable. Alot of research preceeds the decision.
>
> An invester deciding whether to purchase a convience store, for example, must investigate issues such as loans, lisences, taxes, and consummor demand. One principle question for a store in a residentail nieghborhood concerns the sale of liquer. People who come into the store might be asked to compleat a brief questionaire that inquires about there wishes.
>
> Undoubtably, the issue of personal is signifigant as well. Trustworthy employes are indispensible. The investigator can check employmant patterns at the store and in the area overall.
>
> After the study is complete, the investigator will guage the results and make a judgement. Some times the facts are ambigous. Intuition and willingness to take risk will influence the decision. If the project looks feasible, the recomendation will be to procede with the purchase.

2. Which of the spelling errors in the paragraphs in application 1 would a computerized spelling checker *not* catch? Why?

3. From the lists of frequently misspelled and misused words in this chapter, or from your own list of difficult words, select three, and develop mnemonics for remembering their spelling or use. Alternatively, analyze the spelling of these words by analyzing their roots.

4. Get acquainted with your desk dictionary as an editorial resource. What information beyond correct spelling does it provide an editor? Check the front matter and appendixes. Also look for help within the word lists on measurements and on symbols and signs. Name your dictionary, and cite at least three types of information other than spelling and usage. Compare your dictionary with a classmate's to get a sense of how different dictionaries may help an editor.

5. Investigate one or more of the online resources listed above, and report on its value for editors. Look for another Internet resource that could also help editors with their spelling, capitalization, and abbreviation questions.

6. Correct the following sentences for spelling, misused words, capitalization, and abbreviation. Consult a dictionary and style manual when you are uncertain. Use markup symbols.

   a. Development will continue in the Northern part of the city.

   b. Do not spill the Hydrochloric Acid on your clothes.

   c. Do swedish meatballs go well with French fries?

   d. A research project at the sight of a major city landfill has shown how slowly plastic decomposes; i.e., a plastic bottle takes 100–400 years to decompose.

   e. The Society for Technical Communication has planned feild trips to three Hi Tech company's. Later in the year some students will attend the Annual Conference of the STC. All of these plans for travelling requie some extensive fund raising this Fall. The sponsers have proposed the establishment of an Editing Service. Student edtors would aquire jobs thru the Service and return 15 per cent of there earnings to the group.

   f. A minor in computor science in combination with a major in english can make a student an attractive canditate for a job in technical communication.

   g. Capitol investmant in the company will raise in the next physical year.

   h. A prevous employe was the defendent in an embarasing and costly law suit charging sexual harrasment.

   i. Only 10 m. seperates her house from mine. I wish it were further.

10 Grammar and Usage

The subject of grammar makes people feel defensive and resistant, and no wonder: all of us have done something "wrong" grammatically at some point, and perhaps someone has pounced on us for our mistakes. The rules are so numerous that they seem impossible to learn. You can appreciate the anxiety some writers may feel when an editor looks over their work.

Here's another way to think of grammar: it is about meaning, and its rules are less about error than about possibilities for revealing meaning. Grammar rules describe words in relationship to each other. Facts and ideas become meaningful in relationship, not in isolation, and so do words. Assembling sentences according to the conventions of grammar clarifies meaning by revealing relationships. Once you understand these relationships, the rules become almost intuitive. Once writers understand that you are not criticizing but helping to clarify meaning, they will think of you as their best friend.

Grammar refers to relationships of words in sentences: subjects to verbs, modifiers to items modified, pronouns to antecedents. Grammar describes conventional and expected ways for the words to relate. For example, subject and verb agree in number. Grammar changes over time, but the conventions are used widely enough to be described as "rules."

Usage refers to acceptable ways of using words and phrases. Matters of usage include, for example, whether to use a singular or plural verb with *data,* and when to use *fewer* and *less.* Some dictionaries contain usage notes. Professional associations may also state their own conventions of usage.

Grammar has a coherence that long lists of rules and an error orientation do not reveal. This chapter aims to show you the architecture of sentences, structured around parts of speech functioning in conventional and predictable ways. After reviewing parts of speech and sentence patterns, the chapter discusses the common rules for the relationships of words in sentences. Chapter 11 reviews other grammar rules that relate to punctuation. When you can articulate principles of grammar, you will be able to discuss options for sentence construction and punctuation with writers in a professional way. If you have not recently studied grammar, you may find it helpful to refer to one of the books in Further Reading for more details and examples.

Key Concepts

Editing sentences according to the patterns that grammar describes will make the document easier to understand because the patterns will clarify relationships of words. Knowing how different types of words function in sentences will speed up editing. Words can be defined by their parts of speech (noun, verb). The parts of speech have different roles in sentences (subject, modifier). Errors result when parts of speech are asked to do things in sentences that they cannot do or that don't make sense.

Parts of Speech

To understand relationships of parts within the sentence, you need to understand the functions of different types of words. Words are categorized according to the parts of speech. Table 10.1 names and defines the parts of speech. Dictionaries identify the parts of speech of words. (Some words can be different parts of speech, such as noun or adjective, depending on how they are used.) The parts of speech are used in sentences in different ways. For example, the subject of the sentence is usually a noun or pronoun. Adjectives, adverbs, prepositional phrases (preposition plus object), and participles are modifiers. Verbs especially affect the sentence because the four types of verbs are all followed by different material in the predicate.

Sentence Structure

In the previous section, the topic was parts of speech, with a focus on different types of words. Now the topic is parts of sentences, with a focus on sentence architecture. These topics are related because different parts of speech have different functions in sentences.

Sentences have two main parts: subject and predicate. The subject states the topic of the sentence. The predicate comments about the subject. Each part contributes necessary meaning.

       subject   predicate
The new computer has increased productivity by 40%.

The subject of the sentence identifies the agent or recipient of the action in a sentence. The simple subject is the subject without its modifiers. In the example sentence, "computer" (a noun) is the simple subject. Modifiers are the article "The" and the adjective "new."

The predicate tells what the subject does or what happens to the subject. It makes a statement or asks a question about the subject. The predicate always includes a verb. The simple predicate is the verb or verb phrase. In the example sentence, "has increased" is the simple predicate.

TABLE 10.1	**Parts of Speech**
Term	Definition, function, and examples
Noun	A noun names a person, place, thing, or idea. *Roberto, Canada, machine, life,* and *Marxism* are all nouns. A noun functions in a sentence as subject, object of a verb or preposition, complement, appositive, or modifier.
Verb	A verb denotes action or state of being. A verb tells what the subject of the sentence does or what is done to it.
Adjective	An adjective modifies (describes or limits) a noun or pronoun. *Contaminated, careful, flexible, thorough, clean,* and *strong* are adjectives. Adjectives show comparisons with the suffixes *-er* and *-est* (*strong, stronger, strongest*). The articles *a, an,* and *the* are adjectives.
Adverb	An adverb modifies a verb, adjective, other adverb, or the entire sentence. Adverbs are frequently formed by adding the suffix *-ly* to an adjective. The surgeon stitched *carefully*. ["Carefully" modifies (describes) "stitched."] *Surprisingly*, the rates declined. ["Surprisingly" modifies the entire sentence.] The team played *very well*. [Both "very" and "well" are adverbs; "well" modifies "played," and "very" modifies "well."]
Pronoun	A pronoun substitutes for a noun. A relative pronoun makes a clause dependent.
personal	*I, you, he, she, they*
relative	*who, whoever, which, that* (relative pronouns make clauses dependent)
indefinite	*each, one, neither, either, someone*
Preposition	A preposition links its object (noun) to another word in the sentence. Common prepositions are *about, after, among, at, below, between, from, in, of, on, since,* and *with*. The word following the preposition is its object. The preposition plus its object and modifiers form a prepositional phrase. A prepositional phrase is a modifier, either adjectival or adverbial.
Conjunction	A conjunction joins words, phrases, or clauses.
coordinating conjunctions	Coordinating conjunctions join elements of equal value. *and, but, or, for, yet, nor, so*
subordinating conjunctions	Subordinating conjunctions make words that follow "subordinate" to the main clause. A subordinating conjunction makes a clause dependent. *after, although, as, because, if, once, since, that, though, till, unless, until, when, whenever, where, wherever, while* (etc.)
Conjunctive adverb	Conjunctive adverbs modify an entire clause; unlike subordinating conjunctions, they do not make the clause dependent. *consequently, however, moreover, besides, nevertheless, on the other hand, in fact, therefore, thus* (etc.)
Verbal	Verbals convert verbs into words that function as nouns or adjectives, have implied subjects, and may take objects or complements.
gerund	A gerund functions as a noun. It is formed by the addition of *-ing* to the verb. *Editing* on the computer saves time.
participle	A participle functions as an adjective. It is formed by the addition of *-ing* or *-ed* to the verb. In the example, "Planning" describes "she." *Planning* to edit on the computer, she established document templates before the writers began their work.

Verbs and Sentence Patterns

The predicate in the sample sentence includes more words than the verb phrase. These words form the *complement,* words that "complete" the comment on the subject. (Note the *e* in *complement,* which distinguishes this word from *compliment* and should help you remember its role in completing information on the subject of the sentence.) The verb determines the relationship between the subject and the rest of the predicate, or the type of complement. Verbs are *transitive, intransitive, linking,* or *to be.* These categories suggest functions: a transitive verb "carries over" to a direct object, but an intransitive verb does not. Linking verbs "link" the subject and predicate, and *to be* verbs suggest a kind of equation. The following paragraphs will explain these functions further, but for now you should understand that part of the function of the verb is to establish the relationship between the subject and the comment on the subject in the predicate. These relationships form the basis of some important grammar rules.

Transitive Verbs

Transitive verbs "carry over" to a direct object, one type of complement. The object answers the implicit question of the verb: "What?" or "Who?" The sentence would be incomplete without the answer. Most verbs are transitive.

transitive verb
The computer has increased productivity.
subject   direct object

The sentence would not make sense unless it answered the question, "What has the computer increased?"

Intransitive Verbs

An intransitive verb forms a complete predicate and does not require a complement. It does not carry over to a direct object.

The hard drive crashed.
The car crashed into the fence.

In neither case does the sentence raise the question "What?" or "Who?" Nothing more is necessary in the sentence to complete the meaning of the verb. The prepositional phrase "into the fence" in the second sentence is not necessary to complete the meaning but rather is an adverbial phrase that answers the question "Where?"

Linking Verbs and To Be Verbs

Linking verbs (such as *look, seem, appear,* and *become*) and *to be* verbs *(is, are, was, were, am, have been)* require a complement that reflects back on the subject. The complement that follows a linking verb is a subject complement. This complement is either an adjective that describes the subject or a noun (or noun phrase) that could substitute for the subject. The complement can substitute for the subject if it is in the same class of things. (All items in a class share some common feature, such as biological life. A class of people might include students, males, and Tim, with the order of this series ranging from abstract to specific.)

TABLE 10.2	Sentence Structure		

Subject	Predicate		
Noun	Verb	Complement	Comment
	transitive	direct object	The direct object answers "What?" or "Who?"
		object complement	The object complement describes or identifies the direct object.
	linking, *to be*	subject complement	The subject complement describes or identifies the subject.
	intransitive		An intransitive verb takes no complement, but the predicate may include an adverb.

The equipment looks imposing.

"Imposing" is an adjective that describes "equipment."

The new hard drive is reliable.

"Reliable" is an adjective that describes "hard drive."

The supervisor is an expert on computer graphics.

"Expert" is a noun that can stand in the place of "supervisor" because both words are in the category of "people at work."

Dictionaries identify verbs in two classes, transitive and intransitive, based on whether the verb takes a direct object. Thus, dictionaries classify linking verbs as intransitive. The distinction between linking and intransitive verbs is that the linking verb is followed by a complement that says something about the subject (either an adjective or a noun) whereas the intransitive verb is followed, if at all, by an adverb (not something that reflects on the subject).

Table 10.2 summarizes these main sentence patterns, and Table 10.3 provides examples. All of the sentences include a subject and predicate, and all but the sentence with the intransitive verb have a complement in the predicate. To test your understanding of these sentence patterns and verb types, try application 3 in Discussion and Application.

Adjectives, Adverbs, and Modifying Phrases

The simple subject and simple predicate form the core of the sentence, but most sentences include modifiers as well. Modifiers "modify" the meaning of the subject, verb, or complement by describing it and thereby limiting the number of features that it can have. For example, in the phrase *feasibility report*, "feasibility" is an adjective that establishes the subject as one particular type of report.

TABLE 10.3	**Examples of Sentence Patterns with Verb Types**	
transitive verb, direct object	Our supervisor created our web page.	"our web page" answers, "Created what?"
transitive verb, direct object, object complement	The new chip makes our computer obsolete.	"our computer" is the direct object of "makes" (makes what?); "obsolete" is the object complement describing "computer"
to be verb, subject complement	The web page is colorful.	"colorful" is an adjective modifying "web page"
intransitive verb	The page downloads slowly.	"slowly" is an adverb describing "downloads"

Modifiers may be adjectives and adverbs, but they may also be phrases. A phrase is a group of words that functions as a grammatical unit. The phrase may function as a noun (subject, object, or subject complement), verb, adjective, or adverb. For example, in the previous sentence, the prepositional phrase "as a noun" functions as an adverb modifying "functions." Phrases that function as adjectives or adverbs are modifying phrases. These phrases may be prepositional phrases, participial phrases, infinitive phrases, and appositive phrases.

Phrases are identified by their headword. For example, a prepositional phrase is introduced by a preposition, such as *by, from, to, for,* or *behind.* A noun phrase includes a noun and modifiers. An infinitive phrase includes the infinitive form of a verb, identified by *to.* A participial phrase includes a participle (a verb with an *-ing* or *-ed* ending that functions as a modifier). An appositive phrase renames the term just mentioned. Some common types of phrases and their functions in sentences are illustrated here:

Prepositional Phrase
adverbial: modifies "arrived"
The package arrived from New York City.

Noun Phrase
direct object of "obtained"
We obtained the required papers.

Infinitive Phrase
direct object of "will try"
We will try to complete the proposal tomorrow.

Participial Phrase
adjectival: modifies "technician"
The technician taking readings discovered a faulty procedure yesterday.

Appositive Phrase
adjectival: modifies "OKT3"
OKT3, a monoclonal antibody, helps to prevent a rejection crisis in kidney transplants.

Prepositional phrases may be adjectival as well as adverbial, and noun phrases may be subjects as well as objects.

A phrase never contains both a subject and a verb. A group of words with both a subject and verb is a clause. You will read more about clauses in the next chapter.

Relationships of Words in Sentences

Grammar describes the accepted ways of using words in relationship to one another to form sentences. The aim of grammar is clarity. Although structuring sentences according to the rules will not ensure that the resulting sentences can be understood, breaking these rules will almost certainly hinder understanding.

In this section, the principles of relationship are defined. Common errors that violate these grammatical principles are summarized in Table 10.4.

Subjects and Predicates

The most basic rule for forming sentences is that each sentence must contain a subject and a verb. A group of words that lacks both a subject and a verb is only a phrase. A phrase used to substitute for a sentence is a sentence fragment. Fragments are generally more difficult to understand than complete sentences because some essential information is missing. Subjects and verbs must agree with one another in number, and they must make sense together.

Subjects and Verbs Agree in Number

A sentence will be easier to understand if the subject of the sentence agrees in number with the verb. A singular subject takes a singular verb; a plural subject takes a plural verb.

TABLE 10.4	Common Grammatical Errors
Error	**Definition**
Subject-verb agreement	The subject and verb do not agree in number. A singular subject is used with a plural verb, or a plural subject is used with a singular verb.
Faulty predication	The predicate does not comment logically on the subject. Either the subject cannot complete the action that the verb defines, or the subject complement cannot be identified with the subject.
Dangling modifier	The modifier (participle, gerund, or infinitive phrase) defines an action that the subject of the sentence cannot perform.
Misplaced modifier	Misreading is likely because the modifier is separated from the item it is intended to modify. The solution is to move the modifier.
Pronoun-antecedent agreement error	The pronoun does not agree in number with the antecedent.
Ambiguous pronoun referent	It is unclear which of two previously named items in the sentence or a previous sentence the pronoun might refer to.
Pronoun case error	A pronoun in the subject case is used in the object position or vice versa.
Tense error	The tense does not accurately represent the time of the action.
Tense sequence error	The times of actions as implied by the verb tenses contradict logic, or the time shifts arbitrarily.

- Singular subject, singular verb:

  This store enforces the minimum age requirement for the purchase of alcohol.
  Adolescent drinking is sometimes an expression of rebellion against the parental
  authority figure.

- Plural subject, plural verb:

  Age limits on drinking encourage clandestine drinking by young people.
  Correlations between stress and alcoholism are high for all age groups.

An agreement error mixes a singular subject with a plural verb or vice versa.

The identity of words as singular and plural is not always obvious. Editors need to read closely, using the signals of sentence structure as well as knowledge of meanings of words.

- Collective nouns, such as *staff* and *committee,* may be regarded as either singular or plural, depending on whether you think of the group as a single entity or as a collection of individual members.

  The committee is prepared to give its report.
  All staff are responsible for completing their sick leave forms.

- The subject of the sentence, not the modifier closer to the verb, determines whether the verb is singular or plural. In the following sentence, it would be easy to read "fees" as the subject because it is next to the verb.

  An increase in parking fees was recommended.
  ["Increase" is the subject of the sentence.]

  NOT: An increase in parking fees were recommended.

- When *there* functions as a pronoun and begins a sentence, the verb agrees in number with the noun that follows the verb.

  There have been thefts in this location.
  There has been talk of hiring an all-night guard.

Subjects and Predicates Must Make Sense Together

Subjects and predicates must work together logically as well as grammatically. When your ear and intuition tell you that the sentence doesn't make sense, consider whether the subject can do what the verb says or if the subject complement can be identified with the subject. If not, the error is faulty predication.

These observations have concluded that 70% of new employees will be promoted within three years.

The sentence doesn't make sense because an observation cannot conclude; only a person can. The writer meant that the observations have shown or that, on the basis of the observations, researchers have concluded.

Faulty predication occurs when a human agent is required but missing from the subject. This statement does not mean that all subjects or agents of action must be human. Inanimate subjects can be agents of some actions.

These studies have revealed a relationship between diet and disease.

The report explains the marketing problems.

The sentence makes sense.

Types of documents and studies are sometimes agents of action because they are elliptical references to the people who wrote the document or conducted the study. Thus, a report presents, a study shows, and a committee recommends. But rules and principles cannot always explain when inanimate subjects can perform specific actions. One learns the conventions of a language by experience with it.

A Subject Complement Must Reflect on the Subject

Linking verbs and *to be* verbs link the subject of the sentence to a subject complement in the predicate that is identified with the subject. The subject complement must either modify the subject or be able to substitute for it.

Faulty predication occurs when the modifier or noun in the predicate cannot modify or substitute for the noun in the subject.

The application of DDT is one of the best chemicals to eradicate the rootworms.

"Application" is the subject, and "one of the best chemicals" is the complement. However, "application" is not a chemical, so the complement is faulty. A simple revision eliminates the redundant "application."

DDT is one of the best chemicals to eradicate rootworms.

Now the complement can be identified with the subject because DDT is a chemical.

Another revision eliminates "chemicals." It also changes the structure of the sentence by using a transitive verb ("eradicate") and an object rather than a *to be* verb and a subject complement.

The application of DDT could eradicate rootworms.

Faulty predication is about semantic relationships between the subject and predicate: between subject and transitive verb or between subject and complement with a linking or *to be* verb.

Verb Tense and Sequence

Verbs communicate relationships of time. The *tense* of a verb refers to the time when the action takes place—present, past, or future. This information helps readers create a sense of chronology. The six tenses in English are present, past, future, future perfect, past perfect, and future perfect.

Tense	Examples	Use
Present	I edit	Indicates action that takes place now or usually
	I am editing	Indicates continuous action in the present
Past	I edited	Indicates an action that has been completed
	I was editing	Indicates continuous action in the past

Tense	Examples	Use
Future	I shall edit	Places the action in the future
Present perfect	I have edited	Indicates a recently completed action
Past perfect	I had edited	Places the action before another action in the past
Future perfect	I shall have edited	Indicates an action that will be complete before some other event in the future

The Verb Tense Indicates the Time of the Action

Use the present tense for action that occurs in the present or for timeless actions. Most descriptive writing is in present tense.

> The page layout program includes functions for HTML markup.

The past tense is identified by the past participle of the verbs (formed by adding -ed to the verb unless the verb is irregular). It can also be identified by the helping verb *was*. Use the past tense only for action that occurred at a specific time in the past.

> The accident occurred last night.
>
> BUT: Newton illustrates the concept of gravity with the example of an apple falling from a tree.

The future tense is identified by the helping verbs *will* or *shall*. Use the future tense for action that will take place in the future

> When the new equipment arrives, we will convert to desktop publishing.

The "perfect" tenses (past, present, future) place action at a point of time in relationship to other action. In the sample sentence, both the editing and the revision occurred in the past, but the editing occurred before the revision.

> I had already edited the chapter when the writer unexpectedly sent me a revision.
>   past perfect tense   past tense

Shift Tenses Only for Good Reasons

Unnecessary shifts in tense disorient readers by requiring them to move mentally from the present to the past or future.

On the other hand, tenses may be mixed in a document when it describes actions that occur at different points in time. For example, a trip report may include statements in the past tense that recount specific events on the trip and also statements in the present tense that describe processes, equipment, or ongoing events.

> In our meeting, President Schober told us that STC plans a new recruiting drive this fall. The organization presently attracts technical communicators who are well established in their jobs. The recruiting drive will target new technical communicators.

The paragraph includes verbs in the present, past, and future tenses, but the tenses all identify accurately when the action occurs. A confusing shift is an arbitrary shift among tenses.

Modifiers

Modifiers are adjectives, adverbs, and modifying phrases. Readers must be able to perceive the relationship of the modifier to what it modifies. Some uses of modifiers obscure this relationship. Common misuses are called *misplaced* modifiers and *dangling* modifiers.

Misplaced Modifiers

A modifier, in English, generally precedes the noun it modifies. Readers would expect to read "tensile strength," for example, but would be startled and confused by "strength tensile." Adverbs generally follow the verbs or adverbs they modify.

If modifiers are placed too far from the term they modify, the sense of their meaning may be lost, and they may seem to modify something other than what is intended.

> We all felt an impending sense of disaster.

Is the sense impending or is the disaster impending? "Impending" (to hover menacingly) describes the anticipated disaster, not the sense.

> We all felt a sense of impending disaster.

Only and *both* are frequently misplaced. The restaurant owner who posted the sign, "Only three tacos for $2.89," really meant to say, "Three tacos for only $2.89." In the following two sentences, "both" serves first as an adjective modifying "cancer cells" and as an adverb modifying "to kill . . . and to carry." Placement of the modifier determines its function. The first sentence inadvertently claims that there are only two cancer cells.

> The monoclonal antibodies serve to kill both cancer cells and to carry information.

> The monoclonal antibodies serve both to kill cancer cells and to carry information.

Dangling Modifiers

A modifier, by definition, attaches to something. If the thing it is intended to modify does not appear in the sentence, the modifier is said to dangle. Instead of attaching to what it should modify, it attaches to another noun—the subject of the sentence—sometimes with ridiculous results.

> By taking a course and paying court fees, the charge for DWI is dismissed.

Only a person can take a course and pay a fee, but no person is in the sentence. The modifier, "By taking a course and paying court fees," dangles. The phrase modifies "charge," the subject of the sentence. The sentence says that the charge will take a course and pay a fee.

A modifier that dangles may be a verb form: a participle, gerund, or infinitive. (Participles, which function as adjectives, are formed by the *-ing* and *-ed* endings of verbs: for example, *studying, studied.* Gerunds, which function as nouns, are formed by the *-ing* ending. Infinitives are verb plus *to:* for example, *to study.*) These verb forms, called *verbals,* can function as adjectives, nouns, and adverbs, but they maintain some properties of verbs. They have implied subjects and may take objects. The implied subject of a verbal in the modifying phrase should be the same as the subject of the sentence that the phrase modifies; that is, the subject of the sentence should be able to perform the action implied in the verbal of the modifier.

Readers of the following examples, expecting that the subject of the verbal is the subject of the sentence, will be surprised to learn that a hose can change oil or a tax return can complete an application.

> <u>Changing</u> the oil, a worn radiator hose was discovered.   [dangling modifier with a participle]

> <u>To complete</u> the application, the tax return must be attached.   [dangling modifier with an infinitive]

Dangling modifiers often appear at the beginning of sentences, but shifting positions will not eliminate the problem because the implied subject of the verbal will still not be in the subject position of the sentence. Rearrangement can only disguise the problem.

> The charge for DWI is dismissed by taking a course and paying court fees.   [The modifier still dangles.]

One way to correct a dangling modifier is to insert the implied subject of the verbal into the sentence.

> By taking a course and paying court fees, a DWI offender may have the charge dismissed.

An alternative is to rephrase the sentence altogether to eliminate the modifying phrase (that is, to replace the phrase with a clause that contains its own subject and verb).

> If a DWI offender takes a course and pays court fees, the court will dismiss a DWI charge.

One frequent cause of dangling modifiers is a modifying phrase in combination with the passive voice. *Passive voice* means that the subject of the sentence receives the action of the verb rather than performing the action. "Charge" in "The charge may be dismissed" receives the action of dismissal; the sentence is in the passive voice. If the subject is receiving rather than performing action, it cannot perform the action implied by the verbal in the modifier. One way to reduce the possibility of dangling modifiers is to use active voice rather than passive voice.

Pronouns

A pronoun substitutes for a noun that has already been named or is clear from the context. Pronouns relate to other words in the sentence by number, case, and reference.

Pronouns Agree in Number with Their Antecedents

The noun that precedes the pronoun is the antecedent, meaning that it "comes before" the pronoun. Pronouns may be singular (*he, she, it, him, her, one*) or plural (*they, their*). If the antecedent is *people*, the pronoun will be *they*; but if the antecedent is *person*, the pronoun will be *he, she, he or she, she/he*, or *he/she* (depending on whether you know the gender of the person). The pronouns with "one" in them, including *everyone, anyone, none*, and *no one*, are singular. So are pronouns with "one" implied, such as *each, either*, and *neither*. In formal usage, they should be used with other singular antecedents and verbs.

> Every<u>one</u> should complete <u>his or her</u> report by Monday. [singular possessive pronoun with singular antecedent]

> NOT: Everyone should complete <u>their</u> report by Monday. [plural possessive pronoun with singular antecedent]

> Neither of the options is satisfactory. [singular verb with singular subject, "neither"]

> None of the students was prepared.

Informally, however, the singular pronouns are interpreted as plural. *Their* is increasingly accepted when following a pronoun with "one" in it, especially as a way of avoiding pronouns of gender. Pronouns with "one" in them are interpreted as referring to groups of people rather than individuals. Unless you know your readers and co-workers approve of this usage, however, it is safer to be conservative—or to avoid the issue altogether by using plural subjects.

> All applicants should submit their transcripts. [Better than "Each applicant should submit their transcript"]

Relative pronouns (*who, which, that*) used as subjects take the number of their antecedent.

> Applicants who send samples of their writing will have the best chance to be called for interviews.

> The editor who is my supervisor has helped me learn how to estimate time for various tasks.

Choose Pronoun Case According to the Pronoun's Role in the Sentence

Pronouns may be subjects of clauses (*I, we, you, he, she, they, anyone, each*); objects of prepositions (*us, me, him, her, them*); or possessives (*my, mine, your, yours, their, theirs*). Some pronouns can serve as either subjects or objects (*it, everyone, anyone, one*). Their *case* describes their role in the sentence: subject, object, or possessive.

If the pronoun is the subject of a sentence, use a pronoun that serves as a subject; if the pronoun is the object of a verb or preposition, use a pronoun that serves as an object. You may have been chastised as a child for saying, "Her and me are going out," using pronouns in the objective case for the subject. "She and I are going out" uses pronouns in the subjective case for the subject. Likewise, saying, "The coach picked he and I," uses pronouns in the subjective case as objects. The transitive verb requires objects: "The coach picked him and me." People seem self-conscious about using *I* and *me*. "Carlos and myself will conduct the study" errs because the personal pronoun in the subjective case is "I." The pronouns containing *-self* only appear when the noun or pronoun they refer to has already been named in the sentence: "Carlos made photocopies for me and for himself." In the subjective case, myself is used only for emphasis: "I myself will clean up the mess."

Who and *whom* are also confusing. *Who* is a subject; *whom* is an object. If a person does what the verb says, use *who*. If the person receives the action of the verb, use *whom*.

> The person who trained the new network administrator was sent by the computer company.

> The person whom we trained can now perform all the necessary page layout functions.

The Pronoun Must Refer Clearly to the Noun It Represents

Pronouns with more than one possible antecedent are ambiguous and may confuse readers.

> Cytomegalovirus (CMV) is a danger to those whose natural immunity has been impaired, such as persons with transplanted organs. Their immunity is deliberately weakened to prevent it from attacking the donor tissue.

Is "it" the virus or the immunity? Rereading will aid interpretation, but readers, more conditioned to thinking of viruses attacking tissue than immunity attacking tissue, may read "it" as "virus." The sentence demonstrates an error in pronoun reference.

Table 10.4 (page 158) summarizes some of the more common grammatical errors that arise when the principles discussed above are violated.

Conventions of Usage

Expressions in common use may technically violate grammar rules but still be acceptable. Likewise, some expressions in common use are not acceptable in professional writing. A few conventions that technical communicators use frequently are reviewed here: expression of quantities and amounts, use of relative pronouns, and idiomatic expressions.

Quantities and Amounts

The adjectives *fewer* and *less* and the nouns *number* and *amount* have specific uses and are not interchangeable. *Fewer* and *number* refer to measurable quantities, such as pounds, dollars, or liters.

> Fewer people attended the trade show this year than last, but the number of products sold was greater.

Less and *amount* refer to indefinite (not measurable) amounts, such as weight, money, and milk.

> The amount of gasoline consumed rose 50%.

> This shipment of apples shows less contamination from pesticides than the shipment in April.

Relative Pronouns

Who refers to persons; *which* and *that* refer to things. It would be preferable to say, "The people who came to the meeting brought tape recorders," rather than "The people that came. . . ."

In formal usage, *which* and *that* are distinguished according to whether the clause they begin is restrictive or nonrestrictive. A restrictive clause limits the noun's meaning and is necessary to the sentence. A nonrestrictive clause provides additional information about the noun, but it does not restrict the meaning of the noun. Some usage experts specify the use of *that* with a restrictive clause and *which* with a nonrestrictive clause. (See also Chapter 11.)

Restrictive clause, use *that*	The supplies <u>that</u> we ordered yesterday are out of stock. [Only those supplies ordered yesterday, not all supplies, are out of stock.]
Nonrestrictive clause, use *which*	Murray Lake, <u>which</u> was dry only two months ago, has flooded. [Even without the information about the dryness of the lake, readers would know exactly which lake the sentence names.]

Idiomatic Expressions

Some expressions in English cannot be explained by rules alone. In fact, their usage may contradict logic or other conventions. These expressions are idiomatic. It would be difficult, for example, to explain why we say *by* accident but *on* purpose, or why we can be excited *about* a project, bored *with* it, or sick *of* it. The use of prepositions, in particular, is idiomatic in English. One simply learns the idioms of the language through reading and conversation. Idioms are especially difficult for writers for whom English is a second language. These writers may be perfectly competent in English grammar but have too little experience with the language to know all its idioms. A copyeditor quietly helps.

Using Your Knowledge

No one memorizes all the rules of grammar that handbooks define. Neither do editors have time to consult a handbook each time they are uncertain about grammar. A systematic approach to editing for grammar will improve your efficiency and effectiveness.

1. **Edit top down**, from the largest structures in the sentence to the smallest—from clauses to phrases to words.

   a. Begin with the subject and predicate (the sentence core). Assess whether subjects and verbs agree in number and whether they work together logically. If the verb is a linking verb, assess whether the subject complement can be identified with the subject.

b. Check phrases for accurate modification and parallelism. If the sentence contains participles or infinitive phrases, make sure that the subject of the sentence can perform the action the modifier implies.

c. Check individual words—pronouns, modifiers, numbers.

2. **Increase your intuition about grammar by studying sentence structures.** Grammar rules describe the use of words and phrases within sentence structures. When the structures are familiar, grammar will not seem like hundreds of discrete rules, mostly focused on error, but rather like a coherent description of ways in which the parts of the sentence relate. Knowing the patterns will help you learn and remember the details.

3. **Edit for sentence structures, not just to apply rules.** Aim to clarify relationships between words in the sentence, not just to be "correct."

4. **Use a handbook for reference.** To use it efficiently, learn enough of the vocabulary of grammar to know what to look up.

5. **Respect the writer.** Don't add to the writer's anxiety about grammar by being unnecessarily critical. Just get it right for the sake of readers.

Further Reading

Beason, Larry, and Mark Lester. 2002. *A Commonsense Guide to Grammar and Usage.* 3rd ed. St. Martin's. An introduction.

Kolln, Martha. 2003. *Rhetorical Grammar: Grammatical Choices, Rhetorical Effects.* 4th ed. Allyn & Bacon. A sophisticated explanation of language as it works grammatically and rhetorically.

Smith, Edward L., and Stephen A. Bernhardt. 1997. *Writing at Work: Professional Writing Skills for People on the Job.* McGraw-Hill. Offers lively and readable explanations with workplace examples.

Also see grammar and usage handbooks published by major publishing companies.

Online Resources

Capital Community College. *Guide to Grammar & Writing.* www.ccc.commnet .edu/grammar. From the "Word and Sentence Level" menu, choose the topic for which you need more information. Examples are cases of nouns and pronouns, sentence variety and types, subject-verb agreement, and verbs and verbals.

Lynch, Jack. *Guide to Grammar and Style.* andromeda.rutgers.edu/~jlynch/Writing. Handbook arranged alphabetically.

Discussion and Application

1. *Parts of speech.* Identify the parts of speech of all the underlined words in the following sentences. If necessary, consult a dictionary or a grammar and usage handbook.

> [1]A genome is all the genetic material in the chromosomes of an organism. [2]The human genome includes about three billion base pairs that make up human DNA. [3]The Human Genome Project, completed in 2003, was a 13-year, international effort to identify the 20,000–25,000 human genes and make them accessible for further biological study. [4]This research has catalyzed biotechnology.

2. *Sentence parts.* For each of the sentences in application 1:
   a. Identify the simple subject, complete subject, complete predicate, and modifiers.
   b. Identify the type of verb (transitive, intransitive, linking, *to be*) and complement, if any.

3. *Sentence patterns.* Describe the predicate in each of the following sentences. Identify whether the verb is transitive, intransitive, linking, or *to be*. If necessary, consult a dictionary.

   Identify whether the remainder of the predicate consists of a direct object, a subject complement, or an adverb (or adverbial phrase). When you find subject complements, determine whether they are nouns or adjectives. Note that some verbs can be either transitive or intransitive.
   a. The Dow Jones Industrial Average fell by 53 points.
   b. I will lie down.
   c. She laid the newspaper down.
   d. A virus is a pathogen.
   e. The accident occurred yesterday.
   f. The program works well.
   g. The students worked the algebra problems.
   h. The marketing strategy seems successful.
   i. All members are present.

4. *Transitive and intransitive verbs.* The sentences in a and b are both common, but which is "correct"?
   a. She graduated from college.
   b. She graduated college.

Is the verb "graduated" transitive or intransitive? (Check your dictionary.) How does your answer define what should follow the verb? How does "from college" function in sentence a? How does "college" function in sentence b? How would an editor determine the better choice of sentences?

5. *Tense*

   a. For the sentences in application 3, identify whether the verb tense is present, past, or future.

   b. Edit this paragraph with particular attention to verb tense.

   > We take great pleasure in welcoming you to our staff. We hope our relationship is one of mutual understanding and support. The owners had many years of experience in the operation of successful and profitable businesses. We were fortunate in the past with our choices for our staff, and we sincerely hope that you will follow in this path.

6. *Subject, verb, and complement.* The following sentences contain errors in the verb or complement. For each sentence, first identify the simple subject and then the simple verb. (Remember that the grammatical subject may differ from the topic of the sentence.) If the verb is a linking or *to be* verb, identify the complement. Then identify and correct errors in subject-verb agreement or faulty predication.

   a. The overall condition of the facilities are good to very good.

   b. The resources dedicated to repair is minimal.

   c. The record of all courses attempted and completed appear on the transcript.

   d. A wide range of extracurricular activities are available to students.

   e. Shipment of factory sealed cartons are made from our warehouse via the cheapest and fastest way.

   f. The benefit of the annuity to the investor will be a source of additional retirement income.

g. The income from the annuity does not indicate that it will offset living expenses.

h. The agriculture industry is susceptible to pest problems.

i. Overloading on the library floor has moved approximately 20,000 volumes off campus to avoid structural damage to the building.

j. The interaction between the chemical mechanism and the dynamic mechanism appears to be the two important factors behind the depletion of the ozone layer.

k. The purpose of this section of the report is to increase the fatigue strength of an already welded joint.

7. *Dangling modifiers.* The following sentences contain dangling modifiers. Edit the sentences by inserting missing subjects into sentences or by converting modifiers to clauses. You may create two sentences from one if necessary for clarity. Which of the original sentences are written with passive voice verbs?

a. When preparing copy for the typesetter or when correcting errors on the screen, the cursor can be easily moved with the mouse.

b. Rather than make marks on the copy, the change can be placed in the computer for a faster and neater job.

c. The cost of production can be reduced by purchasing software and hardware for desktop publishing.

d. A mosquito bit Lord Carnarvon on his left cheek five months after entering King Tut's tomb.

e. Growing up to five feet long and weighing over 600 pounds, natives on the Moluccas Archipelago use the shells of the giant man-eating clam as children's bathtubs.

f. By using lead-free gasoline, harmful lead oxides and lead chlorides and bromides are not released into the atmosphere as is the case with leaded (regular) gasoline.

8. *Misplaced modifiers*. The following sentences contain misplaced modifiers. Edit to show where the modifier should go. In your own words, explain the difference between a misplaced modifier and a dangling modifier.

a. Only smoke in the break room.

b. Racquets with safety thongs and bumpers are only allowed.

c. Journalists must be able to operate equipment used to produce the stories such as computers.

9. *Pronouns: case, number, antecedent*. The following sentences include pronoun errors of various types. Identify the error and edit to correct the sentence.

a. A positive attitude allows the waitperson to laugh at oneself and learn from their mistakes.

b. If an investor wants to sell their shares of stock, they are sold at the market price at the time of sale.

c. He and myself will conduct a workshop on investments.

10. *Usage*
   a. What issues of usage do sentences 1 and 2 raise?
      (1) Use the express lane if you have less than ten items.
      (2) The project to edit the employee handbook could be divided between several students.

b. Consult your dictionary and handbook to determine what advice they may offer about usage on the following two issues:

(1) *Hopefully:* Is the adverb *hopefully* misused as a substitute for "it is hoped" in the sentence "Hopefully, we will finish before Friday"? Define the grounds on which the use of *hopefully* could be considered a matter of usage rather than of grammar.

(2) "The reason is because . . .": Check *The American Heritage Dictionary* under "because" for a usage note. Remembering sentence patterns with linking verbs and guidelines for subject complements, explain why the structure might be considered grammatically incorrect. Discuss how usage rather than grammar may determine whether a structure is acceptable.

11 Punctuation

Punctuation corresponds to sentence structure. In the previous chapter you reviewed the sentence patterns that follow from the use of transitive, intransitive, and linking verbs. By signaling the structures in sentences, punctuation provides readers with clues to meaning.

The chapter illustrates four sentence patterns and their punctuation as well as the punctuation of phrases. It aims to help you make good editorial decisions about punctuation by defining key terms: *clause* (independent or main and dependent or subordinate), *phrase, conjunction* (coordinating and subordinating), *relative pronoun, restrictive and nonrestrictive modifier,* and *parallel structure.*

Key Concepts

> Punctuation helps people to read a sentence accurately. Internal marks— commas, semicolons, dashes—as well as end punctuation reveal sentence patterns and relationships of clauses and phrases. Incorrect punctuation may create document noise and confusion. Knowing where to punctuate and why contributes to accurate editing.

Clauses, Conjunctions, and Relative Pronouns

As Chapter 10 explains, each sentence contains a subject and predicate. These components create the core of a sentence, the *clause*, which always includes both a subject and a verb. Sentences may contain more than one clause. The correct joining of the clauses in sentences depends on whether the clauses are independent or dependent and on the joining words, called *conjunctions*.

This chapter continues the analysis of sentence patterns by considering ways in which clauses may be joined in single sentences. Specific punctuation marks show the different ways to join clauses. The terms that describe structural parts of the sentence, including *clause, conjunction,* and *relative pronoun,* are defined and illustrated in this section.

Independent and Dependent Clauses

A *clause* is a group of words containing a subject and a predicate. The subject is a noun or noun substitute; the predicate tells what is said about the subject and must contain a verb. The clause is the fundamental unit of a sentence—all sentences must contain a clause. (Advertising copy and informal writing may use fragments, but most technical writing follows conventional rules.) Sentences may contain two or more clauses.

Clauses may be independent or dependent. An *independent clause* (sometimes called the main clause) includes the subject and predicate and can stand alone as a sentence.

A *dependent clause* (sometimes called a subordinate clause) contains a subject and predicate, but by itself it is a sentence fragment. (It is dependent on or subordinate to an independent clause.) Like an independent clause, it contains a subject and predicate, but it also includes a subordinating conjunction or relative pronoun. The inclusion of a subordinating conjunction or relative pronoun makes the clause dependent on an independent clause. The following list summarizes the definitions and gives examples.

	Structure	**Descriptor**	**Example**
Clause	subject + predicate	an essential part of a sentence	editor proofread
Independent (main) clause	subject + predicate	may stand alone as a sentence	The editor proofread the text.
Dependent clause	subordinating conjunction or relative pronoun + subject + predicate	by itself is a sentence fragment	Although the editor proofread the text, . . .

Subordinating conjunctions and relative pronouns are defined and discussed on the next few pages. For now, note in the example ("Although the editor proofread the text, . . .") that the dependent clause, because of the subordinating conjunction "although," makes readers expect that the thought will be completed elsewhere. This clause is dependent on the independent clause that presumably will follow.

When you are determining whether a group of words is a clause, make sure the verb is a verb and not another part of speech formed from a verb. Gerunds and participles, for example, are formed from verbs, but they function as nouns and modifiers. Thus, "The experiment being finished" is not a clause because it contains no verb. To turn this group of words into a clause, an editor could substitute a verb, such as *is* or *was*, for the participle "being."

The subject of a sentence in the imperative mood—phrased as a command—is understood to be "you." "Turn on the computer" is an independent clause because it means "[You] turn on the computer."

In order to punctuate sentences correctly, you must be able to distinguish a clause from a phrase. A phrase is a group of related words, but it does not contain a subject and predicate. Phrases function in a sentence as the subject or predicate, as modifiers, or as objects or complements. Because they do not contain both a subject and verb, however, they cannot be punctuated as independent clauses (or as sentences).

Conjunctions

When a sentence contains more than one clause, the clauses are often joined by a conjunction. As the root word *junction* suggests, the purpose of the conjunction is "to join." The type of conjunction, coordinating or subordinating, determines whether a clause is dependent or independent.

Conjunctions may be *coordinating,* joining independent clauses or other sentence elements (such as the terms of a compound subject) of equal grammatical value. Only seven words can function as coordinating conjunctions, all with three or fewer letters. Three of the words rhyme. You can memorize them.

> **Coordinating conjunctions**   *and, but, or, for, nor, yet, so*

Conjunctions may also be *subordinating,* joining a dependent clause to an independent clause. They establish that one clause is less important than another, or subordinate to it. If a clause contains a subordinating conjunction, it is a dependent or subordinate clause. You probably won't memorize this list, but being familiar with it will help you make punctuation decisions accurately and quickly.

> **Subordinating conjunctions**   *after, although, as, because, if, once, since, that, though, till, unless, until, when, whenever, where, wherever, while* (etc.)

Recognizing conjunctions and their type will help you determine whether the clause you are punctuating is independent or dependent.

Relative Pronouns

Relative pronouns relate to a noun already named in the sentence. They introduce dependent clauses (dependent because readers must know the noun already named in the sentence to understand the meaning of the clause). Like subordinating conjunctions, relative pronouns make clauses dependent.

> **Relative pronouns**   *that, what, which, who, whoever, whom, whomever, whose*

Clauses beginning with relative pronouns may be the subject of the sentence, or they may be modifiers.

> <u>Whoever comes first</u> will get the best seat.   [The dependent clause is the subject of the main clause.]

> The library book <u>that Joe checked out</u> is overdue.   [The dependent clause, with the subject "that," modifies "book."]

These groups of words introduced by relative pronouns are called clauses, not phrases, because they contain subjects and predicates (the pronoun is the subject). They are dependent clauses because the relative pronoun must relate to something in another clause.

To summarize, a clause contains a subject and verb, and each sentence must contain at least one. Subordinating conjunctions and relative pronouns make clauses dependent. A dependent clause must be attached to an independent clause to form a complete sentence. A phrase, like a clause, is a group of words, but a phrase does not contain both a subject and a verb. To punctuate sentences accurately, you will need to identify clauses as independent or dependent.

Sentence Types and Punctuation

The number and type of clauses in a sentence determine its type and its internal punctuation. The four types of sentences are simple, compound, complex, and compound-complex.

Sentence type	Structure	Example
Simple	one independent clause (subject + predicate)	Rain fell.
Compound	two independent clauses	Acid rain fell, and the lake was polluted.
Complex	one independent clause plus one dependent clause	Because the factory violated emission standards, its owners were fined.
Compound-Complex	two independent clauses plus a dependent clause	The factory that violated emission standards was fined; therefore, the owners authorized changes in procedures and equipment.

Each type of sentence uses particular punctuation marks to signal the clauses. The punctuation patterns are summarized in Figure 11.1 and explained here.

Punctuating Simple Sentences: Don't Separate the Subject and Verb with a Single Comma

Simple sentences contain one independent clause. Simple sentences have end punctuation only unless there are introductory phrases or interrupting elements within the sentence. No single commas should separate the subject from the predicate because separation interferes with perception of the predicate as a comment on the subject.

Paradoxically, although a single comma separates, a pair of commas does not—the comma pair functions like parentheses. A pair of commas may enclose an interrupting modifier or an appositive phrase.

> WRONG: Some student editors with good grades in English, assume they know grammar well enough to edit without consulting a handbook.   [The comma separates subject and verb.]

> CORRECT: Leigh Gambrell, a good student in English, edits effectively for grammar. [The pair of commas sets off the modifying phrase.]

Simple sentences may contain a compound subject, verb, or object. Nevertheless, they are still simple sentences (because they contain only one subject), and they generally should be punctuated as simple sentences. If a comma is inserted before a second verb appears, it separates this verb from its subject.

The following sentences use an unnecessary comma that separates the second verb from the subject:

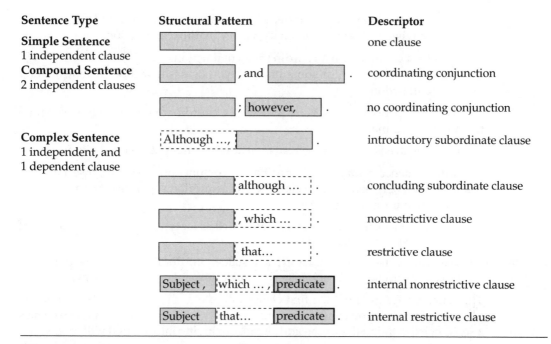

Figure 11.1   Punctuation Options for Sentence Types

The shaded boxes with solid borders represent independent clauses. The clear boxes with dotted borders represent dependent clauses. Compound-complex sentences, not represented here, combine the patterns for compound and complex sentences. Note that the semicolon appears only in the compound sentence that lacks a coordinating conjunction.

MISLEADING PUNCTUATION: This draft prunes away some wordiness, and attempts to present the remaining information more efficiently. [single clause, compound verb]

MISLEADING PUNCTUATION: The writer prepared the draft of the text on the computer, but drew the visuals by hand.

Perhaps the commas in the sentences appeared because the writer expected to breathe after the long beginning. Perhaps "and" or "but" seemed to begin a clause rather than the second part of a compound verb or object. However, it is illogical to separate parts of the sentence that belong together. The sentence type is simple; the comma should be deleted, or a pronoun should be added before the second verb, which would make the sentence compound.

Punctuating Compound Sentences: Determine Whether There Is a Coordinating Conjunction

Compound sentences contain two independent clauses. Internal sentence punctuation (comma, semicolon, colon, or dash) signals the end of one clause and the beginning of another. These punctuation marks cause readers to anticipate a new subject and a new verb.

There are two ways to join independent clauses in a compound sentence depending on whether they are joined by a coordinating conjunction:

1. Use a comma with a coordinating conjunction.
2. Use a semicolon (or sometimes a colon or dash) without a coordinating conjunction.

If the independent clauses are joined by a coordinating conjunction, separate them with a comma.

Joe catches typos when he proofreads, but he is less careful about word-division errors.

If the sentence contains no coordinating conjunction, however, insert a semicolon or colon between the two clauses. The semicolon separates statements relatively equal in importance.

Good proofreaders check for omissions as well as for typos; they also note details of typeface, type style, and spacing.

The dynamic mechanism theory explains sudden and drastic concentration changes of ozone; the chemical mechanism theory cannot account for these huge depletions.

The colon is appropriate if the first clause introduces the second, if the second explains the first, or if you could insert "namely" after the first clause. A colon creates a sense of expectation that some additional, qualifying information will follow.

One fact remains clear: XML provides a flexible way to identify information.

Comprehensive editing is rhetorical: it begins with an editor's understanding of the document's audience, purpose, and context of use.

The Conjunctive Adverb in the Compound Sentence

Clauses in compound sentences are often linked by the conjunctive adverbs *consequently, however, moreover, besides, nevertheless, on the other hand, in fact, therefore,* and *thus.* Conjunctive adverbs both join and modify. They are often mistaken for subordinating conjunctions and thus invite punctuation errors. *However* and *although* are easily confused because they both signal contrasting information, but they are different parts of speech and affect clauses differently. If you use only a comma between clauses in a compound sentence whose second clause begins with a conjunctive adverb, you will have an error called a comma fault. A semicolon must be used with a conjunctive adverb in a compound sentence following the pattern of separating independent clauses with a semicolon unless they are joined by a coordinating conjunction.

COMMA FAULT: We tag our documents with XML, therefore, we can easily customize information for different users.

CORRECT PUNCTUATION: We tag our documents with XML; therefore, we can easily customize information for different users.

The first sentence in the following pair consists of two independent clauses (a compound sentence). Because the sentence has no coordinating conjunction, the clauses are separated with a semicolon. The second sentence in the pair consists of an independent plus a dependent clause introduced by the subordinating

conjunction *although*. Because the sentence pattern is complex rather than compound, it requires no internal punctuation.

> We have no leather recliners in stock; however, we do have vinyl ones.
>
> We have no leather recliners in stock although we do have vinyl ones.

> If *however* appears in a single clause, it is set off in the sentence with commas.

> We have no leather recliners in stock. We do, however, have vinyl ones.

The second sentence is a simple sentence (with one independent clause) and
therefore requires no semicolon.

You will make the right decision about punctuating clauses within a sentence
if you can identify which clauses are independent and which are dependent. As
the example of *however* illustrates, it is smarter to punctuate clauses than to try to
punctuate words. (*However* sometimes follows a semicolon and sometimes follows a comma.) Although you do not need to memorize the entire list of subordinating conjunctions, you may need to memorize the common conjunctive adverbs
and recognize that they are *not* subordinating conjunctions.

Punctuating Complex Sentences

Complex sentences contain both an independent and a dependent clause. There
may be no punctuation between the clauses, or they may be separated by a
comma. Semicolons and colons should not be used to separate independent from
dependent clauses.

An introductory clause with a subordinating conjunction is followed by a
comma. A comma is usually not necessary if the dependent clause follows the independent clause.

> Although we use spelling and grammar checkers, we also depend on editors to read
> for meaning.
>
> We depend on editors although we also use a spelling checker.
>
> We depend on editors because spelling checkers cannot determine meaning.

A comma before the concluding dependent clause may sometimes be necessary to clarify meaning, especially if the main clause includes "not." Compare the
following two sentences to see how a comma can change meaning.

> I did not take the job because I want to live in a warm climate.   [Presumably, the writer took
> the job for reasons other than the opportunity to live in a warm climate.]

> I did not take the job, because I want to live in a warm climate.   [Presumably, the unattrac-
> tive climate was the reason why the writer did not take the job.]

Dependent clauses formed by relative pronouns (who, which, that) are modifiers. Whether a modifying clause should be separated from the main clause
with a comma or not depends on whether the modifying clause is restrictive or
nonrestrictive.

Restrictive Clause (Essential to the Definition)

A *restrictive clause* restricts the meaning of the term it modifies. The term may suggest a class of objects, but the restrictive modifier clarifies that the term in this sentence refers only to a subclass.

> Concrete that has been reinforced by polypropylene fibers is less brittle than unreinforced concrete.

In this sentence, the term *concrete* refers to all concrete. The clause "that has been reinforced by polypropylene fibers" clarifies that the sentence is not about all concrete but just the concrete that has been reinforced. The clause restricts the meaning of concrete as it is used in the sentence.

To illustrate visually, the circle below represents all concrete. The shaded segment represents the subject of the sentence—the concrete that has been reinforced. The information in the clause has restricted the meaning of concrete.

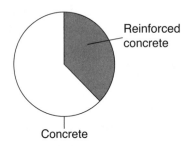

Because the restrictive modifier must be understood in conjunction with the term it modifies, it would be illogical to separate it from the term with a comma. If the clause is restrictive, do not use a comma.

Nonrestrictive Clause (Meaningful but Not Essential)

A *nonrestrictive clause* adds additional information, but it does not restrict the meaning of the term it modifies. Although removing the clause from the sentence would cost the sentence some meaning, a reader would still understand the meaning of the term. Because the term can be identified on its own, it does not need to be joined to its modifier. Use a comma to separate a nonrestrictive clause from the term it modifies. Use a pair of commas if the dependent clause falls within the sentence rather than at the beginning or end.

> Research in the reinforcement of concrete with polypropylene fibers led Shell Chemicals to develop their "caracrete" material, which consists of polypropylene fibers and concrete.
>
> Caracrete, which consists of polypropylene fibers and concrete, can be used in . . .

Usually a proper name or specific term will be followed only by a nonrestrictive clause. "Caracrete" defines a material that is not further subclassified. The nonrestrictive clause that follows tells readers something about the material, but it does not define just one type of caracrete. On the basis of the information in the restrictive clause, we could not draw a circle representing "caracrete" and identify a pie slice that represented just one type of caracrete.

In formal usage, a restrictive clause begins with *that,* while a nonrestrictive clause begins with *which.* Because most writers do not know or do not apply that distinction, it is an unreliable clue about whether the clause is restrictive or nonrestrictive. Readers will depend on the comma or the lack of it rather than on discriminating between *that* and *which.* Usually you can determine whether the clause is restrictive or nonrestrictive by close reading of the sentence on its own and in context. If you cannot, query the writer to clarify meaning.

When the relative pronoun in a restrictive or nonrestrictive clause refers to people rather than to objects, use the pronoun *who* rather than *that.* Never use *which* to refer to people.

> The proposed computer center would benefit walk-in customers, who would receive inexpensive and prompt service.   [nonrestrictive clause, relative pronoun referring to people]

The table below summarizes the distinction between restrictive and nonrestrictive clauses.

	Function	Punctuation	Preferred Usage
Restrictive Clause	essential to the definition	no comma	that
Nonrestrictive Clause	meaningful but not essential	comma	which

Punctuating Compound-Complex Sentences

To punctuate sentences that contain two independent clauses and a dependent clause, you apply the rules both for punctuating compound sentences and for punctuating complex sentences. Begin with the biggest structures, the two independent clauses. If the clauses are joined by a coordinating conjunction, separate them with a comma preceding the conjunction; otherwise, use a semicolon (or possibly a colon or dash). Next, look at the dependent clause. If the dependent clause introduces the independent clause, place a comma after it. If the dependent clause modifies the subject, determine whether it is restrictive or nonrestrictive, and punctuate accordingly.

Punctuating Phrases

Phrases often require punctuation within the sentence, usually commas. The commas mark the beginnings and endings of phrases and signal whether they are restrictive or nonrestrictive modifiers.

Series Comma and Semicolon

A series is a list within a sentence of three or more items such as terms, phrases, or clauses. Parallel structure means that items with related meanings share a common grammatical form. The items in the list may be a series of prepositional phrases ("by planning, by scheduling tasks, and by evaluating"), or nouns, adjectives, infinitive phrases, clauses, or other structural units that may appear in a series. The structural similarity helps to identify the items in the list and to establish that they are related.

The items in the series are generally separated by commas, with the conjunction *and* preceding the final item. The commas signal to readers where one item ends and another begins. In fact, conventions about whether to place a comma before the final item in the series have changed over time and differ, to some extent, among disciplines. However, most handbooks (except in journalism) recommend the comma. The comma is especially important for clarification when one item in the series includes a conjunction. A coordinating conjunction may signal that one of the items in the series has two components as well as signal the end of the series. The comma establishes with certainty that the series is ending.

Some readers may stumble over the following sentence:

> Good hygiene, such as sneezing into a tissue, washing hands before handling food and cleaning utensils and cutting boards thoroughly, eliminates most of the risk of transmitting bacteria in food handling.

On first reading, a reader may regard "food and cleaning utensils" as objects of "handling." Rereading and reasoning will confirm that the writer is not advising washing hands before handling cleaning utensils, but reading time has been wasted. One problem is that the series is built on *-ing* words ("sneezing," "washing," "cleaning"), but other *-ing* words ("handling," "transmitting") interfere with the recognition of their relationships. A comma before the final item could eliminate the need to reread.

> Good hygiene, such as sneezing into a tissue, washing hands before handling food, and cleaning utensils and cutting boards thoroughly, eliminates most of the risk of transmitting bacteria in food handling.

If you are editing according to journalistic style, you will not use the comma. Appreciation of the possible confusion resulting from its absence, however, should encourage you to restructure sentences to minimize the possibility of misreading.

When items in a series contain subdivisions separated by commas, semicolons may be used to separate the main items.

> The editing course covers sentence-level issues such as grammar and spelling; whole-document issues such as style, organization, and format; and design and production issues such as layout and electronic distribution.

When you edit a structurally complex series, punctuate the main divisions first and then punctuate the smaller divisions.

Commas with a Series of Adjectives (Coordinate Adjectives)

If adjectives in a series both or all modify the noun, use a comma between them but not after the final one.

The older, slower models will be replaced. [The models are both older and slower.]

Sometimes one adjective must be understood as part of the term modified. The phrase *theoretical research* identifies a type of research that differs from *empirical research.* The phrase *recent theoretical research* does not require a comma between the two adjectives, "recent" and "theoretical," because "recent" modifies the whole phrase *theoretical research* rather than *research* alone.

Parallelism

Parallelism means that related items share a grammatical structure. Faulty parallelism results when the structure of items in a series shifts. Faulty parallelism creates confusion for readers because the structural signals contradict the meaning of the words. Because of faulty parallelism, readers may not interpret the sentence correctly. In the following sentence, the shift from the noun form, "setting," to the verb form, "remove" and "install," reveals a shift from description to instruction, inappropriate in a list of steps.

> FAULTY PARALLELISM: The major steps in changing spark plugs are setting the gap on the new plugs, remove the spark plug wire, and install the new plugs.

The noun form is the correct one to follow the *to be* verb, because the information that follows such a verb either modifies or substitutes for the subject of the sentence.

> The major steps in changing spark plugs are setting the gap on the new plugs, removing the spark plug wire, and installing the new plugs.

What appears as faulty parallelism is often incorrect punctuation. You may recognize faulty parallelism first but edit the sentence by restructuring. For example, the punctuation in the following sentence signals three research tasks, but analysis reveals that the punctuation is misleading.

> MISLEADING PUNCTUATION: In researching the career of an agricultural scientist, I consulted the *Occupational Outlook Handbook,* Standard and Poor's *Index of Corporations,* and interviewed a research associate at the Agricultural Experiment Station.

As you analyze the structure of sentences in order to punctuate, it may help to list key terms in outline form. The outline, which preserves the three-part series, reveals the faulty parallelism in the sentence.

> consulted
> Standard and Poor's
> and interviewed

The first and third items are verbs, but the second is the name of a book. Thus, you might try a second outline.

> consulted a and b
> and interviewed . . .

The second outline reveals how the predicate includes a compound verb, not a series of three, with the first verb taking two objects. Two items are joined by a coordinating conjunction, without punctuation.

> In researching the career of an agricultural scientist, I consulted the *Occupational Outlook Handbook* and Standard and Poor's *Index of Corporations* and interviewed a research associate at the Agricultural Experiment Station.

The punctuation identifies the structure of the sentence as a simple sentence (structurally speaking) with a compound verb. The verbs are joined by the coordinating conjunction "and," without punctuation. However, because of the complexity (compound object and compound verb), it will help readers if you create two independent clauses here by repeating the "I."

> In researching the career of an agricultural scientist, I consulted the *Occupational Outlook Handbook* and Standard and Poor's *Index of Corporations,* and I interviewed a research associate at the Agricultural Experiment Station.

A series of three is often easier to understand than a double compound. If you can preserve the meaning accurately by converting a double compound to a series of three, choose the series of three.

Original (faulty parallelism)	Thousands of openings will occur each year resulting from growth, experienced workers transferring, or retiring.
Double compound	Thousands of openings will occur each year resulting from growth and from experienced workers transferring or retiring.
Series of three	Thousands of openings will occur each year resulting from growth, transfers, and retirements.

When the items in a series are prepositional or infinitive or noun phrases, the word that introduces the phrase (for example, the preposition *to* or the article *the*) may or may not be repeated for each item. The choice depends on the complexity of the sentence. It may be sufficient to use the introductory word just once, in the first item.

> The accountant is responsible for preparing the balance sheet, analyzing the data, and reporting findings to the manager.

The series is short, and the three responsibilities are evident. In a more complex series, however, the repetition of the introductory word with each item will help readers recognize when they have arrived at subsequent items on the list.

> Damaged O-rings have been tested in order to determine the extent of failure in the material and the temperature at which failure occurred and to evaluate the effect of joint rotation.

Be consistent in your choice.

Introductory and Interrupting Phrases

Introductory phrases are usually followed by a comma unless they are quite short. The comma signals that one main part of the sentence, the subject, is about to begin. Phrases at the ends of sentences usually are not preceded by commas. Those phrases are part of the predicate rather than a separate division.

In the appendix to our style guide, we have listed the correct spellings for specialized terms that are frequently used in our publications. [introductory phrase followed by a comma]

Interrupting phrases (and words) are set off by a pair of commas or none, but not by a single comma. A single comma would interrupt another structural part. A pair of commas acts like parentheses. You "close" the commas just as you close parentheses.

Use a pair of commas for an interrupting adverb, for an appositive phrase, and for another modifying phrase that is nonrestrictive.

Natural fibers, such as cotton, silk, and wool, "breathe" better than fibers manufactured from oil byproducts. [A pair of commas sets off the prepositional phrase beginning with "such as."]

On May 6, 2005, we introduced our accounting software at the convention. [A pair of commas sets off the year.]

If the internal phrase, such as an appositive, is restrictive, don't use any commas.

Plants such as poison ivy may cause allergic dermatitis.

"Such as poison ivy" identifies a small category of plants. It restricts the meaning of "plants" to those that are like poison ivy. Other plants do not cause allergic dermatitis.

Punctuation within Words

The punctuation discussed thus far reveals sentence structure. Punctuation within words also gives readers useful information. This category of punctuation includes the apostrophe and the hyphen.

The Apostrophe

Apostrophes show either possession or contraction. The apostrophe before a final *s* distinguishes a possessive noun or pronoun from a plural. For example, *engineer's report* and *bachelor's degree* show that the second term in some way belongs to the first. Whenever you could insert the conjunction *of* in a phrase, the noun or pronoun is likely to be possessive and should have an apostrophe. Although we probably wouldn't say "report of the engineer" or "degree of a bachelor," it would not be wrong to do so, and the implied "of" points to the need for an apostrophe.

Its and other pronouns (*his, hers, theirs, whose*) are exceptions to the use of an apostrophe to show possession. The word *it's* is the contraction for "it is"; *its* is the possessive, as in "its texture" or "its effect." *Who's* means "who is."

Many writers ignore the apostrophe or confuse them with plurals. You may have seen signs like the one in the grocery store announcing "banana's for sale" or the building directory identifying an office for "Veterans Adviser's." A visual mnemonic device to remember whether plurals or possessives require the apostrophe is this: for a *pos*sessive, use an a*pos*trophe (except for pronouns).

Words can be both plural and possessive. In these cases, you form the plural before you show the possessive. For nouns whose plural ends in *s*, the *s* will precede the apostrophe, as in "students' papers" (more than one student) or "pipes' corrosion" (more than one pipe). Irregular plurals are an exception: "women's studies."

The plurals of years or abbreviations do not require an apostrophe (the *s* alone forms a plural):

> the 1920s, the late 1800s
> IQs, YMCAs

Finally, with a compound subject, use the possessive only on the second noun.

> Mark and Steve's report

The Hyphen

The hyphen shows that two words function as a unit. Often new terms introduced into the language are formed from two familiar words. These new terms are compound nouns and compound verbs, such as *cross examination* and *cross-examine*. Compound terms may be hyphenated, "open" or "spaced" (that is, two words, unhyphenated, such as *type style*), or "solid" or "closed" (that is, a single word, such as *typesetting*). Hyphens are also frequent in modifiers when two terms join to modify another, but not all modifiers are hyphenated.

You can often determine which compounds are hyphenated by consulting a dictionary. Check the dictionary, for example, to determine whether to hyphenate "cross" compounds. You will find inconsistencies: *cross section* but *cross-reference* and *crosscurrent; crossbreed* but *cross-index; halftone* but *half note* and *half-truth.* Where the dictionary is incomplete, consult a style guide for general principles.

The same compound may be treated differently depending on the part of speech. For example, "back up" is open as a verb but solid as a noun or adjective.

> We back up our computer files every week. We create a backup copy on a CD.

Noun Forms

Noun forms are more likely to be open or solid than hyphenated. Noun phrases formed from two nouns are usually open or solid. The noun phrase *problem solving*, for example, is not hyphenated; however, the nouns *lovemaking* and *bookkeeping* are closed up as one word. Noun phrases formed from a noun and an adjective are more likely to be hyphenated or solid. Thus, the "self" compounds—for example, *self-service, self-study,* and *self-treatment*—are hyphenated. *Hydrochloric acid,* however, and many other compounds of noun plus adjective, are open. Fractional numbers, such as *one-half* and *two-thirds*, are hyphenated.

Adjective Forms

Compound adjectives are hyphenated if they precede the word modified and if they are formed in the ways listed in the examples that follow:

- Adjective or noun + past participle (if the compound term precedes the noun):

country-smoked ham	BUT: The ham was country smoked.
computer-assisted editing	BUT: The editing was computer assisted.
dark-haired girl	BUT: The girl was dark haired.

- Noun + present participle:

decision-making responsibility	BUT: decision making (noun)
interest-bearing account	BUT: an account that is interest bearing

- Compounds with "all," "half," "high," or "low" (whether they precede or follow the noun):

all-purpose facility, all-or-none reaction, all-out effort, all-around student	
half-raised house	BUT: halfhearted effort, halfway house
high-energy particles, high-grade disks	BUT: high blood pressure

- Compounds with "well" (if they precede a noun):

well-researched report	BUT: The report was well researched.

  Compounds with "well" are not hyphenated in a predicate.

Do not use a hyphen in compound adjectives in these situations:

- Two proper names:

  Latin American countries

- Two nouns (unless there might be some confusion):

  blood pressure level

Unit Modifiers with Numerals

When a numeral forms part of a compound adjective, it should be hyphenated when it precedes the noun it modifies. Thus, we would write "two-unit course" or "six-foot fence." The hyphen is particularly important if two numerals are involved.

three 2-liter bottles

A predicate adjective, however, is not hyphenated.

The fence is six feet tall.

Make sure that the phrase is a modifier and not simply a measure.

a two-semester course	BUT: a course lasting two semesters

If you have a series of unit modifiers, repeat the hyphen after each numeral.

The foundation offers two- and three-year scholarships.

Finally, spelled-out fractions are hyphenated.

three-fourths full, one-half of the samples

Color Terms

If one color term modifies another, do not hyphenate.

> bluish gray paper

If the two color terms are equal in importance, however, insert a hyphen.

> blue-gray paper

Combination terms working together as a unit are hyphenated before the modified term but not in the predicate.

> black-and-white photography   BUT: The photograph is black and white.

Prefixes and Suffixes

Generally, prefixes and suffixes are treated as part of the word and spelled solid although some prefixes are hyphenated to prevent misreading. For example, the prefix *re-* is hyphenated to distinguish it from another word: *re-create* (versus *recreation*). The dictionary will show both hyphenated and solid alternatives for some of these types of words. In that case, you should choose one form and place it on your style sheet so that you can be consistent in future choices.

When in Doubt, Leave It Out

If you judge that readers will read accurately without the hyphen, chances are you don't need one—except for unit modifiers with numerals. This principle also acknowledges a trend in language development away from hyphens. In the history of a word's development, the two familiar words that form a new compound may be hyphenated initially so that readers will see their relationship. When, over time, the compound becomes familiar, the hyphen isn't needed for this information and often disappears. This trend is one reason why *copyediting* is spelled solid in this book, though you will see it spelled elsewhere both as *copy editing* and *copy-editing*. *Proofreading*, an older term, has become a permanent compound spelled solid though it began as *proof reading* and then became *proof-reading*. Increasingly, *email* replaces *e-mail*. Likewise, the words *longterm* and *fulltime* (and even *shortterm* and *parttime*) are often spelled solid. Within a document, you should always hyphenate consistently.

Marks of Punctuation

The discussion in this chapter so far has emphasized punctuation as a clue to meaning. As an editor, you also need to know the conventions of placing the marks of punctuation. This section reviews common uses of quotation marks, parentheses, dashes, colons, and ellipsis points.

Quotation Marks

In editing for a North American audience, place quotation marks outside a comma or period but inside a semicolon or colon.

The prosecutor described the defendant as "defiant and uncooperative."

The defendant was described by the prosecutor as "defiant and uncooperative"; however, the defendant's behavior changed once he took his place on the witness stand.

British usage is to place quotation marks inside the comma and period as well as inside the semicolon and colon.

Do not use quotation marks with a block quotation (one that is indented and set off from the rest of the text). In such a situation, they are redundant, as the block form identifies the material as a quotation. Do retain the quotation marks if you are repeating dialog, however.

Parentheses

If parentheses enclose an entire sentence, include the end-of-sentence punctuation within the parentheses. But if parentheses enclose only part of a sentence, place the period or other punctuation outside the parentheses.

Twelve reports in the library deal with air pollution and acid rain in various ecosystems (e.g., the urban environment, streams, forests, and deserts).
[The period is outside the parenthetical part of the sentence.]

Twelve reports in the library deal with air pollution and acid rain in various ecosystems. (Examples of ecosystems are the urban environment, streams, forests, and deserts.)
[The entire sentence is parenthetical; the period is inside the parentheses.]

Attach parenthetical cross-references and references to visuals to a sentence, or create a separate sentence for the cross-reference within parentheses. Do not let a parenthetical cross-reference "float" in a paragraph without sentence punctuation.

The survey revealed that 40% of the employees favor a profit-sharing plan for retirement (see Table I).

OR: . . . plan for retirement. (See Table 1.)

Dash

An em dash—the length of two hyphens without space around them—can substitute for parentheses, show a break in thought, or provide emphasis. In the sentence you just read, they function as parentheses. Commas could have been used, but the dashes are more emphatic. Dashes also signal additional information at the end of the sentence that helps a reader interpret the significance of the primary information in the sentence.

Some state prison systems apply the policy of risk-group screening for AIDS only to pregnant women—a very small number of inmates.

Dashes have their proper and formal uses, but overuse will diminish their power to emphasize, and constant breaks in thought will make the writer seem immature or disorganized. There are no rules about how many are too many; you will have to trust your hunches and common sense.

Colon

The colon can introduce a list after a clause, but it is often used superfluously and therefore incorrectly. Not every list needs to be introduced with a colon. Do not use a colon between a preposition and its object nor between a verb and its complement or object. (See Chapter 10 and the glossary for definitions of these terms.)

WRONG: Natural fibers, such as: cotton, silk, and wool, are comfortable next to the skin.

WRONG: Three types of communication are: written, oral, and graphic.

CORRECT: A technical communicator is competent at three types of communication: written, oral, and graphic.

In the sentences labeled "wrong," the colon separates the preposition from its object or the *to be* verb from its subject complement. The preposition requires the object, and the *to be* verb requires a complement. Punctuation should not separate necessary parts of sentences. In the sentence labeled "correct," the verb and complement are complete before the colon.

Ellipsis Points

Ellipsis points (three spaced periods) indicate that some words have been omitted from a quotation. They are rarely used at the beginning or end of a quotation. If the ellipsis comes at the end of a sentence, insert a fourth period before the ellipses.

Typing Marks of Punctuation to Emulate Typesetting

Some typing habits linger from the days of typewriters, but the characters produced by word processing emulate professional typesetting. For example, to create the em dash, you can press certain key combinations to create a solid line of em dash length rather than the broken line created by two hyphens. Likewise, you can set your options to create curly "smart quotes" rather than straight quotes, and ellipsis points that won't break at the ends of lines. (In Microsoft Word, use Tools > AutoCorrect > AutoFormat.)

Although many people type two spaces after a period or colon (a remnant of typewriter days), professional typesetters use only one space. Before you submit an electronic file to the typesetter, use the edit/replace function in the word processing program to replace the double spaces with single spaces.

Using Your Knowledge

The strategies that work in editing for grammar also work in editing for punctuation. As in editing for grammar, aim to clarify meaning; do not apply rules arbitrarily.

1. **Edit top down**, from the largest structures in the sentence to the smallest—from clauses to phrases to words.

   a. Look for clauses and determine the sentence type. Determine whether any clauses are dependent (because they include a subordinating conjunction or relative pronoun). Punctuate the clauses according to Figure 11.1.
   b. Identify introductory or internal phrases and words or phrases in a series. Use punctuation in pairs around nonrestrictive internal phrases.
   c. Identify words with hyphens and apostrophes. Mark them according to the dictionary and your style sheet.

2. **Use a dictionary.** If you are not sure of a part of speech and how a word is affecting a sentence pattern, look it up.

3. **Take pride in editing on the basis of knowledge.** You are a technical expert on punctuation just as the person whose work you edit may be a technical expert on computer hardware or biotechnology. Editing and speaking like the expert you are encourages others to respect you and encourages you to edit expertly. Thus, you will not advise punctuation on the basis of "breathing" and "pausing" but rather according to sentence patterns.

If you learn the basic principles and terms in this chapter, read closely for meaning when you edit, and consult a handbook when you have questions, you will offer substantial value to the work of writers and to the ease by which readers can read and comprehend.

Further Reading

Alred, Gerald J., Charles T. Brusaw, and Walter E. Oliu. 2003. *Handbook of Technical Writing*. 7th ed. St. Martin's.

Truss, Lynne. 2003. *Eats, Shoots & Leaves: The Zero Tolerance Approach to Punctuation*. Gotham Books. The witty, narrative style of this book makes learning about punctuation fun.

Also see grammar and usage handbooks published by major publishing companies.

Online Resources

Capital Community College. *Guide to Grammar & Writing*. www.ccc.commnet .edu/grammar. From the "Word and Sentence Level" menu, choose the topic for which you need more information. Examples are clauses and compounds.

Conjunctions and conjunctive adverbs: www.ablongman.com/rude (textbook website).

Discussion and Application

1. *Phrases and clauses.* Determine which of the following groups of words are phrases, which are clauses, and which combine a phrase with a clause.

   genetic instructions

   have recently determined

   researchers have determined the cause

   on the bottom

   in the event that the program crashes

2. *Sentence types.* Identify the sentence type—simple, compound, complex, or compound-complex—for each of the following sentences.

   [1]Since 1974, thousands of papers have addressed the subject of ozone depletion. [2]The first hard evidence that proved a problem existed did not surface until 1985; in that year, Dr. Joe Farman of the British Antarctic Survey team reported finding a hole in the ozone directly over Antarctica. [3]In order to learn more about this Antarctic ozone phenomenon, scientists around the world have joined forces to create the National Ozone Expedition (NOZE). [4]Each year since 1986, these scientists have braved the cold of Antarctica in order to study the ozone depletion patterns that occur there every spring.

3. *Sentence types and punctuation.* Punctuate the sentences to clarify the structure of clauses and phrases within them. Articulate the reason for each mark of punctuation you insert or the reason for deleting a mark.

   a. I had hoped to find a summer job in the city however two weeks of job hunting convinced me that it was impossible.

   b. We came to work today in the rain; although we all preferred to stay in bed.

   c. Most of our studies have focused on the response of cultured mammalian cells to toxic inorganics such as cadmium, these in vitro studies benefit from the ease and definition with which cultured cells may be manipulated and from the absence of complicating secondary interactions that occur in vivo. Perhaps the greatest advantage of working with cultured cells however is that it is often possible to derive populations that vary in their response to the agents in question. Because the mechanisms involved in cell damage or protection are often altered specifically in such cells; they provide tools invaluable to identification of these mechanisms and to a definition of their importance in the overall response of the cell

   d. *This intentional fragment appears in a college catalog course description. Explain why the colon won't work, and suggest alternative punctuation.*

   The study of contemporary techniques of music: modes, synthetic scales, serialism, vertical structures, with a term project.

e. Since computer skills, including page design have become essential to professional writers, college writing courses include a computer component.

f. Other substances inhibit digestion of insects, alter insect reproduction to make it less efficient or interfere with insect development.

4. *Restrictive and nonrestrictive modifiers.* In each of the following sentences, identify the dependent clause that begins with a relative pronoun. Determine whether the clause functions as a restrictive or nonrestrictive modifier, and punctuate it accordingly. Change "which" to "that" when the modifying clause is restrictive. When you cannot tell for sure, explain the difference in meaning if the modifier is restrictive rather than nonrestrictive.

a. Adding polypropylene fibers to concrete increases its resistance to dynamic loading which is characterized by high strain rates that result from explosive impact or earthquake loading.

b. The death of 3,000 white-tailed deer in 1962 resulted from heavy overpopulation and range abuse which led to malnutrition and its various side effects.

c. Americans admire intelligence, which has practical aims, but intelligence which ponders, wonders, theorizes, and imagines is suspect.

d. The photoconductive cells measure the amount of radiant energy and convert it to electrical energy which is then interpreted by the computer and displayed on the meter.

e. Present lab equipment allows pulse energy experiments, which require 300,000 kw or less of electric power. New equipment would increase the potential of the lab.

f. A high deer population which continuously feeds on the seedlings of a desired tree species can severely retard the propagation of that species.

g. Pulse welding can join many metals which are impossible to join by conventional welding methods.

5. *Parallelism, series, and compounds.* Each of the following sentences contains faulty parallelism or a double compound punctuated as a series of three. The present punctuation and item structure do not accurately reveal the overall sentence structure. Punctuate and use parallel structure to clarify the relationship of the items. If you cannot determine the writer's meaning well enough to punctuate, write a query to the writer to elicit the necessary information.

a. The rehabilitation center is raising funds to purchase Hydro-Therapy bathing equipment, wheelchairs, and to renovate the Hydro-Therapy swimming pool.

b. The disease is typified by a delay in motor development, by self-destructive behavior, and it leads to death.

c. The virus may cause AIDS, cancer, or even kill.

d. The literature component of the professional writing major allows students to develop their sensitivity to language, texts, and their ability to read critically.

e. This anthology is for psychologists, students of psychology, and of other related fields. The articles deal mostly with contemporary problems in psychology, minority, cultural, and other underrepresented groups.

f. These instructions are written for automobile owners who do their own minor repairs, know the primary parts of an engine, and the use of tools.

6. *Parallel structure.* For each of the following, make sure all related items in a series have parallel structure.

a. The old copiers produce copies of inconsistent quality, sometimes too light and at other times a black smudge of toner.

b. Some qualities that are needed are the ability to communicate orally and written, and possess good judgment and tact.

c. We have investigated ways of raising capital for building the Ronald McDonald House. We have spoken to banks about loans, to a foundation about getting grants, as well as fundraisers.

d. Please send information on the following:
   - the national office's program for low-interest loans
   - the matching funds program
   - any other helpful advice on funding

e. The ten-year cost is $5,000 for renting and $3,864 if they were to buy.

f. *Be careful, this one is tricky. Hints: A verb + "-ing" can form a participle (adjective) or gerund (noun substitute); "fall" is an intransitive verb, but "skip" and "lose" are transitive.*

   A teenager may reveal a drug habit by skipping classes, falling grades, and losing friends.

7. *Plurals and possessives.* Which of these phrases require an apostrophe to show possession?

a. bears paw

b. two years experience

c. two years ago

d. jobs requirements

e. requirements for the job

f. its colors

g. joints rotation

h. Steves Café

i. Masters degree

      j. two months allowance

      k. policies cash value

      l. employees cafeteria

8. *Internal sentence punctuation.* Correct the punctuation in the following sentences, and explain why each change was necessary. Some sentences require more than one change. If the sentence is punctuated correctly, cite the principle that verifies its correctness.

    a. The copy center, like the check cashing service and the convenience store would be open 24 hours each day.

    b. Harris / 3M offers a 36 month leasing plan. For the proposed copiers for the center, the 6055 and the 6213. Lease costs would be $702 per month which includes maintenance.

    c. If the target goal of 36,000 copies per month billed at $.045 is met the equipment would pay for itself in 2.5 years sooner if the monthly allowance is exceeded.

    d. The fixed costs involved with this project; electricity, ventilation and floor space are not considered.

    e. There are two types of ultraviolet (UV) radiation; UV-A and UV-B. UV-A radiation which is frequently used for tanning beds, is lower in energy (longer in wavelength) than UV-B therefore it is considered safer than UV-B radiation. Some experimenters; however, believe that UV-A is just as damaging as UV-B; although higher doses of UV-A are required.

    f. Many theories about the depletion of ozone have been proposed but at present, two main hypotheses are widely accepted; the chemical mechanism and the dynamic mechanism.

    g. Chlorofluorocarbons (CFCs) such as the refrigerant freon, are very inert compounds, however, when CFCs drift into the upper atmosphere ultraviolet light will activate them.

9. *Compound terms and hyphenation.* Determine whether the compound terms and modifiers in the following sentences should be hyphenated, solid, or open.

    a. While studying day to day read outs from the Total Ozone Mapping Spectrometer, Dr. Wilson discovered that drastic depletions of ozone in a short period of time are not at all out of the ordinary.

    b. The manual includes step by step instructions.

    c. We took a multiple choice exam.

    d. The grant requires a semi-annual progress report.

    e. Type the sub-headings left justified.

    f. The left justified headings are not recognizable as level one headings.

    g. The pipe is two meters long.

    h. The valves in the two, three, and four meter pipes have corroded.

    i. Cross fertilization joins gametes from different individuals. The parents may be different varieties or species.

    j. This text book also serves as a reference book.

    k. The architect designed a multi purpose cafeteria for the school.

10. *Punctuation and meaning*. The following pairs of sentences differ only in punctuation. Explain the difference in meaning that results.

    a. A style sheet is a list of the general style—spelling, capitalization, abbreviation, hyphens—and unfamiliar or specialized terms.

    A style sheet is a list of the general style—spelling, capitalization, abbreviation, hyphens, and unfamiliar or specialized terms.

    b. When you make an entry on your style sheet, write the page number of the first occurrence (or every occurrence if you think you may change the style later). You will then be able to check your choice in its context.

    When you make an entry on your style sheet, write the page number of the first occurrence, or every occurrence, if you think you may change the style later. You will then be able to check your choice in its context.

11. *Your own writing*. Examine a sample of your own writing. Determine

    a. sentence patterns: what types of sentences do you most use?

    b. punctuation: is your work punctuated correctly?

    Prepare some goals for yourself for structuring and punctuating sentences.

12. *Word play*. Explain why these terms are named as they are.

    a. Why is there a "junction" in a conjunction?

    b. Why are some pronouns "relative"?

    c. Why are some clauses "subordinate"?

    d. How is "punctuation" related to "punctuality"?

    e. How does a "period" relate to time?

    f. How is a pronoun "pro" the noun?

    g. What does a restrictive modifier "restrict"?

    h. How does a modifier "modify"?

    i. What is "semi" about a semicolon?

    j. How does a "dash" dash?

12 Quantitative and Technical Material

Mathematical, statistical, and technical material has a language that shares many features with language constructed of words. The vocabulary of numbers and symbols may be unfamiliar, but even equations have subjects and verbs, and to use statistics is to make an argument. An editor, even one who does not fully understand the meaning of symbols and formulas, offers the knowledge of language structure, the ability to perceive patterns, and the awareness of resources for checking what is unfamiliar.

The editor establishes grammar, punctuation, consistency, and accuracy. In addition, the editor prepares the text for production by marking for italics and capitalization, distinguishing letters from numbers, and indicating spacing and other typographic elements.

This chapter introduces some principles for using numbers in math and statistics as well as basic standards of measurement and some scientific symbols. It also introduces guidelines for displaying and marking mathematical material in equations and in tables. The chapter will not make you an expert math and statistics editor, but if you learn the principles outlined here and check the applicable style manuals as you work, you can be competent in and confident about marking this type of text, and you will have a start on developing expertise. You will increase your value as an editor if you develop your ability to edit quantitative and technical material.

Key Concepts

Your knowledge about punctuation and grammar, about checking style manuals, and about marking type are pertinent to copyediting material full of numbers and symbols. You do not have to be an expert in the subject matter to do well at editing statistical and mathematical material, but you must be diligent about checking details, cautious about marking changes, and willing to inquire when in doubt. Style manuals offer an abundance of help.

Using Numbers

The conventions for the treatment of numbers in technical texts differ from those in humanities texts, and they may differ among the technical disciplines. The conventions concern questions of style such as whether to spell out a number or use a figure or whether to use the metric, British, or U.S. system of measurement. A style manual will provide specific guidelines within a discipline, but the manuals are mostly consistent on the following guidelines. Check a comprehensive or discipline style manual for specific directions on the treatment of numbers in dates, money, and time.

1. **Use figures for all quantifiable units of measure,** no matter how small, rather than spelling out the numbers. Also, use a figure whenever you abbreviate the measure.

2 m	1 ½ in	0.3 cm
12 hours	$1 million	18 liters

   Figures aid readers in comprehending and in calculating. In documents that are not scientific or technical, however, spelling out both the number and the measure is common if the number is lower than 10 or possibly lower than 100, depending on the style guide.

2. **Do not begin a sentence with a figure.** Either rearrange the sentence to avoid the figure at the beginning or spell out the number.

3. **Don't mix systems of measurement.** For example, do not describe some objects in inches and related objects in centimeters. The dual usage may confuse or mislead readers, and they may calculate incorrectly.

4. **Set decimal fractions of less than 1.0 with an initial zero** (0.25, 0.4). An exception is a quantity that never equals 1.0, such as a probability (P) or a correlation coefficient (r).

5. **Convert treatment of numbers in a translation** according to usage in the country where the document will be used. There are cultural variations for dates, time, and money. See Chapter 20.

Measurement

Three systems of measurement are widely used: the U.S. Customary System, the British Imperial System, and the International (metric) System. In the U.S. and British systems, the standard measures are the yard and the pound, but these systems vary in some measures and in the expression of them. For example, in France and Germany, and formerly in Great Britain, a billion equals a million million, whereas in the U.S. system it equals a thousand million. The metric system is used for most scientific and technical work.

The International System of Units, an extension of the metric system, is a standard system of units for all physical measurements. Its units are called SI units (for *Système International*, in French). The International System has seven fundamental units, listed here.

Quantity	Unit	Abbreviation
length	meter	m
mass	kilogram	kg
time	second	s
electric current	ampere	A
temperature	kelvin	K

luminous intensity	candela	cd
amount of substance	mole	mol

Other measures, based on these fundamental units, are derived units; that is, they are multiples or parts of the fundamental units. Because the metric system is a decimal system, all the derived units are multiples of 10. The prefix indicates the multiple: a kilometer equals 1,000 meters, and a centimeter equals 0.01 meter. Use numbers only between 0.1 and 1,000 in expressing the quantity of any SI unit; use the derived units for smaller or larger numbers. For example, 10,000 m equals 10 km. SI units established for other physical quantities are used primarily in science and engineering. The following list presents a few of these units. The SI units and symbols pertinent to a given field are likely to be listed in that field's handbook.

Quantity	Unit	Symbol or Abbreviation
acceleration	meter per second squared	m/s^2
electric resistance	ohm	Ω
frequency	hertz	Hz
power	watt	W
pressure	newton per square meter	N/m^2
velocity	meter per second	m/s

Whereas the units are lowercase, even when they represent names, some abbreviations include capital letters, especially if they represent proper nouns. For example, *watt* and *newton* are lowercased as units but capitalized as abbreviations. The units and abbreviations are set in roman type and without periods.

The abbreviations are never made into plurals. Thus, six watts would be written as "6 W," not "6 Ws." However, if the measures are written in prose, the words are formed into plurals according to the same rules that govern other words. Thus, "6 watts" is correct in a sentence. Fractions of units are always expressed in the singular whether they are expressed in symbols or in prose: 0.3 m and 0.3 meter.

A hyphen is used between the measure and the unit when they form a modifier but not when they simply define a measure. Thus, both "The trial lasted 10 seconds" and "the 10-second trial" are correct.

Marking Mathematical Material

Documents that include fractions, equations, or other mathematical expressions require close copyediting attention. To save space, built-up fractions may have to be converted for inline presentation. Equations may have to be displayed, numbered, or broken. In addition, markup must clarify characters that could be confusing (for example, the number *1* and the letter *l*, the number *0* and the capital letter *O*, the unknown quantity *x* and the multiplication sign) and indicate the position of subscripts and superscripts on the page, as well as noting italics and capitalization. Fortunately computer programs generate equations and other mathematical material in the appropriate format, reducing the complexity of markup.

Fractions

Fractions may be set "built up" or "solid" (inline). The inline version substitutes a slanted line, or *solidus*, for the horizontal line of the built-up fraction.

Built up	**Solid (Inline)**
$\dfrac{1 + (x - 3)}{y + 2}$	$[1 + (x - 3)]/(y + 2)$

The built-up version is preferable for comprehension because it conveys visually the relationship of the numbers. However, it takes more space in typesetting. If a fraction appears in a sentence, it may have to be converted to inline form in order to avoid awkward line breaks and extra space between the lines. Complexity in fractions, however, will justify the extra space for building up.

When fractions are converted to inline form, parentheses may be necessary to clarify not only the numerator and denominator but also the order of operations, which will affect the result obtained. (Operations within parentheses are completed before other operations.) In the previous example, the denominator is the quantity (or sum) of $y + 2$. Thus, $y + 2$ will be added before the sum is used as a divisor; otherwise, the numerator of the fraction would be divided by y and then 2 would be added to the total. The entire numerator is enclosed in brackets to establish that those operations must be performed before division. Both the square brackets and the parentheses, as well as braces, are signs of aggregation or fences. The preferred order for these signs is parentheses, square brackets, and braces:

$$\{[()]\}$$

Fractions containing square root signs can be set inline if the sign is converted to the exponent ½. The square root sign may also be set without the top bar, thereby allowing it to fit within a normal line of type. All three of the following expressions say the same thing.

$$\frac{a + b}{\sqrt{\dfrac{2a - 12}{6}}} \qquad (a + b)/[(2a - 12)/6]^{1/2} \qquad \frac{a + b}{\sqrt{[(2a - 12)/6]}}$$

Mark a built-up fraction for inline presentation by creating a line break mark using the line in the fraction.

$$\frac{1}{2} \quad \text{marked} \quad {}^{1}\!/_{2} \quad \text{becomes} \quad \text{½}$$

Equations

Equations are statements that one group of figures and operations equals another. Equations can contain known quantities (expressed as numbers), variables (often expressed as a, b, and c), and unknowns (often expressed as x, y, and z). Equations always include an equal sign.

Displaying and Numbering Equations

Equations may be set inline, or they may be displayed (set on a separate line). If multiple equations in a text are displayed, they should be either centered or indented a standard amount from the left margin.

Displayed equations may be numbered if they will be referred to again in the document—the numbers provide convenient cross-references. If an equation is to be numbered, it must be displayed, but not all displayed equations are numbered. Generally, the equation number appears in parentheses to the right of the equation.

Equations are numbered sequentially either through the work or through a section of it. Double numeration (chapter number followed by a period and the equation number) can save time if an equation number must later be changed (perhaps because one equation was deleted). With double numeration, only the equation numbers in that chapter must be changed. The possibility that equation numbers may change during copyediting and revision is also a good reason for numbering only the equations that must be referred to, not all those that are displayed.

Let there begin the value of the $y_0, y_1, \ldots, y_n$ of the function $y = f(x)$ at the $(n + 1)$ points $x_0, x_1, \ldots, x_n$. A unique polynomial $P(x)$ whose degree does not exceed n is given by the equation

$$P(x) = a_n x^n + a_{n-1} x^{n-1} + \ldots + a_0, \tag{2.2}$$

for which $P_n(x)$ at x_i must be satisfied by

$$P_n(x_i) = f(x_i) = y_i, i = 0,1, \ldots, n. \tag{2.3}$$

The conditions in (2.3) lead to the system of $n + 1$ linear equations in the a_1:

$$a_0 + a_1 x_i + \ldots + a_n x^n = y_i, i = 0,1, \ldots, n. \tag{2.4}$$

Equation (2.3) is referred to again by its number. The parentheses identify the figures as an equation number. Some professional associations, such as the American Psychological Association, prefer that the text read "Equation 2.3" rather than (2.3).

Breaking Equations

Equations that are too long to fit on one line should be broken before an operational sign ($+$, $-$, $\times$, $\div$) or a sign indicating relations ($=$, $\leq$, $\geq$, $<$, $>$). However, you should never break terms in parentheses. Use the line break mark to indicate a break in the equation.

$$a_0 + a_1 x_i + \ldots + a_n x^n = y_i, i = 0,1, \ldots, n.$$

In a document with a series of related equations, the equations should be aligned on the equal sign, regardless of the amount of material to the right and left of the equal sign.

The mathematical formula for computation of the mean is

$$\overline{X} = \frac{\Sigma X}{N}$$

where

$\overline{X}$ = mean of the scores,

ΣX = sum of the scores, and

N = number of scores.

Punctuating Equations

Equations are read as sentences when they appear in prose paragraphs, with the operational signs taking the place of verbs, conjunctions, and adjectives. Thus, $a + b \geq c$, where $b = 2$, would be read aloud as *a plus b is greater than or equal to c, where b equals 2.*

Practice in punctuating equations varies. Some publishers omit punctuation such as commas and periods on the grounds that the marks are potentially confusing if read as part of the equation rather than as punctuation. They also argue that when the equation is displayed, the space around it signals its beginning and end, making punctuation redundant. Other publishers, however, follow the same rules for punctuating equations as for punctuating sentences. Displayed equations are punctuated as though they appear in sentences. Thus, they may be followed by commas or periods as the structure of the sentence requires. In the following examples, no colon appears after "by" in the first equation, but a colon appears after "as follows" in the second equation. The words that introduce the displayed equation are not followed by a colon unless they would be followed by a colon in the same circumstances if all the text were prose. The displayed equations do not have end punctuation; the space takes the place of commas and periods.

Let the polynomial be given by

$$P(x) = a_0 + a_1 x + a_2 x^2 + \ldots + a_n x_n$$

where the coefficients a_1 and a_2 are to be determined.

If the center of a circle is at the origin and the radius is r, the formula can be reduced as follows:

$$x^2 + y^2 = r^2$$

Grammar and Punctuation

Mathematical copy, like prose copy, should follow accepted rules of grammar: subjects and verbs must agree, nouns take articles, and clauses contain subjects and verbs. You do not need to understand the following equation to edit the subject-verb agreement error, to place a comma before the nonrestrictive clause, and to conclude the sentence with a period.

There exists a unique polynomial $P(x)$ whose degree does not exceed n which is given by $P(x) = a_n x^n + a_{n-1} x^{n-1} + \ldots + a_0$

A common grammatical error in mathematical copy is the dangling participle. Dangling participles occur in mathematical text for the same reason they occur in prose: the subject to be modified is absent from the sentence, often because the passive voice omits the agent. To revise, you insert the agent or delete the modifier.

Constructing interpolation polynomials, the inverse can be explicitly calculated.

By solving $x = 12 - y$, a contradiction can be obtained.

SOLUTIONS:

The inverse can be explicitly calculated with the construction of interpolation polynomials.

Solving $x = 12 - y$, we obtain a contradiction.

Solving $x = 12 - y$ yields a contradiction.

Markup for Typesetting

Markup clarifies type style (italic, roman, or bold), spacing, subscripts and superscripts, and ambiguous characters. Unknowns and variables in expressions and equations are set as lowercase italic letters. These are usually letters in the English alphabet, but they may be Greek letters. If they are not typed in italics on the paper copy, they should be marked with an underline to denote italics. A slash through a letter indicates lowercase when the letter is typed as a capital. Numbers, symbols, and signs are set in roman type. Vectors are set in bold. (Vectors are quantities with direction as well as magnitude. They are sometimes set with an arrow over them.)

Operational signs ($+$, $-$, $\times$, $\div$) and signs of relation ($=$, $\leq$, $\geq$, $<$, $>$) are set with space on either side. The sign for multiplication is roman, not italic as for the unknown x.

Markup should clarify the intent when characters are ambiguous. For example, a short horizontal line may be a minus sign, hyphen, en or em dash. It could even indicate polarity (a negative charge). Some, but not all, possible ambiguities are listed here.

0	zero	β	Greek beta
O	capital "oh"	B	capital "bee"
o	lowercase "oh"	e	exponent
°	degree sign	e	charge of an electron
1	number one	Σ	summation
l	lowercase "el"	$\in$	element of
$\times$	multiplication sign	m	abbreviation for meter
x	unknown quantity	m	unknown quantity
χ	Greek chi		

If the text contains just a few of these ambiguities, they may be marked as in any other markup. A clarifying statement may be circled next to the character. For example, you might write and circle "el" next to the letter that could be misread as the number one. Mark unknown quantities as italic if they are not typed in italic.

Even when the typist has been careful to use symbols, italics, and spacing correctly, marks can clarify the intent for the compositor or production specialist and leave no doubt as to how the elements are to be set. For example, you might mark the superscripts and subscripts, even if they are typed correctly, to confirm that the positions should be retained in typesetting, and you might write and circle "minus sign" next to what should set as a minus sign so that it will not be mistaken for an en dash or a hyphen. The extent of your marking for clarification will depend, in part, on your company's practice and on your production specialist's expectations, but it is better to err on the side of overclarification than to leave interpretation of elements up to the production specialist and spend time—and money—correcting errors after typesetting. Check with the writer if you cannot tell what characters were intended.

Statistics

Statistics is a system of procedures for interpreting numerical data. The science of statistics governs the collection of data (as by sampling), the organization of data, and the analysis of data by mathematical formulas. Using statistics, researchers learn the significance of "raw" data, or figures that have not been analyzed. Researchers use the numbers to make predictions. Statistics is widely used in any empirical research, or research that relies on observation and experiment. Thus, it is used in the life sciences, physical sciences, social sciences, and engineering. You may edit material based on statistical analysis if you work in any of these fields and if you edit research proposals and reports. The results of product usability testing may also be analyzed statistically.

Two usage notes: the word *statistics* is singular if it refers to the system of interpreting numerical data; the word is plural if it refers to numbers that are collected. "Statistics is a science" is the correct usage because the reference is to the science. "The statistics are misleading" is correct as a reference to specific numbers. Also, the sciences (though usually not engineering) regard the word *data* as plural; thus, "data are" is the correct usage in science, in spite of what your ear may tell you.

If you work extensively with statistical material, you should learn more about it than this chapter will tell you. But this chapter will introduce some of the more common terms and symbols to enable you to copyedit. A fuller, though not overwhelming, introduction is available at the website by Robert Niles listed in Further Reading.

Letters used as statistical symbols are italicized, whether they appear in prose or in tables, unless they are Greek letters. If they are not italicized or underlined in the typescript, they will have to be marked. Some common symbols and abbreviations and their meanings follow:

ANOVA	analysis of variance
r	correlation
df	degrees of freedom
F	F-ratio
μ (Greek letter mu)	mean
n	number of subjects
N	number of test results
P or ϕ (Greek letter phi)	probability
SD	standard deviation
$S; Ss$	subject; subjects
t-test	test of differences between two means

Equations are placed on the page according to the same rules that govern mathematical material. The following guidelines are basic ones:

- Align related equations on the equal sign.
- Set operational or relational signs with space on either side (for example, $\mu = 92.55$, not $\mu=92.55$).
- Place a zero before a decimal in a quantity of less than 1, except for correlation coefficients *(r)* and probabilities *(P),* which are always less than 1.
- Capitalize *experiment* or *trial* when these words refer to specific tests. Thus, "The results of Experiment II revealed that . . ." but "the experiment."

Tables

Tables represent an efficient way to present quantitative and verbal data. An editor should be alert to misspellings and other errors; inconsistencies in capitalization, spelling, and abbreviation; clarity of identifying information, such as whether the numbers represent percents or totals and whether the measures are meters or feet; correctness of arithmetic, such as whether totals are correct and whether percentages total 100. Apparent inconsistencies in numbers or other data may signal inaccuracy and should be checked.

In addition, the editor should check readability of the table, such as the amount of space between rows and columns and variations in type style (bold, italic, capitalization) to indicate different levels of headings. A reference to the table in the text should precede the table. Tables should be numbered sequentially in the chapter, and each table number should match the number used in cross-references. The title should accurately reflect the contents. The structure of the table should match the structure of information in the text; that is, topics should appear in the same order in both places. The wording of headings should also match.

General Guidelines

Tables should be easy to read and understand. You can find guidelines to aid you in editing tables in textbooks on technical writing and in some handbooks and style manuals. The following points briefly summarize these guidelines.

- Illustrations that are tabular are called tables; other illustrations, such as line drawings, graphs, and photographs, are called figures. The information in the rows and columns of tables may be quantitative, verbal, or even pictorial. The table title and number are identified at the top of the table, although informal tables that lack titles and numbers are also permissible. Some disciplines specify roman numbers for table numbers.
- Items compared in a table are listed down the stub (left) column, while points of comparison are listed across the top. This arrangement allows easy comparison of related items.
- Vertical and horizontal lines (rules) separating rows and columns are discouraged except for highly complex tables because the lines clutter the table with visual noise. White space is the preferred method for separating rows and columns. Figure 12.1 uses a few lines to distinguish different

levels of information, but most of the columns are separated by space rather than by lines. Too much space between columns, however, may result in inaccurate reading because the eye may skip to a lower or higher row. The table width can be narrower than the width of the paragraphs around it to control the amount of space between columns.

■ To increase accuracy of reading across the rows of long tables, a space may be used between groups of five or so rows, or alternate groups of five rows may be lightly shaded.

■ Columns of numbers should be aligned on the decimal.

■ Headings for columns should identify the measure ($, %, m) that is represented by the numbers in the column. The tables are cleaner (less noisy) if the abbreviation is not repeated for each entry. Another method is to insert the symbol or abbreviation before or after the first number in each column. This method reduces clutter in column heads.

No. 379. Toxic Chemical Releases by Industry: 2001

[In millions of pounds (6,158.0 represents 6,158,000,000), except as indicated. "Original Industries" include owners and operators. Covers facilities that are classified within Standard Classification Code groups 20 through 39, 10, 12, 49, 5169, 5171, and 4953/7169 that have 10 or more full-time employees, and that manufacture, process, or otherwise uses any listed toxic chemical in quantities greater than the established threshold in the course of a calendar year are covered and required to report]

Industry	1987 SIC[1] code	Total facilities (number)	Total on and off-site releases	On-site release				Off-site releases/ transfers to disposal
				Total	Air emissions	Surface water discharges	Other [2]	
Total [3]	(X)	24,896	6,158.0	5,580.3	1,679.4	220.8	3,680.1	577.7
Metal mining	10	89	2,782.6	2,782.0	2.9	0.4	2,778.7	0.5
Coal mining	12	88	16.1	16.1	0.8	0.8	14.6	-
Food and kindred products	20	1,688	125.1	118.9	56.1	55.2	7.6	6.2
Tobacco products	21	31	3.6	3.2	2.5	0.5	0.2	0.3
Textile mill products	22	289	7.0	6.2	5.7	0.2	0.3	0.7
Apparel and other textile products	23	16	0.4	0.3	0.3	-	-	0.1
Lumber and wood products	24	1,006	31.4	30.9	30.5	-	0.4	0.5
Furniture and fixtures	25	282	8.0	7.8	7.8	-	-	0.2
Paper and allied products	26	507	195.7	189.9	157.2	16.5	16.2	5.8
Printing and publishing	27	231	19.7	19.3	19.3	-	-	0.4
Chemical and allied products	28	3,618	582.6	501.3	227.8	57.6	215.9	81.3
Petroleum and coal products	29	542	71.4	68.1	48.2	17.1	2.8	3.3
Rubber and misc. plastic products	30	1,822	88.5	78.1	77.1	0.1	0.9	10.5
Leather and leather products	31	60	2.6	1.3	1.2	0.1	-	1.3
Stone, clay, glass products	32	1,027	40.5	35.4	31.3	0.2	4.0	5.1
Primary metal industries	33	1,941	558.6	286.8	57.6	44.7	184.5	271.8
Fabricated metals products	34	2,959	64.0	42.8	40.4	1.7	0.6	21.2
Industrial machinery and equipment	35	1,143	15.4	10.7	8.3	-	2.5	4.6
Electronic, electric equipment	36	1,831	23.9	16.4	12.7	2.9	0.7	7.6
Transportation equipment	37	1,348	80.6	67.7	66.7	0.2	0.8	13.0
Instruments and related products	38	375	9.4	8.6	7.2	1.4	-	0.8
Miscellaneous	39	312	8.4	6.8	6.8	-	-	1.6
Electric utilities	49	732	1,062.2	989.2	717.6	3.5	268.1	73.1
Chemical wholesalers	5169	475	1.5	1.3	1.3	-	-	0.2
Petroleum bulk terminals	5171	596	21.3	21.2	21.2	-	-	0.2
RCRA/solvent recovery	4953/7369	223	219.9	168.4	1.0	-	167.4	51.4

- Represents or rounds to zero.   X Not applicable.   [1] Standard Industrial Classification, see text, Section 12.   [2] Includes underground injection for Class I and Class II to V wells and land releases.   [3] Includes industries with no specific industry identified, not shown separately.

Source of Tables 378 and 379: U.S. Environmental Protection Agency, *Toxics Release Inventory*, annual.

Figure 12.1 Table with Footnotes and Multiple Levels of Headings

Source: U.S. Census Bureau. *Statistical Abstract of the United States: 2003.* 236.

- Notes to clarify information or to cite sources may be necessary, just as they are in prose text (see Figure 12.1). When the table as a whole comes from a single source, a footnote number often appears on the table title, and a note that cites author, title, and publication data appears at the bottom. The form of the note corresponds to the form used for the reference list.
- All tables must be referred to in the text.

Application: Editing a Table

Marking right on the table is sufficient for simple corrections. You can specify capitalization, deletions, and insertions on the table and supplement these marks with marginal instructions. When the changes are extensive, however, you may have to reconstruct the table electronically.

Figure 12.2 illustrates an editor's marking of a table for correction before printing. The table was created for a report on AIDS in correctional facilities. It shows the results of screening to determine how many inmates test positive for the HIV antibody, which would indicate that they have been exposed to AIDS. At the time of the screening, only four states had mass screening programs.

The editor has marked several types of changes, aiming to make the text correct, consistent, accurate, and complete and also to increase visual readability.

- **Correctness.** A check for correctness reveals several problems. The illustration was inappropriately labeled as a figure, but because the material is tabular, its title is changed to *table*. Also, the numbers in columns 2 and 4 are marked to align on the decimal, and the misspelling in the fifth row under "Mass Screening" is corrected.
- **Consistency.** The column headings are marked for flush left. Capitalization in headings and in the third column is made consistent. And, because "number" is spelled out in the heading for column 4, the "percent" is also spelled out in the heading for column 5.
- **Accuracy/completeness.** Placing "HIV" at the beginning of the title gives a more accurate view of the subject matter than the generic "results." The selection of states might raise questions about completeness, but the prose part of the text verifies that only a few jurisdictions have screening programs. Also, the numbers are consistent enough to suggest that they are accurate.

  An editor may question the middle category, "Epidemiological Studies," because the title names only the first and third categories, and the middle category contains only one study. The table may in fact be complete without this information. However, a deletion would be a content change and therefore beyond the expectations for copyediting. The editor should query the writer before such a deletion, perhaps suggesting that the information on that single study be included within the text of the report where the table is discussed. Another alternative would be to change the title of the table to "Results of Screening Programs for the HIV Antibody in Inmates," which does not imply just two categories. However, a title change would also have to be approved by the writer, and all references to the table in the text and in preliminary pages would have to be changed accordingly.

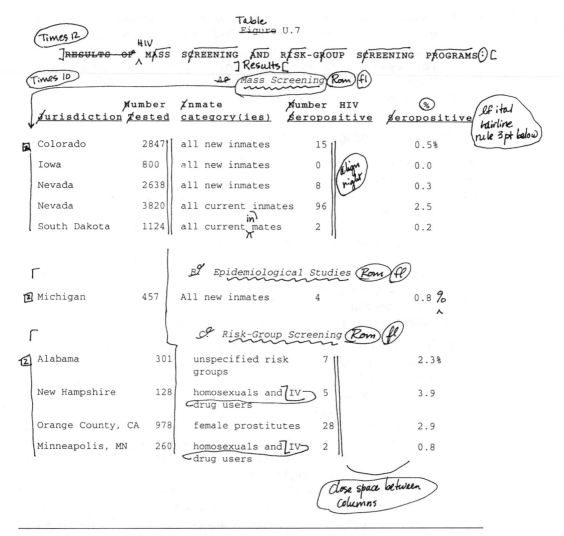

Figure 12.2 Edited Table

- **Visual design.** The editor's biggest task is using space and type to make the table easy to read.

Title: initial caps and lowercase

The title is marked for initial capital letters with lowercase letters rather than all caps. Presumably, the titles of other tables in the report use caps and lowercase letters, a style that is easier to read than all caps.

Headings

Headings for the three major categories are aligned on a left margin that protrudes farther into the left margin than other data. Such placement is conventional for tables as well as logical for left-right reading. The headings are printed in boldface type rather than italicized for greater typographic prominence. The letters "A," "B," and "C" are deleted as irrelevant because

Table U.7

HIV Mass Screening and Risk-Group Screening Programs: Results

jurisdiction	number tested	inmate category(ies)	number HIV seropositive	percent seropositive
Mass Screening				
Colorado	2847	all new inmates	15	0.5%
Iowa	800	all new inmates	0	0.0
Nevada	2638	all new inmates	8	0.3
Nevada	3820	all current inmates	96	2.5
South Dakota	1124	all current inmates	2	0.2
Epidemiological Studies				
Michigan	457	all new inmates	4	0.8%
Risk-Group Screening				
Alabama	301	unspecified risk groups	7	2.3%
New Hampshire	128	homosexuals and IV drug users	5	3.9
Orange County, CA	978	female prostitutes	28	2.9
Minneapolis, MN	260	homosexuals and IV drug users	2	0.8

Figure 12.3   Printed Version of the Edited Table in Figure 12.2

they are not used in the text references and because the three side headings establish visually that there are three categories of information.

Column headings   Column headings are placed above the heading for the first category because they head the columns for all three categories. They are marked to set lightface italics rather than boldface. The rule distinguishes the column headings from the body of the table.

Alignment   Numbers are marked to align on the imagined decimal. Columns are marked to maintain consistent indentations.

Column spacing   The space between columns 4 and 5 is reduced so that the eye will not have to leap so far to get from one item to the next. In this case, the specific amount of space is left up to the compositor and will be based partly on how much space the column headings require.

Typeface   The table is set in the same typeface as the rest of the text.

Figure 12.3 shows how the table will look when printed as marked by the editor.

Standards and Specifications

Standards and specifications describe products and processes according to measures of quality and intended uses. They affect the work of a technical editor in several ways: they may describe the details of type, structure, layout, and coding of the materials that the editor prepares and even the process of developing those materials; they may describe a product yet to be developed and serve as a tool for planning and designing the product's documentation; they may be documents to edit.

Standards for technical materials and processes are developed and published by government, industry, and national and international standards organizations. Some prominent developers of standards are the United States Department of Defense, the American National Standards Institute (ANSI), the Institute of Electrical and Electronics Engineers (IEEE), and the International Organization for Standardization (ISO). These standards cover many kinds of equipment and processes, including aircraft, computers, and quality assurance procedures. They may specify, for example, the size and brightness of screens for visual display terminals and the interfaces for microprocessor operating systems.

In addition to standards for equipment and processes, a number of standards, particularly in the military, govern publications. These specify standards for abbreviations, format, parts lists, readability, abstracts, and reports. In the United States, military specifications are identified in labels with the prefix MIL- and are commonly referred to as MIL-SPECS. Documents for all government agencies except the Department of Defense are governed by Federal Information Publication Standards. These specifications for publications will be particularly important if you work for the military, the government, or military and government contractors. The standards for SGML (Standard Generalized Markup Language—see Chapter 5) are defined in an ISO standard. The standards aim for a uniform quality and consistency in products that may be produced by various organizations. Products that meet those standards give consumers and other users some confidence in their quality. An editor checks publications for conformity to applicable specifications.

One set of standards particularly important to technical communicators is ISO 9000. This standard requires that all companies doing business in the European Economic Community maintain and follow a quality management system, documented in a quality manual. The quality management system encourages a systematic process of document development that requires planning and review, a welcome requirement for editors who work in organizations that ask the editors to perform miracles at the end of document development when it is too late to do more than surface cleanup. In addition, editors may be responsible for maintaining the quality manual that documents the policies and procedures in a company.

In addition to the standards that apply across organizations, a product to be developed within a company will probably be described by specifications that establish product users, environment of use, purposes, requirements, and functions as well as details such as weight, size, materials, parts from other products, and interfaces. These specifications may also define resources for development, documentation needed, and a development schedule. Product documentation often begins on the basis of such specifications—and sometimes before the specifications become detailed. By developing or reviewing product specifications, editors can participate early in designing the documentation—contents, structure, publication medium, and other high-level features that determine usability and usefulness. The first versions of this documentation may even help to determine the way the product itself develops.

Because specifications precede the product, they may be vague or incomplete. The editor has a chance to check for completeness and accuracy of the information about the product to be developed. These checks support good planning at the beginning—much more efficient than efforts to correct problems at the end of the document development cycle.

Editing specifications also requires the ability to discover ambiguities in phrasing that might provide a loophole for a contractor trying to save money. John Oriel notes, for example, that the phrase "stainless steel nuts and bolts" requires only that the nuts be of stainless steel.[1]

Using Your Knowledge

A style manual is a great resource for any editor but especially for an editor developing expertise in editing material with numbers and statistics. Refer to comprehensive and specialized manuals often as you edit, and spend some time browsing the manuals to learn about the editorial issues. Respect the limits of your knowledge, and check with the author when you are in doubt.

Further Reading

Fisher, Barry. 1995. Documenting an ISO quality system. *Technical Communication* 42: 482–491.

Knuth, Donald E., Tracy Larrabee, and Paul M. Roberts. *Mathematical Writing*. math.stanford.edu/~rubin/110/mathwriting.pdf. Offers twenty-seven useful notes on writing and editing mathematics.

Niles, Robert. *Statistics Every Writer Should Know*. www.robertniles.com/stats.

Swanson, Ellen, Arlene O'Sean, and Antoinette Schleyer. 1999. *Mathematics into Type* (updated ed.). American Mathematical Society. The authority of the professional society makes this reference the first choice of a mathematics editor.

Tufte, Edward R. 1997. Visual and statistical thinking: Displays of evidence for decision making. Graphics Press. Separately bound Ch. 2 of *Visual Explanations: Images and Quantities, Evidence and Narrative*. Reasonably priced and a good supplement for the editing course.

University of Chicago Press. 2003. *The Chicago Manual of Style*. 15th ed. Ch. 9: Numbers; Ch. 14: Mathematics in type.

---

[1]Oriel, John. 1992. Editing engineering specifications for clarity. *Conference Record: IPCC 92–Santa Fe.* Institute of Electrical and Electronics Engineers, Inc.: 168–173.

Discussion and Application

1. Edit the sentences below to show the correct use of numbers, symbols, and abbreviations. You may need to refer to the discussion of abbreviations in Chapter 9 and to a list of abbreviations and a style guide. Prepare a style sheet if there are options.[2]

   a. When the electrode is fully in the spinal cord tissue, the resistance shoots up to 1000-ohms or more.

   b. There are two methods of applying a coagulating current. One uses a fixed time, eg. 30 sec, and varies the power applied, eg, 5-mA, then 10mA, 15mA, etc, up to a limit given by the manufacturer. The other method fixes the power, eg at 30 mA, and varies the time, eg 5 sec, then 10 seconds, 15 sec, etc.

   c. The spinal cord at the C1-C2 level is about 15-MM across and 10- to 12-mm from front to back, so the maximum lesion needed is 6 mm x 4 mm. Furthermore, a cylindrical electrode with a 2-mm uninsulated tip will provide a lesion somewhat barrel shaped; an exposed tip of 3-mm. is also used and will provide a lesion of about 4.5 x 3.0 mm.

2. Edit the following sentences and equations to clarify punctuation, spacing, italicization of letters that represent unknowns, and grammar.

   a. Consequently an integer p is even if p^2 is even and is odd if p^2 is odd.

   b. Then, squaring we have:

   $$p^2 = 2q^2$$

   c. Hence: since 2 is rational, both p and q are even integers.

---

[2] Sentences in 1a, b, and c are adapted from "Neurolytic Blocks around the Head" (first draft), by Samson Lipton, M.D. Used with permission of the author.

d. Thus,

$$|4| = 4, \qquad |-4| = -(-4) = 4 \qquad |0| = 0.$$

e. Using the *less than*, *equal to*, and *greater than* symbols, these three possibilities are written:

a<b   a=b   a>b

3. Convert these built-up fractions for inline presentation.

a. $\dfrac{1}{16}$

b. $\dfrac{3}{a + b}$

c. $\dfrac{x + 2}{2y}$

4. Mark this equation to show an appropriate line break.

$$H(-x) = -(-x)^4 + 3(-x)^2 + 4 = -x^4 + 3x^2 + 4 = H(x).$$

5. Mark these equations to align them on the equal sign. Clarify other spacing as necessary.

$$(2x^2 - 2)(x^2 - x - 12) = 0$$

$$2(x - 1)(x + 1)(x - 4)(x + 3) = 0$$

$$x = -1, 1, 4, -3$$

6. Mark these sentences from a statistical analysis to show italics, spacing, and capitalization.

a. The analysis revealed a significant effect of method in both experiments. In experiment I, F = 8.09 and p = .001. In experiment II, F = 8.58 and p =.0007.

b. Results of a two-tailed t test (t=.728, p<.20) indicate that the performance difference between the two groups is not significant.

13 Proofreading

An edited text requires attention before publication or online launch. After the editing, the production staff may need to insert illustrations at the points marked, apply styles to the type, lay out the pages, and correct any errors that the editor noted. Before the document can be printed or launched online, a proofreader checks that the document incorporates the editing and graphic design directions.

This chapter distinguishes proofreading from copyediting. It then identifies goals of proofreading, symbols and their placement, and strategies for effective proofreading.

Key Concepts

Proofreading provides a check of the document after copyediting but before printing or online launch. An editor provides directions for preparing the document for publication; a proofreader verifies that those directions have been followed and that the published document will meet the standards of content and appearance that have been set for it.

Distinguishing Proofreading from Copyediting

The term *proofreading* is sometimes used loosely as a synonym for *copyediting*. Both processes share symbols and the goal of correctness. Electronic manuscripts and editing blur the distinctions between copyediting and proofreading because many of the editorial directions can be incorporated at the time of editing, not in a separate procedure. The use of styles and templates ensures visual consistency. However, a final check of the document, separate from copyediting, provides good quality control. The two processes differ in the purposes they serve, in the stage of production during which they occur, and in the placement of marks on the page.

Copyediting	Proofreading
■ prepares the text for publication in print or online; makes style choices	■ verifies that the copyediting specifications have been implemented
■ takes place early in production	■ occurs toward the end of production
■ when the typescript is double-spaced, uses mostly interlinear marks	■ uses mostly marginal marks

Figures 13.3 and 13.4, later in the chapter, compare the copyediting and proofreading of the same document.

Copyeditors establish spellings and mechanics when more than one option is available and also impose correct grammar and punctuation. Copyeditors or graphic designers establish type style and size, typeface, and spacing. Proofreaders do not change these choices but verify that they have been incorporated into the document. Proofreaders introduce change only if the copyeditor has overlooked errors.

Proofreaders compare a current version of a document with the earlier version to establish that marked corrections have been made and instructions followed. Proofreading may occur at any point in production when the document is prepared in a new form, as from the author's original to page proofs. A proofreader confirms that errors previously noted have been corrected and checks that new procedures (such as page breaks or the placement of illustrations) have been completed correctly.

As each version of the document is replaced by a newer version, it becomes "dead" copy. When page proofs are being proofread, the author's original is the dead copy. The dead copy indicates the established text. Proofreading marks show where the proofs differ from the dead copy and what corrections need to be made before the next stage in production.

Editors often proofread, though large publishing companies may hire people with the job title of proofreader. Editors who supervise proofreaders can better schedule and support the task if they understand the time and skill required to proofread well.

The Value and Goals of Proofreading

Three justifications for careful proofreading are accuracy, dollars, and credibility. Errors in numbers and names can cause confusion about budgets, orders, sales, and management. For example, is the part number 03262 or 03226? Have 20 or 200 parts been ordered? Is Andersen or Anderson in charge of arranging the meeting? Should the pressure be set to 22 psi or 32? Is the claim for $38 million or $3.8 million?

The cost of fixing errors increases at each stage of production, especially with full-scale printing. In offset lithography, reburning a metal plate for printing, the final stage of production before printing, can cost more than $100. If errors are recognized only then, the cost of the document escalates. But even with electronic publishing, it is cheaper to fix problems early. If a paragraph is added or deleted after the pages have been made up, for example, new spacing could throw off the placement of illustrations or require new work on smart page breaks. Thus, while a proofreader's main job may be at the end of production, getting the document right as it develops saves time for everyone.

Figure 13.1 illustrates some failures of proofreading. They may also be failures of copyediting, but because the texts have appeared in print and the proofreader has responsibility before print, they represent lapses in proofreading. A reader who recognizes these errors will probably laugh, gloat, and feel superior—not exactly the response a writer, publisher, or company might like. Spotting the one error that remains in a printed document, the reader will not stop to think about the fifty errors that the proofreader found. See how quickly you can spot the errors in the examples in Figure 13.1.

Readers may forgive an error or two, but if errors are too common, readers may dismiss the document as carelessly prepared and lose trust in the information. One editorial services company distributed a brochure advertising experi-

from a physician's advertisement	**LAURA WARD, M.D.** *Obstetrician & Gynecologist* *Diplomate American Board of Obstetrics and Gynecology* *Fellow American, College of Obstetricians & Gynecology* *Medical Director for Planned Parenthood*
from an investment firm's newsletter	Mr. Alberts rises early every weekday for a busy day of investing. After a 35-minute drive into town, Mr. Alberts starts his day pouring over the morning papers at a nearby coffee shop.
from a newspaper report of the Chernobyl disaster in the Soviet Union	The first detailed Soviet description of the Chernobyl disaster and its aftermath was given by Moscow Communist Party chief Boris Yeltsin.... "We are undertaking measures to make sure this doesn't happen again," said Yeltsin, who was attending a Communist Party in Hamburg.
from a newspaper ad for a series of workshops for women	Sexual Harassment - Petticost Wars - Time Management Men's Rections to The Women's Movement - And More

Figure 13.1 Typos and Punctuation Errors in Print

From an annotated bibliography

> The metric system is increasingly the norm in U.S. indus- try. The system is economical...

From a technical report

> A well-constructed message prototype for an emergency is important to the quick dissemination of information. The system and content of a message can have a dramatic effect on public response. Enough research has been conducted to discern a poor message from a good one.

From the printed minutes of an annual meeting

> 1. **Report of the Secretary/ Treasurer**
>   The treasurer reported a balance of $9,978.90 in the checking account, but observed that the expenses exceeded income in this fiscal year.
> 2. **Report of the   Journal editor**
>   The editor presented a budget of $11,002 for 1998 for the
>   journal. The Board advised her to investigate the feasibility of new
>   page layout software.

From the headings for a project summary

> **PROJECT:** Multi-Enzyme Reactor Design
> **SPONSOR:** State of Texas
> **PRIMARY INVESTIGATOR:** F. Senatore
> **Graduate Students:** A. Lokapur, P. Zuniga, T. Jih

From a book on word processing. The spacing error— no indention for a new paragraph— indicates a content omission

> ...This is easy and the results far more convincing than simply adding
> one's name to someone else's style.
> the whole essay into the computer in the first place. Some enterprising...

Figure 13.2   Spacing and Typography Errors in Print

enced "proffers" for proofreading. It is hard to imagine that the company got much proofreading business.

Readers may not notice errors in spacing, type, and capitalization as readily as they see typos, but such errors also detract from the quality of the publication. Shifts in spatial patterns and typography can distract readers or even give false signals about meaning. A spacing error may signal an omission. Figure 13.2 shows spacing and typography errors.

Most of us have spotted proofreading failures in printed or online documents and may therefore feel confident about our proofreading ability. Yet even the best of proofreaders will miss some errors. Proofreaders may get caught up in content and overlook surface errors. Reading quickly, we are likely to identify words by their shapes rather than by their individual letters, yet the word shape will remain essentially the same with a simple typo, particularly one that involves a narrow letter such as *i* or a missing member of repeated letters, such as two *m*'s. In the example that follows, the word shapes are identical for both the correct spelling of evaluation and its misspelling. "Evalvation" would be easy to overlook in proofreading.

evaluation   evalvation

Proofreading Marks and Placement on the Page

Some proofreading can be paperless, especially for online documents, but when you work on paper, you use marks like those you use in copyediting. Table 13.1 identifies proofreading marks for typos, punctuation, typography, and spacing. A comparison of these marks with those for copyediting (Tables 4.1–4.3) will reveal the similarity. However, the marks are placed differently on the page, and proofreading uses some spacing and typography marks that copyediting does not require.

TABLE 13.1	Proofreading Marks—Spelling, Punctuation, Typography, Spacing

Proofreading Marks: Spelling

Margin	In Text	Meaning	Result	Comment
ℓ	wₒords	delete	words	
ℓ	to the the printer		to the printer	
i	edting	insert character	editing	Place the caret at the point in the text where the insertion will go. In the margin, write the characters to be inserted.
the	to printer	insert word	to the printer	
#	inthe	insert space	in the	
⌢	edit ing	close up	editing	
ℓ	typefface	delete and close up	typeface	
(tr)	typescr[ip]t	transpose	typescript	
a	tolerence	substitute letter(s)	tolerance	Write just the letters to be substituted, not the whole word.
(sp)	(%)	spell out	percent	Spell the whole word if the spelling will be in question.

TABLE 13.1	**Proofreading Marks** *(continued)*

 procee~~d~~ | "let it stand"; revert to the unmarked copy | proceed | Use dots under the words in the text that should remain the same and write "stet" in the margin.

Proofreading Marks: Punctuation

Mark all punctuation insertions in the text with an insertion symbol (caret) at the point where the punctuation occurs. The mark of punctuation appears in the margin as noted here. To replace one mark with another, show the insertion only, not the deletion.

Margin	Meaning		Margin	Meaning		Margin	Meaning
⋏,	comma		:		colon		V	quotation marks
⊙	period		|=|	hyphen		V	apostrophe
set ?	question mark		M	em dash		⧘ ⧙	brackets
;|	semicolon		N	en dash		⧙ ⧘	parentheses

Proofreading Marks: Typography

Margin	In Text	Meaning	Result	Comment
lc	T̲ITLE	set in lowercase	title	
cap	t̲i̲t̲l̲e̲	set in capital letters	TITLE	Note three underscores.
ulc	T̲HE T̲ITLE	set in upper- and lowercase letters	The Title	If only the first word is to be capitalized, write "init cap" for "initial cap." "ULC" means that all words are capitalized.
sc	d̲o̲c̲u̲m̲e̲n̲t̲	set in small caps	DOCUMENT	Small caps have the shape of regular caps. They are not lowercase. Their most frequent use is in the AM/PM abbreviations.
ital	document	set in italic type	*document*	
rom	*document*	set in roman type	document	
bf	document	set in boldface	**document**	
lf	**document**	set in lightface	document	
wf	helvetica	wrong font	helvetica	Specify the font if you know it.
x	bro̷ken	reset a broken letter	broken	

(continued)

TABLE 13.1	**Proofreading Marks** *(continued)*				
⤻	m2⤻	set as superscript	m²	Visualize the inverted caret pushing up.	
⌃2	H2O	set as subscript	H$_2$O	Visualize this caret or "doghouse" pushing down.	

Proofreading Marks: Spacing

Margin	**In Text**	**Meaning**	**Result**	**Comment**
¶	sentence. ¶ A new...	start paragraph	sentence. A new...	
(run in)	sentence. ⌐ ⌐ A new...	run lines together	sentence. A new...	
—————⎤	end.———⎤ The line is	fill line	end. The line is	Use with awkward gaps at the ends of lines, especially with right justification.
(eq #)	to ✓a ✓book	equal space between words	to a book	Checks within the text mark the spaces that should be equal.
⌐⌐	the ⌐first book	move type up	the first book	
⌊⌋	the second book	push type down	the second book	Use for words or letters within words.
↓	⌐Headings ↳No space after.	push the line of type down	**Headings** no space after.	Use for whole lines of type.
( ) or (close up)	extra space (between lines)	close up the space between lines	extra space between lines	
⌉	wor⌉d hangs out. ⌊Move it in.	move right	word hangs out. Move it in.	
⌊	⌐paragraph	move left	paragraph	
⌉⌊	⌉Title⌊	center	Title	
‖	‖margin is ‖ not straight ‖ on the left	align horizontally	margin is not straight on the left	
═	base⌊ine	align vertically	baseline	

To affirm that the proofs are correct and that production may proceed, the proofreader must review such visual features as alignment, spacing, and type size. Thus, the proofreading marks include marks for alignment, space between letters and between lines, and shifts in type. Editors may have tagged the text for its visual features, but the proofreader examines whether the tags have been applied correctly.

Figure 13.4, the typeset version of the copyedited document in Figure 13.3, shows that most of the proofreading marks appear in the margin. The marginal marks alert the keyboard operator to check specific lines. The marks in the body of the text only show where to make the correction requested in the margin. In Figure 13.4, the caret in the second line signals that the marginal word "calcium" should be inserted at that point.

When a line requires multiple corrections, they may be listed, left to right—in the order in which they appear—with a slash between the individual corrections. The title in Figure 13.4 contains both a transposition and a capitalization error. Both are listed in the margin with a slash between the two circled instructions.

Corrections may be placed in either margin, but the marginal marks will be easier to match up with the body marks if they are in the margin closer to the error. Don't confuse the person making the corrections by placing in the left margin a note about an error on the right side of the line while placing in the right margin a note about an error on the left side of the line.

Where the corrections within the body of the text might be confusing, particularly when there are multiple corrections within the same area, you may write out the correct version in the margin, as well as marking within the body. Otherwise, it is sufficient to write just a single letter when there is an insertion or substitution. To write more than is necessary will make you look like an inexperienced proofreader and will waste valuable time.

Usually, the publication specialist or compositor will be able to distinguish instructions from modifications of the text, but if there can be doubt, circle your instructions. The circled "run in" that appears beside paragraphs 2 and 4 is an instruction to the compositor.

If there is more than one way to mark an error (for example, by transposing or substituting letters), choose the mark that will be easier to understand. Also try to anticipate how the compositor will apply the direction: if the probable method will be to delete an entire word and retype it, then your mark may reflect that process; if only one or two characters need to be replaced, mark single characters.

Use a colored pencil or pen to mark paper copy when proofreading so that the marks will be easy to see. Legible marks also increase the likelihood that the compositor will be able to follow your proofreading instructions. When your company contracts with someone outside the company for printing or web design services, you may need to identify the responsibility for each correction. The buyer of print or web design services should not be charged for the printer's errors, nor should the publishing company absorb the expense of excessive changes by the writer after the page proofs have been made. In these cases, mark each correction with a circled identifier: "pe" for printer's error, "ea" for editorial alteration, "da" for designer's alteration, and "aa" for author's alteration.

Generally, you will proofread to make the new version conform to the previous version; that is, you will not edit—because of the expense and because substantial changes will slow down production. However, if errors remain in the proof copy that were uncorrected in the dead copy, you will, of course, correct them rather than making conformity to the dead copy a higher aim than an accurate publication.

MILK: NOT QUITE THE PERFECT FOOD

[handwritten: New Century Schoolbook / BF 12 uc]
[handwritten: New Cent 10/12 x 25]

Many people think that milk is nearly a perfect food. However, while calcium is a necessary mineral in the diet, many adults cannot tolerate even small amounts of milk. A scoop of ice cream or a glass of milk can cause abdominal cramps, gas, and sometimes, diarrhea, for about thirty million Americans. These people lack an enzyme, lactase, which is needed to digest lactose, the sugar in milk. After about the age of two, many people gradually stop producing this enzyme. If is thought that fully 70% of the world's population has this condition. People of Asian, Mediterranean, or African descent are more likely to be lactase-deficient while those of Scandinavian descent rarely lose the enzyme.

When lactose, the sugar found in milk and milk products, is not digested, bacteria in the colon use the sugar to produce hydrogen and carbon dioxide. The result is gas, pain, and a number of other symptoms that are sometimes mistaken for. Lactose intolerance is not hte same thing as an allergy to the protein in milk or irritable bowel syndrome, although the three conditions may produce similiar symptoms. Because hydrogen gas is produced, a test that measures hydrogen in the breath can confirm lactose intolerance is the problem.

Since lactose digestion depends on the ratio of lactase to the amount of lactose-containing food eaten, many people can drink small amounts of milk or eat some ice cream. The type of milk product can also make a difference. While yogurt contains more lactose than milk, it is often better tolerated because the bacteria in live cultures of yogurt break down the milk sugars. The same things appears to happen in hard cheeses such as cheddar or swiss.

Fortunately there are some products on the market that can make life easier for those with lactose intolerance. Some foods are treated with the enzyme to break down the lactose into glucose and galactose. LactAid or Lactrase may be added to food 24 hours before consumption.

Figure 13.3 Copyedited Article for a Health Newsletter

Milk: Not Quiet the perfect Food   ⟨tr⟩/⟨cap⟩

Many people think that milk is nearly a perfect food. However, while is a necessary mineral in the diet, many adults cannot   *calcium* tolerate even small amounts of milk.

A scoop of ice cream or a glass of milk can cause abdominal cramps, gas, and sometimes diarrhea for about 30 million   ⟨eq.#⟩ Americans. These people lack an enzyme, lactase, which is needed to digest lactose, the sugar in milk. After about the age of two, many people gradually stop producing this enzyme. Fully 70 per-   ⟨run in⟩ cent of the worlds population have this condition. People of   ⟨tr⟩ Asian, Mediterranean, or African descent are more likely to be   ⟨tr⟩ lactase deficient while those of Scandinavian descent rarely loose   ⟨fr⟩ the enzyme.

When lactose, the sugar found in milk and milk products, is not digested, bacteria in the colon use the sugar to produce hydrogen and carbon dioxide. The result is gas, pain and other symptoms   ⟨tr⟩/⟨#⟩/⟨,⟩ that are sometimes mistaken for allergy to the protein in milk or   ⟨eq.#⟩/an irritable bowel syndrome. Because hydrogen gas is produced, a test that measures hydrogen in the breath can confirm that lactose intolerance is the problem.

Since lactose digestion depends on the ratio of lactase to the amount of food eaten containing lactose, many people can drink small amounts of milk or eat some ice cream. The type of milk   ⟨rom⟩ product can also make a difference. While yogurt contains more lactose than milk, it is often better tolerated because the bacteria in live cultures of yogurt break down the milk sugars. The same thing appears to happen in hard cheeses, such as Cheddar or Swiss. Fortunately, some products on the market can   ⟨lc⟩/⟨¶⟩ make life easier for those with lactose intolerance. Some foods are treated with the enzyme to break down the lactose into glu-   #/⟨fr⟩ cose and galactase.   ⟨run in⟩ LactAid or Lactrase may be added to food 24 hours before   ⟨wf⟩ consumption.

Figure 13.4   Proofread Version of the Copyedited Document in Figure 13.3

Using Your Knowledge

Following these guidelines will increase your proofreading effectiveness:

1. **Take advantage of the computer.** Electronic transmission of files means that corrections can be inserted throughout development. Your company will need a good system of naming versions so that anyone who works on the document will have a file with the current corrections included. Styles and templates regularize type and spacing for parallel structural parts. Use spelling and grammar checkers, but don't depend on the computer completely for proofreading.

2. **Check the author's original copy and the editor's marks and comments.** Proof copy can make sense even when phrases are omitted or words are substituted. Conversely, a word that looks strange to you may be accurate.

3. **Pay special attention to text other than body copy.** Titles, tables, lists, and nonverbal text may contain just as many errors as body copy, but they are easy to overlook.

   - **Titles and headings**. Typos; capitalization, type style, and typeface; spacing before and after.

   - **Illustrations.** Typos; spacing, alignment, keys, completeness, and color or shading.

   - **Numbers.** Accuracy (reversal of numerals and errors in addition are common); alignment on the decimal; match of numbers with references to them in the text.

   - **Reference lists and bibliographies.** Spelling of authors' names and of titles; spacing, capitalization, and italics as specified by the appropriate style manual.

   - **Names of people and places.** Match with the edited copy; use reference books if the dead copy seems inaccurate.

   - **Punctuation.** Close quote marks and parentheses.

4. **If you find one error in a line or word, go back over the line to look for other problems.** Mistakes cluster, perhaps at points where the production staff became confused or tired. Yet you feel as though you have "finished" a line if you find an error.

5. **Review at least once solely for visual errors.** Review just for spacing, alignment, type style, and typeface without the distraction of the text content. Some proofreaders turn the copy upside down to check spacing; others read right to left.

   - **Typeface, type size, type style.** Type should not change arbitrarily.

   - **Spacing above and below headings.** Measure the type from the baseline of one line of type to the baseline of the next if you don't trust your eyes.

   - **Centering or left or right justification.** Check all same-level headings (first level, second level) together to spot inconsistencies.

■ **Indention.** Check systematically, especially after headings.

■ **Alignment.** Scan the left and right margins and any other aligned material.

6. **Check hyphenated words at the end of lines.** Word divisions should conform to those in the dictionary. Note, for example, the difference between *thera • pist* and *the • rapist.*

7. **Use techniques to force slow reading.** Some errors will pop out when you skim, but most will not, especially because the eye recognizes and accepts word shapes even when some of the letters may be incorrect. Unless you can slow yourself down, you may read for content and miss errors. Some proofreaders read with a partner, with one reading aloud from the edited copy while the other reads proof copy. Some proofreaders read pages out of sequence to focus on the presentation rather than the content.

8. **Use multiple proofreaders and proofreadings.** Ideally several proofreaders, including someone other than the writer and editor, will read the document. Proofreaders who are familiar with the document or its content are likely to miss errors because they see what they expect to see. A single proofreader can increase productivity by reading the document for a specific feature (such as the visual features or word division), then rereading for another feature.

9. **Schedule to proofread when you are alert.** If you are tired, interrupt proofreading to do other tasks and return to the proofreading when refreshed.

Good proofreading helps to ensure a high-quality publication, notable for the apparent care with which it has been produced and the absence of distracting errors. Proofreading is skilled work.

Further Reading

Smith, Peggy. 1997. *Mark My Words: Instruction and Practice in Proofreading.* 3rd ed. Editorial Experts, Inc.

Online Resource

Lyon, Jack M. Paperless proofreading. *Editorium Update* May 14, 2003. lists.topica.com/lists/editorium/read/message.html?mid=1713004126. Describes procedures for using Microsoft Word's reviewing tools to streamline.

Discussion and Application

1. Compare the copyedited newsletter article in Figure 13.3 with the proofread version of the same document in Figure 13.4. In your own words, distinguish the types of corrections made on each document, and explain how and why the marks differ.

2. During the preparation of an annotated bibliography on technical editing, two collaborators missed the typos in the following words after repeated checks of the paper copy. (The spelling checker on the computer identified them.) For each word or phrase, hypothesize why the collaborators overlooked the error; that is, what is it about the spelling, word shape, or other feature of the word that encouraged the collaborators to read it as correct?

gullability	progams	responsibilites
indentification	edtors	comform
labortatory	embarassing	comunication
manuscritps		

3. With regard to application 2, discuss how the fact that the collaborators were proofreading their own document affected the quality of their proofreading.

4. A national insurance company mailed to shareholders a ballot for trustees and a booklet describing the qualifications of twenty-seven nominees. Inside the printed booklet was a slip of paper titled "Errata" that included these comments:

> Robert C. Clark is 45, not 55, as appears on page 5 of the Trustee booklet.
>
> William H. Waltrip is 52, not 62, as appears on page 6 of the Trustee booklet.
>
> Uwe E. Reinhardt's name incorrectly appears on the ballot as Uwe E. Remhardt.

What is the meaning of *errata*? Speculate on why the proofreaders missed these particular errors. When the errors were discovered (after the booklet was printed), what were the company's three options? Why did the company choose to include the errata slip? What were the consequences of the proofreading failures to this company, in terms of money, time, and image?

5. Proofread the version of the document in the right column by comparing it with the version in the left column. Mark the typeset version to make it match the typescript version.

 Times 10

Ivory Trade Continues

Elephants at risk of extinction

Recent studies have established that

ivory poaching has reduced the elephant

populations in East Africa by half in less

than a decade. The same story is

repeated for the rest of Africa except

parts of southern Africa where rigorous

management has actually made it

Ivory Trade Continues

Elephants at risk of extinction
Recent studies have established that ivory paoching has reduced the elephant populations in East Africe by half in less than a decade. The same story is repeated for the rest of Afirca except parts of Southern Africa where rigorous managment has actually made it possible for the elephont numbers to increase. Over much of central

possible for the elephant numbers to increase. Over much of the east and central Africa, elephants are so heavily poached, even in previously secure sanctuaries such as Selous, Tsavo, and the Luangwa Valley, that it will not be long before the elephant is extremely rare or even extinct. There is little doubt that short-term profit-motivated poaching is responsible for the enormous decimation of the large herds of elephants. Conservation organizations of the world are now demanding immediate enlightened action, strong political will, and a high degree of international cooperation to avert a disaster.

Africa, elephants are so heavily poached, even in previously secure sanctuaries such as Selious, Tasvo, and the Luangwa Valley, that it will not be long before the lelphant is extremely rare or even extinct. There is little doubt that short term profit-motivated poaching is re-sponsible for the enormous deci-mation of the large herds   of ele-phands. Conservation organizations of the world are now demanding imediate enlightened action, strong politicial will and a high degree of international coop-eration to avert a disaster.

6. Proofread the right column to match the dead copy on the left; check especially for errors in type style, typeface, and spacing.

Ban on Ivory Imports Established
A moratorium on the importation of African elephant ivory was im-plemented through an announce-ment in the *Federal Register*.

Ban on Ivory Imports Established
A moratorium on the importation of African elephant
ivory was implemented through an announcement in the Federal Register.

A quota system for legal, regulated trade in ivory authorized under the Convention on International Trade in Endangered Species of Wild Fauna and Flora (CITES).

A quota system for legal, regulated trade in ivory was was authorized under the Convention on International Trade in Endangered Species of Wild Fauna and Flora (*CITES*).

The United States, Western Europe, and Japan consume two-thirds of the world's "worked ivory."

The United States, Western Europe, and Japan consume two-thirds of the world's "worked ivory."

7. Correct spacing errors in this paragraph.

If the United States, Western Europe, and Japan   were concerned enough as parties to CITES to agree to the appeal and pro hibit importation ofall ivory without exception,

the present enormous demand for ivory wou ld cease. IN   turn, poached ivory would become less lucrative. CITES, the   one instrument   of international standards available, should impose a world wide ban on the ivory trade to stop the convention   being used to channel hundreds of tons of illegal ivory into legal trade.

14 Comprehensive Editing: Definition and Process

In Part 3 of this book, you learned ways to make a document correct, consistent, accurate, and complete. But a document may achieve all of these standards and still not work—readers may not be able to use it or comprehend it. Usefulness and comprehensibility depend on a document concept that matches the need for the document. Organization, visual design, and style support comprehension and the probable uses of the document by readers.

To make the document functional for its readers, editors consider a document's concept and content, organization, design, and style. Editors must keep the writer's intentions in mind while anticipating the reader's needs and imagining the document in use. Style manuals and handbooks provide little help for decisions on this level. Instead, an editor relies on a good process and on principles of good writing and usability. The process and the principles help editors make good judgments. Comprehensive editing will take more time and effort than basic copyediting, but it gives you more options for making the document work. Other terms for this type of editing are *developmental editing, substantive editing, macro editing, analysis-based editing*, and *heavy editing.*

In this chapter, you will compare the copyediting and the comprehensive editing of a single document to illustrate the differences between the two editing tasks. The chapter will walk you through the process of document analysis and goal setting that precedes comprehensive editing. The chapters that follow in this section review principles of style, organization, visual design, and illustration on which comprehensive editing relies.

Key Concepts

Comprehensive editing should make a document more usable and comprehensible. Because documents and purposes vary, no handbook can establish the best content, organization, visual design, or style. Instead of relying on a manual, an editor uses a systematic process of analysis and applies principles of good writing. The process and principles help editors avoid changes that could make the document worse.

The comprehensive editing process requires the editor to analyze the document's purpose, readers, and uses; to evaluate the document; to establish specific editing objectives; and to consult with the writer about the plans before the editing takes place.

Comprehensive editing precedes editing for grammar, punctuation, and mechanics.

Example: Copyediting versus Comprehensive Editing

To compare the purposes, methods, and results of copyediting and comprehensive editing, let us consider the following scenario. You work in the communications department of a large medical center. Today you are editing a grant proposal, due in five days, worth $500,000 if it is funded. You also have an appointment with the graphics specialist to review the photographs for a slide show for which you are writing the script, to be completed in two weeks. The Director of Maintenance, with whom you have a cordial relationship, drops by with a memo (Figure 14.1) announcing a policy and procedure change. The director wants to distribute the memo in paper copy, through the internal mail system, in order to get the attention of potential users, though he plans to update the policy online as well. He asks you to look it over. The information is time sensitive, and the memo needs to be mailed today. Although you can spare only about 15 minutes, you agree to check the memo.

The punctuation and mechanics of this memo need attention, but the work will be fun. It will offer relief from the intensity of the grant proposal. And you can perform a service for the Director of Maintenance with little effort.

If you edited the document in Figure 14.1 for grammar and mechanics, your edited memo would look much like the marked-up version in Figure 14.2. On most points your editing would agree with the editing shown in Figure 14.2 because editors apply the same rules of grammar and spelling. Some editing for mechanics would depend on your assumptions about titles. For example, is "Maintenance Department" a proper name or a descriptive one? Is "manual" a part of a title or just a descriptor? Answers to these questions affect decisions about capitalization. If you actually worked at this medical center, you would either know the answers to these questions or be able to check them quickly.

Some decisions require more thoughtful analysis. For example, in the last line of the first paragraph, the series ending in "hazards" is not clearly punctuated. "Safety" does not fit in the same class of things as wiring and fire (it is not a problem requiring a service call), so it must modify "hazard." But then, where does the series end, and where should you insert the "and" to conclude the series? You have two options:

. . .wiring, fire, and safety or security hazards.

. . .wiring, and fire, safety, or security hazards.

The second option identifies "fire hazards," not "fire," as the problem requiring a service call. You can check with the maintenance department if necessary, but good sense tells you that fire, as opposed to a fire hazard, would require a call to the fire department.

Apart from these possible variants, however, the results of copyediting by any two competent editors will be similar. Furthermore, the edited document, as shown in Figure 14.3, will be similar in content, form, and style to the original.

TO: All Medical Center personnal

FROM: Tom Barker

DATE: July 13, 2005

SUBJECT: Service Calls

The Maintenance Department will initiate a Service Call system beginning Monday,July 18th, 2005, that is designed to provide an effective response time with equal distribution for all departments for work categorized as Service Calls. Service Calls are defined as urgent minor work requiring immediate attention, for example, loss of heat, air conditioning, water leaks, clogged plumbing, faulty electrical wiring, fire, safety or security hazards.

To initiate a Service call a telephone call to the Maintenance Department, (742-5438, 24 hours a day), is all that is required. A Service Call number will be assigned to the job and furnished to the originator if requested. This number may be used when referring to the status of the Service Call

During the hours of 5:00 p.m. through 8:00 a.m. daily, all day week-ends and holidays, only one Maintenance Mechanic is on duty, who must insure total systems are oprational on his shift, in addition to accomplshing Service Calls. That will limit the amount of calls, but should not deter anyone from calling in a Service Call to insure it is accomplished when the manhours are available. If no one answers, please call again.

For urgent or emergency requirements, the PBX Operator must be notified to page Maintenance. Please do not use the paging system unless your request is justifiable.

Your cooperation in this matter would be greatly appreciated.

Service Call requirements are outlined in the Policies and Procedures Manual. Requests exceding these these requirements will be classified as Job Orders, and are also outlined in the Policeis and Procedure Manual.

Figure 14.1 Unedited Memo on Service Calls

Excess capitalization is distracting.

The punctuation in the list makes the examples confusing. Surely "air conditioning" and "safety" are not problems requiring service calls. ("Loss" and "hazards" are.)

A list of two items (daily hours and weekend hours) requires a conjunction. Use "number" with quantifiable amounts.

TO: All Medical Center personnel

FROM: Tom Barker, *Director of Maintenance*

DATE: July 13, 2005

SUBJECT: Service Calls

The Maintenance Department will initiate a Service Call system beginning Monday, July 18th, 2005, that is designed to provide an effective response time with equal distribution for all departments for work categorized as Service Calls. Service Calls are defined as urgent minor work requiring immediate attention, for example, loss of heat, air conditioning, water leaks, clogged plumbing, faulty electrical wiring, fire, safety, or security hazards.

To initiate a Service call a telephone call to the Maintenance Department, (742-5438, 24 hours a day), is all that is required. A Service Call number will be assigned to the job and furnished to the originator if requested. This number may be used when referring to the status of the Service Call.

During the hours of 5:00 p.m. through 8:00 a.m. daily, and all day weekends and holidays, only one Maintenance Mechanic is on duty, who must ensure total systems are operational on his shift, in addition to accomplishing Service Calls. That will limit the number of calls, but should not deter anyone from calling in a Service Call to ensure it is accomplished when the manhours are available. If no one answers, please call again.

For urgent or emergency requirements, the PBX Operator must be notified to page Maintenance. Please do not use the paging system unless your request is justifiable.

Your cooperation in this matter would be greatly appreciated.

Service Call requirements are outlined in the Policies and Procedures Manual. Requests exceeding these these requirements will be classified as Job Orders, and are also outlined in the Policies and Procedures Manual.

Figure 14.2 Memo from Figure 14.1 Marked to Show Copyediting

TO: All Medical Center personnel

FROM: Tom Barker, Director of Maintenance

DATE: July 13, 2005

SUBJECT: Service Calls

The Maintenance Department will initiate a service call system beginning Monday, July 18, 2005, that is designed to provide an effective response time with equal distribution for all departments for work categorized as service calls. Service calls are defined as urgent minor work requiring immediate attention; for example, loss of heat or air conditioning, water leaks, clogged plumbing, faulty electrical wiring, and fire, safety, or security hazards.

To initiate a service call, a telephone call to the Maintenance Department (742-5438, 24 hours a day) is all that is required. A service call number will be assigned to the job and furnished to the originator if requested. This number may be used when referring to the status of the service call.

During the hours of 5:00 p.m. through 8:00 a.m. daily, and all day on weekends and holidays, only one maintenance mechanic is on duty, who must ensure that total systems are operational on his shift, in addition to accomplishing service calls. That will limit the number of calls, but should not deter anyone from calling in a service call to ensure it is accomplished when the manhours are available. If no one answers, please call again.

For urgent or emergency requirements, the PBX operator must be notified to page Maintenance. Please do not use the paging system unless your request is justifiable.

Your cooperation in this matter would be greatly appreciated.

Service call requirements are outlined in the *Policies and Procedures* manual. Requests exceeding these requirements will be classified as job orders, and are also outlined in the *Policies and Procedures* manual.

Figure 14.3 Copyedited Memo from Figure 14.2

What, then, has copyediting accomplished? First, because the document is correct, readers will take the message more seriously than they would if they recognize errors. Readers, fairly or not, evaluate the message on superficial text characteristics such as spelling and punctuation. Editing has increased the chances that readers will respond to the memo in the intended way rather than being distracted by errors.

Copyediting has also increased clarity. For example, in line 6 of Figure 14.2, the insertion of the conjunction *or* establishes that "loss" refers to "air conditioning" as well as to "heat." The insertion of the conjunction *and* in line 7 clarifies that "safety" modifies "hazards." These insertions clarify the information. The reduction of document noise (in this case excess punctuation and capitalization as well as errors) also makes the message clearer.

From the perspective of your workload, the task has taken little time and energy. You can now return to your more important task—the proposal.

From the perspective of your relationship with the memo's writer, you have performed a service without challenging his competence. If you and he disagree on some specific emendations, you can confirm your choices by pointing to a handbook or style guide, or you can yield on issues when his choice is acceptable (if not preferable). If copyediting is correct, it should create a happy writer.

Comprehensive editing, on the other hand, looks beyond words and sentences to the way in which readers will read and use the document. Editors may make global changes with substantial impact. Instead of reacting line by line to the text, as in basic copyediting, you begin by assessing the document as a whole and interviewing the writer to determine how readers will use the document and what editing objectives to establish. An additional review follows the editing to ensure that editing has not introduced content errors.

The Process of Comprehensive Editing

Comprehensive editing is a multistage process. It begins with analysis to determine how the document as a whole will be read and used and with a plan for revision. The editor may participate in the analysis and planning that take place before the writing begins. In the development of the online tutorial described in Chapter 1, Charlene helps to determine the goals for the document. All the members of the development team understand and share the goals for the document from the start. But some comprehensive editing occurs after documents are developed—the editor enters the project at the end, as in the service call memo. Analysis before editing discourages line-by-line reaction to errors and sentence structure. The line-by-line approach can work for copyediting, but it does not direct the editor's attention to big-picture issues of content, organization, and style nor to the document in use.

As editor, you are more likely to achieve the overall goal of improving the document's usability and comprehensibility if you edit with a plan. Before you begin to mark the page or edit on the computer screen, you should complete a four-part process that will result in a plan.

1. Analyze the document's purpose, readers, and uses to determine what the document should do and the ways it will be used.
2. Evaluate the document's content, organization, visual design, style, and reader accommodations to determine whether the document accomplishes what it should.
3. Establish editing objectives to set forth a specific plan for editing.
4. Review the plan with the writer to work toward consensus.

The steps in comprehensive editing are discussed more fully in the following pages, and the process is then applied to the memo on service calls.

Analyze the Document's Purpose, Readers, and Uses

Analysis begins with questions. These are familiar questions—the ones writers ask before drafting documents:

Purpose

- What is the purpose of this document? What should happen as a result of its use?

Readers

- Who will read the document, and why? Are there cultural variations in readers that will affect comprehension? Will the document be translated? Do some readers have visual or other physical impairments? Do they read well?
- What are their attitudes toward the subject of the document? (Cooperative? Unsure? Hostile?)
- Where and when will they read it? (While doing a task? In good light or poor? Inside or outside?) Will they read it straight through or selectively? If they are reading online, how fast is their connection speed?
- What should readers do or know as a result of reading it? What do they already know about the subject? Should they memorize the contents or use the document for reference?
- What will they do with the document once they have read it? (Bookmark it? Throw it away? File it? Post it? Refer to it again? Print it from its online version?)

Uses

- What constraints of budget, equipment, or time influence the options for the document?
- Will this document be used in multiple forms, perhaps in an online as well as a print version? Is some content reused from other sources? Will this content be reused by another document?

For complex documents, you will need to interview the writer to determine answers to these questions. You can answer a number of the questions, though, by inference and from experience.

Evaluate the Document

While the purpose of analysis is to determine what the document should do, the purpose of evaluation is to determine how well the document does it. An evaluation should systematically review these features as they support (or detract from) the document's purpose and uses:

- *content:* completeness and appropriateness of information
- *organization:* order of information; signals about the order
- *visual design and navigation:* prose paragraphs, lists, or tables; paper size; screen display; links
- *style:* writer's tone or persona; efficiency of sentence structure; concreteness and accuracy of words; cultural bias; grammar, usage, punctuation, spelling, and mechanics
- *illustrations:* type, construction, placement
- *accessibility:* accommodation for readers with visual or other physical impairments
- *reuse:* display in more than one medium; content from another source or to be used in another document

Establish Editing Objectives

Setting objectives may begin with an evaluation of what is ineffective with the original document, but it goes beyond a critical evaluation to identify what you want to accomplish in editing. If you observed in evaluation, for example, that the writer's style is full of nouns and weak verbs, the editing objective may be to substitute strong verbs for weak ones. If you determined that readers would be likely to read selectively but that the document provided inadequate identifiers of sections, your goal might be to insert headings for major divisions or better navigation links for online documents. Thus, you convert evaluative statements to goal statements, from what's wrong to what needs to be done. These goals give focus to your editing.

Review Your Editing Plans with the Writer

After you have analyzed the document and established editing objectives, consult with the writer about the editing plans. You may save yourself the time of misdirected editing if you have misinterpreted editing needs or if the writer can offer additional suggestions before you begin. Furthermore, the review encourages cooperation and support. You will avoid surprising the writer and thereby avoid the negative response that surprise can provoke. The more substantial the changes, the more important is the review.

Complete the Editing

Experienced editors work mostly "top down," first considering the most comprehensive document features, the ones that will affect others. These features include content, organization, and visual design. Bottom-up editing, beginning with sentences, can waste time and divert your attention from the comprehensive goals. You could start by correcting sentences for grammar and editing them for style and then proceed to content; however, you might ultimately decide to delete many of those edited sentences. On the other hand, if the surface errors distract you from the content, you might clean up the document first by correcting spelling and other errors. With clean copy, you can pay closer attention to more comprehensive document features. A strategy for perceiving the document structure may be to make the headings consistent—a bottom-up task that supports top-down editing.

If the needs are complex, you may need to work through the document more than once to achieve all your objectives. On your first pass, you might work on the content, organization, and visual design. On the second pass, you can edit for style. A third time through the manuscript, you may refine the editing and attend to consistency. As you see the document take shape, you may revise some editorial objectives.

Evaluate the Outcome

What determines that the editing is good or right? With comprehensive editing, the criteria include not just the textual features but also the ways in which the document will be used. These criteria are outside the text. The editor needs to consider the *context*. As Chapter 2 explains, the best decisions about textual features reflect not just handbook standards but also what will happen when the document is distributed. For example, a textbook on editing may include practice documents, but a general book on editing would not. Different contexts of use (the classroom, a workplace) help to define what features might be appropriate. In addition to purpose and readers, the context includes storage and disposal, budgets, and related documents.

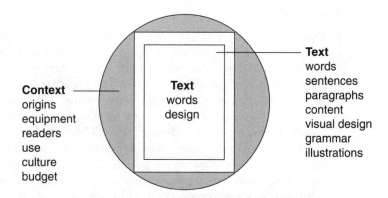

Comprehensive editing is good if it has produced a document that achieves the goals established in document analysis. The document is not simply different. The editing has been based on a thoughtful analysis of document function and use and applies known principles of good writing. The emendations are thoughtful and functional, not arbitrary. The editor can give a reason for each emendation, showing the logic of his or her decisions. The results can be tested, either objectively (as in a usability test with sample users) or by informed judgment.

Review the Edited Document with the Writer or Product Team

Both as a courtesy to the writer and as a check on the accuracy of your editing, you should give the writer a chance to review the edited document and to approve it or to suggest further emendations. You may need the writer to clarify ambiguous points. The writer may also have questions, and you need to be able to answer them intelligently. (See Chapters 3 and 25 for suggestions on how to confer productively with a writer.) The outcome of the review should be either the writer's approval or identification of additional editing tasks.

Application: The Service Call Memo

Comprehensive editing of the service call memo will result in a document that differs from the copyedited memo. Figures 14.4 and 14.5, which appear later in the chapter, illustrate comprehensive editing. This discussion will show how those versions of the document emerged from the various stages of the editing process—analysis, evaluation of the content, establishment of editing goals, and evaluation of the outcome.

Analysis

Even this short memo raises questions that may require some input from the writer. However, common sense will let us make some initial judgments about purpose, readers, and uses.

- **Purpose.** The primary purpose is to enable the readers to act according to the policy. Thus, the style and design of the document should make it easy for readers to determine what to do. A secondary purpose is to encourage cooperation and goodwill. Perhaps the procedure has changed because the previous one alienated some departments. Whatever the situation, the procedure will work best if readers want to and can cooperate.
- **Readers.** This memo is addressed to "All Medical Center personnel." That label, however, only generalizes about readers. More specifically, the readers who will use the memo are people who will actually place service calls: secretaries and nursing supervisors. Secondary readers may be people who need to know the procedure but who will not actually place the calls,

such as department managers who will direct a staff person to place the call. Readers will be receptive to the instructions, because they want to do their jobs correctly; they will be even more receptive if they can see how the new system will benefit them. But if they have had trouble in the past with the maintenance department or if they are constantly bombarded with procedure changes, they may be impatient or unreceptive.

- **Uses.** Readers will not use the instructions immediately upon receiving them by mail. Rather, they will skim the memo to get its gist and then file it for reference when they need to place a service call. Thus, the document should be designed for filing, easy reference, and selective reading. Some readers will look for the policy online, and one option would be to skip the print version, announcing the change in policy by email. The edited memo should be suited for both print and online publication.

Evaluation

A critical evaluation of the memo should yield the following observations:

- Specific information (such as the hours in which the policies are in effect) is difficult to find and interpret. Nothing attracts attention to the subject of the memo.
- The memo shows the value to the maintenance department of the change, but the benefit to users is not explicit. The memo could motivate readers to cooperate if it stressed the service itself rather than the justification for it. The proposed date for implementing the new policy may occur before readers get the memo, which may exasperate them.
- The sentences and words are unnecessarily long.
- The instructions are buried in the passive voice.
- Excessive capitalization creates distracting document noise.
- There is no obvious plan for making the document available online as well as in print.

Editing Objectives

Based on the evaluation, the following objectives would be reasonable for the service call memo.

- **Content.** The information seems complete, but emphasize some information and delete some information to fit the memo on one page. Propose a later date of implementation.
- **Organization.** Begin with an overview of the process; then give specific instructions. The information at the beginning will establish the purpose and significance of the instructions.
- **Visual design.** First, restrict the memo to one page to prevent some instructions from getting lost and to force simplification of the instructions. Second, if necessary, create a memo of transmittal to explain the document, but place the instructions that will be filed on a separate single page.

Third, hole-punch the left side of the instructions on the assumption that readers will file the instructions in the *Policies and Procedures* manual and that the manual is in a three-ring binder. Finally, use white space, boldface headings, and lists to make information accessible.

- **Style.** Use the language of instructions—verbs in the imperative mood.
- **Illustrations.** Do not add any illustrations. Because no equipment (other than the telephone) is necessary and no forms will be filled out, illustrations will not clarify the procedure.
- **Mechanics.** Establish consistent capitalization style for "service call," "maintenance department," and the manual title.
- **Document reuse.** Develop a version for online distribution that fits onto one screen.

The Outcome of Editing

Figure 14.4 shows the results of comprehensive editing based on the established objectives. Although the message is the same as the message in Figure 14.3, this document will enable readers to read selectively and to follow instructions. The edited version shows a new date of implementation, but the editor must confer with the writer before such a change is made. The editor does not have liberty to change a policy, but thinking about the document from the reader's perspective, the editor can recommend such a change to maintain goodwill with readers.

This version shows some emendations that the objectives did not specify. Because the memo fits on one page, a separate memo of transmittal is unnecessary. Some new information has been added after all, following "How do I place a service call?" Besides placing the call and requesting the number, the caller will have to identify the problem and location. The definition of service calls that previously appeared at the end is now incorporated in the introduction with other definition material. Finally, in the last section, the statement that service calls could be placed 24 hours a day contradicted the purpose of the memo to encourage users to place their calls during normal working hours. The sentence has been restructured.

This version can be adapted for online presentation at the website of the maintenance department. The website version can begin with the section, "What is a service call?" The announcement of the change in policy will soon become dated. The *Policies and Procedures* manual, in both its print and online versions, may need to be updated to reflect the new procedure.

The editing plan did not anticipate each emendation, but it provided direction to make editing efficient and purposeful. Other goals became apparent as the editing progressed. These goals could be incorporated into the overall plan.

You may have imagined a different document. You can probably think of ways to improve the document in Figure 14.4. Or you may believe the version in Figure 14.5 would be easier to read. Differences in editing result from different initial assumptions. If you have assumed (in contrast to the analysis outlined here) that readers will read primarily for comprehension and that they will read most thoroughly when they remove the memo from the envelope, you will

Shorter sentences make comprehension easier. The procedure is introduced in terms of benefits to readers.

Headings in question form aid selective reading.

List form lets readers skim the examples. It also clarifies that "fire" and "safety" modify "hazards" rather than being problems that require service calls.

The reference to the manual relates to the definition of a service call. Reorganization groups related information.

Verbs in imperative mood direct a reader to act.

A new piece of information, the request for identifying information, clarifies the three-step procedure.

Space sets off and calls attention to the second-level heading, "for emergencies."

TO: All Medical Center personnel

FROM: Tom Barker, Director of Maintenance

DATE: July 13, 2005

SUBJECT: **Service calls: New procedure**

Beginning Monday, July 25, 2005, a new procedure for placing service calls will take effect. This procedure will allow the maintenance department to respond to requests from all departments quickly and effectively.

What is a service call?

Service calls are for minor work requiring immediate attention. Examples:

>   loss of heat or air conditioning
>   water leaks or clogged plumbing
>   faulty electrical wiring
>   fire, safety, and security hazards

See the *Policies and Procedures* manual (page 18) for service call requirements. Requests exceeding these requirements are classified as job orders.

How do I place a service call?

Call 742-5438.

Identify the nature of the problem and the location.

Request your service call number. Use this number if you need to check on the status of your service call.

For emergencies

Ask the PBX operator to page maintenance. Please request paging only when you have a true emergency.

When may I call?

You may place service calls 24 hours a day.

From 5:00 p.m. through 8:00 a.m. weekdays and all day weekends and holidays, only one maintenance mechanic is on duty. He or she can respond to a limited number of calls during those times. However, you can still place a service call on nights or weekends to be completed when the mechanic is available.

Thank you for your cooperation.

Figure 14.4   The Memo after Comprehensive Editing

The information beginning with the heading "What is a service call?" can be the text of an online policy statement. The reference to the manual can be linked to the appropriate place.

TO: All Medical Center personnel

FROM: Tom Barker, Director of Maintenance

DATE: July 13, 2005

SUBJECT: Service calls: New procedure

On Monday, July 25, 2005, the Maintenance Department will begin a new service call system. This system will let us respond to all departments quickly and effectively.

Service calls are for minor work requiring immediate attention, such as loss of heat or air conditioning, water leaks, clogged plumbing, faulty electrical wiring, and fire, safety or security hazards.

To place a service call, telephone the Maintenance Department (742-5438), 24 hours a day. A service call number will be assigned to the job. You may request this number and use it when checking on the status of your service call.

From 5:00 p.m. through 8:00 a.m. daily, and all day on weekends and holidays, only one maintenance mechanic is on duty. He or she can respond to a limited number of service calls during these times. However, you can still place a service call on nights or weekends to be completed when the mechanic is available. If no one answers, please call again.

In an emergency, notify the PBX operator to page Maintenance. Please do not use the paging system unless your request is justifiable. We will appreciate your cooperation.

Service call requirements are outlined in the *Policies and Procedures* manual (page 18). Requests exceeding these requirements will be classified as job orders, which are also outlined in the *Policies and Procedures* manual.

Imperative verbs and "you" in this paragraph help readers identify themselves as actors. Elimination of passive voice clarifies who is to do what.

Use of "we" and "you" throughout personalizes the memo and is consistent with the reality that this procedure involves people working together.

Figure 14.5 Memo Edited for Mechanics and Style but Not for Format and Organization

probably have planned for paragraphs rather than the highly formatted version in Figure 14.4. The highly formatted version assumes selective reading at the time a service call is placed. It further assumes that the main response will be an action (placing the call). Because so many judgments have been involved in this editing and nothing comparable to a style guide tells what form a memo giving instructions on placing service calls should take, any two editors might edit in different ways. Careful analysis will help keep judgments sound. Consultation with the writer will reveal assumptions that are not apparent from the document itself.

Determining Whether Comprehensive Editing Is Warranted

Comprehensive editing requires more commitment to a project than does basic copyediting. The choice to edit comprehensively has significant consequences for the editor's time as well as for the document. Furthermore, the greater the editorial intervention, the greater the risk of changing the message. The choice about the level of editing will depend on several criteria.

- **Limits in your job description.** You may not have a choice. Your supervisor or the writer may set limits on the extent of editing you can do. If you are charged only with correcting grammar and spelling and making mechanics consistent, you will stop there even if you can see how different organization or visual design might improve it. You can always present an argument for a higher level of editing, but sometimes you must be satisfied with doing less than you could.

  An exception to this general guideline to edit according to your job description is a situation where there is potential danger to someone's health and welfare or violation of laws or ethics. You have an ethical responsibility to make sure the document will not result in harm to a reader. If you are only a copyeditor but can see that a safety warning ought to be added or that a warning about risks should be emphasized, you should make the recommendation anyway.

- **Time.** If you are pressed by more important projects and the task has not been previously scheduled, you will have to stop when you run out of time.

- **Importance of the document.** Comprehensive editing is more appropriate for important documents than casual ones. The memo on service calls is important because safety is an issue. It may be important as well if productivity in the maintenance department is low because callers are abusing the service call system. Other criteria of importance are money and number of users. If the document will be used by thousands of people or if it accompanies a product that accounts for a large percentage of your company's revenue, it's worth comprehensive editing. If it is an internal newsletter, basic copyediting may be satisfactory.

- **Document anonymity.** Messages that represent an anonymous voice of the company—whose author doesn't particularly matter—can generally be edited more comprehensively than can personal statements, such as editorials or essays. The memo on service calls, though it is signed, is essentially anonymous; it is not a personal statement but rather a procedure. On the other hand, an article for a professional journal reflects an individual's point of view and requires more cautious editing. A middle ground between copyediting and comprehensive editing is editing for style but not for content, organization, or visual design. The memo on service calls, for example, would be easier to follow in its original paragraph form if it used verbs in the imperative mood rather than passive voice. Figure 14.5 is an example of a document edited for style but not for form, content, or organization. It accomplishes the objectives of shortening sentences and words, making the instructions friendlier and more reader oriented, and clarifying the information. Its editor has paid less attention to the objective of facilitating selective reading.

Be sure to determine what outside criteria may limit your right to edit comprehensively, either before you begin or after you have evaluated the document and set objectives. And don't exceed the limits without permission or other good cause related to safety and ethics.

Using Your Knowledge

Comprehensive editing offers more options for improving a document than basic copyediting, but it also enables an editor to do more damage. To add value to the document and to minimize the chance of doing harm, follow these steps:

1. Get a sense of the needs of the whole document before you begin to edit. Analyze, evaluate, plan, and review the plans with the writer.
2. Edit according to known principles of what makes documents work. Take pride in sharing the knowledge and expertise of writers and editors.
3. Query changes of content, even word changes, to verify that you understand. Likewise, alert the writer to reorganization and to substantial changes in sentence structure.

Further Reading

Bay Area Editor's Forum. *Editorial Services Guide.* www.editorsforum.org/what_do.html. Defines light, medium, and heavy editing.

Grove, Laurel. 1994. When the basics aren't enough: Finding a comprehensive editor. *IEEE Transactions on Professional Communication* 37.3: 171–174.

Weber, Jean Hollis. Working with a technical editor. www.techwrl.com/techwhirl/magazine/writing/technicaleditor.html. Identifies editing tasks and distinguishes "rule-based editing" from "analysis-based editing."

Discussion and Application

1. The following definition of open heart surgery is prepared for patients and their families. It is intended to answer a frequent question in a way that permits patients to study and ponder the answer. It may save the surgeon time, and it may answer questions that patients neglect to ask during consultation. The definition is printed on an 8½ × 11-inch page and folded in half for a two-page look.

Analyze the document's readers and possible uses in more depth, evaluate the document, and establish editing objectives for content, organization, style, visual design, and possible use of illustrations. Do not edit. As you analyze, focus on readers and the document rather than on the writer. Use "readers" and "documents" rather than "the writer" as the subjects of your analytical statements. Try to anticipate what questions the readers may ask and the ways in which they will use the brochure.

If your instructor directs you to, write a letter to the writer proposing the editorial emendations. Indicate the concept of the emendations you propose, and request the writer's response and suggestions. Refer to Chapters 3 and 25 for suggestions about communicating with authors.

What Is "Open Heart Surgery"?

By Donald L. Bricker, M.D.

This question is often asked perhaps more of patients who have experienced "open heart surgery" than of physicians. It is even posed in an argumentative fashion in some circumstances, and the author has been called more than once to arbitrate as to whether a given surgical procedure was or was not really "open heart surgery." The confusion surrounding the use of the term is quite understandable, since the term "open heart surgery" was coined over two decades ago and is very vague today when applied to the large area of cardiac surgery which it may be used to describe.

Perhaps the term was coined originally because the heart surgeon was concerned with methodology which would allow him to correct congenital heart defects which actually entailed opening the cardiac chambers for repair. Yet, this nosological consideration for procedures performed within cardiac chambers overlooked the true area of common ground which set cardiac surgery apart in terms of magnitude and risk. This area of common ground was simply the need to relieve the heart of its physiological burden while operating on it. In other words, heart operations of great magnitude are best grouped together by the necessity of providing an external mechanical support system to substitute for the function of the heart and lungs in pumping and oxygenating blood. Any heart operation, therefore, which requires that the heart either be stopped for the procedure, or undergo such manipulation that it cannot perform its designated function, would fit this classification. The external mechanical support system referred to is, of course, the "heart-lung machine." What is today implied by the term "open heart surgery" in common medical parlance, then, is any cardiac operation requiring the use of the "heart-lung machine."

What does the "heart-lung machine" do? First, let us exchange that term for "cardiopulmonary bypass" to aid in our understanding. Basically, cardiopulmonary bypass removes the heart and lungs as a unit from the circulatory system and temporarily bypasses them while performing their function. To accomplish this, blood is diverted from the heart by tapping into the great veins delivering blood from the upper and lower extremities. This blood flows by gravity into an oxygenating device which performs the lungs' function of adding oxygen and dissipating carbon dioxide. This blood is then pumped back into the circulatory system through a convenient artery, usually the aorta, the great artery that comes immediately from the heart. It can be seen then, that with this system functioning, appropriately placed surgical clamps on the venous and arterial sides of the heart and lungs would totally isolate them. This allows the surgeon to stop the heart if he wishes and perform his operation in a precise and unhurried fashion. Of course, the cardiopulmonary bypass unit pump-oxygenator, or if you insist, "heart-lung machine," cannot do this job indefinitely, but sufficient time is safely at hand with today's equipment that the surgeon need not hold the concern he once did for the time factor. Improvement in this equipment has, more than any other factor, led to the successes we routinely enjoy today. One question frequently asked is about the blood supply to the heart and lungs themselves during this period of "bypass." Their blood supply is markedly reduced since they are removed from the circulatory system, but since they are at rest, oxygen requirements are minimal, and for the duration of most procedures no problems are posed.

In conclusion, "open heart surgery" has come to be a less than literal term and in common usage implies a cardiac surgical procedure requiring use of cardiopulmonary bypass. It is this factor which sets the operation apart from other operations on the heart. If your operative procedure was or is to be done under cardiopulmonary bypass, rest assured you have undergone or will undergo "open heart surgery."

Reprinted by permission of Donald L. Bricker, M.D.

2. Organizations publish policies, such as the service call policy (page 243), online. Assuming that the medical center will publish its policies online, discuss how you might advise its publication from an editor's perspective.

a. Should the online version replicate the print version, or should it be designed for the Internet? A PDF version could replicate print. A separate online version would be coded with HTML or XML (see Chapter 5). What would be some criteria for the decision? What would be advantages and disadvantages of publication in both versions (PDF and HTML/XML)?

b. Might the policy be published entirely online? If yes, how would "all medical center personnel" be notified of the change in policy? What document naming policies would increase the odds of a successful search for the policy? In what situations might a print version have value?

    c. Develop an email notification of the policy change, using the information in the memo in the email text, but abbreviated.

    d. Design an online version of the policy.

3. Locate a brief document or section of a document that may benefit from comprehensive editing. This document could be a letter, short instructions, flyer, brochure, announcement, or chapter from a text. Assume that you have been assigned to edit the document. Using the procedure for comprehensive editing described in this chapter, analyze, evaluate, and set objectives for editing your document. If your instructor requests, bring the document to class and share your analysis, evaluation, and objectives with the class orally.

4. Referring to Chapter 6, discuss ways in which editing with the computer might facilitate or limit comprehensive editing.

Writers have many choices about words and their arrangements. Sentences may be long or short; words may take form as nouns, verbs, or modifiers; verbs may be expressed in the active or the passive voice. The rules of grammar govern these choices, and more like them, in only limited ways. The choices are matters of style. Style in this sense differs from mechanical style—(choices about capitalization, hyphenation, and numbers)—and from computer styles (format).

People use subjective terms in describing style. We may say that a document is dense, clear, or wordy (terms that reflect the reader's response to the content); or we may call the style formal, stuffy, pretentious, informed, casual, or warm (terms that reflect the image the writer projects). These impressions are created by specific components of language—words and sentence structures. Style in documents is analogous to style in clothing or music. For example, in America the western style in dress is the cumulative effect of components—denims, plaid shirts, boots, and belts with big buckles. Different styles in music, such as classical and jazz, are created by choices of notes, rhythms, and instruments. As with dress and music, styles in writing are appropriate according to the situation.

Creation (and modification) of a style, whether in dress, music, or language, depends on knowing component parts and options for arrangement. To make a document less stuffy or pretentious or to make it informal or casual, an editor may shorten sentences and word length and convert verbs in passive voice to active voice.

This chapter continues with a fuller definition of style. It then presents guidelines about sentence structure. Chapter 16, also on style, emphasizes words. This chapter will be more meaningful to you if you have reviewed the components of sentence structure (noun, verb, complement, clause) in Chapters 10 and 11.

Key Concepts

Style is not just a decoration but rather is a matter of substance. Style affects comprehension and a reader's attitude toward the document. The sentence core—the subject and verb—announces the sentence content and reveals what is important. Subordinate and parallel structures signal relationships of ideas. The beginning and end of the sentence help readers connect one sentence and its content to the next. Reading carefully and knowing the context for use can help an editor apply these guidelines.

Definition of Style

Style in language is the cumulative effect of choices about words, their forms, and their arrangement in sentences. These choices create a writer's persona (voice and image) and affect comprehension. Although the purpose of editing for style is to increase overall effectiveness, the work is done at the word and sentence levels.

Writer's Persona and Tone

The choices about words and sentence structures project an image of the writer—a persona—that influences a reader's response to the content of the document. Even an image of an objective and detached person has a corresponding style. The writer's persona reveals his or her attitude toward the subject and readers. The writer may seem to be serious, superior, disdainful, indifferent, or concerned. This attitude creates a document's tone, analogous to a speaker's tone of voice. An inappropriate tone can make readers resistant. For example, a condescending tone in a user manual discourages readers working on a difficult task.

Style and Comprehension

Style affects comprehension. Certain structures and word choices are easier to comprehend than others. Readers usually understand subject-verb-object structures better than inversions, concrete terms more readily than abstract terms, and positive expressions more readily than negative ones. Editors aim for reader comprehension because a universal purpose of technical communication is to inform. Editing for style is editing for meaning.

Editing for style gives editors significant power in clarifying meaning, but editors need to use this power wisely. Editors can be arbitrary or wrong; they can distort meaning. They can't depend on rules, on the ear (what sounds better), nor even entirely on guidelines. Inverted structures, abstract terms, and negative expressions may usually interfere with comprehension, but sometimes they have their own meaning and should be left alone. This chapter and the next will help you develop knowledge to edit wisely for style.

Example: Analysis of Style

The effectiveness of a document depends not just on style but also on accuracy and completeness of information, organization, visual design, correctness, and consistency. To distinguish the effect of style, let us look at two versions of the same paragraph that differ in style but not in any other features. The paragraphs illustrate the ways in which words and structures create a style and affect a reader's response and probable action.

The example is the first paragraph of a proposal for a grant to fund some research to study the effect of job complexity in sheltered workshops. Like all proposals, this one tries to persuade the funding agency that the research will be worthwhile and that the person who proposes it is capable of completing the project. To understand the paragraph apart from the whole proposal, you should know that sheltered workshops are places where people with handicaps complete tasks comparable to those in any industry but in settings where the pressures are not so great. The assumption has been that simplifying their jobs will improve their performance. The question this proposal will raise is whether the tedium that results from simplification hurts performance.

Original Paragraph

The field of job complexity in industry has a varied history encompassing as it does philosophy, economics, social theory, psychology, sociology, and a myriad of other disciplines. In the rehabilitation field, the job simplification model has been the mainstay in sheltered workshops for retarded persons almost since their inception. As in industry, the reasons for adopting the job simplification model are varied, but it is the author's contention that in the workshop the reasons are philosophical and technological rather than economic.

In evaluating the paragraph for possible editing, the editor might ask: Does the idea seem important? Does the writer project competence and other desirable attributes? Does the style seem right for the audience and purpose? If the answer to these questions is "yes," the sentences should not require editing for style.

Some readers have described this writer as windy and inflated. That image would be counterproductive in a proposal. Inflation gives the impression that a writer has to expand a trivial idea to make it seem important. If the writer invites a negative impression through style, the proposal concept and method will have to be proportionally stronger to persuade the people at the funding agency to approve this project rather than a competing one.

To emend an inflated style, an editor looks for stylistic features that create it. This writer uses redundant categories, words that restate a concrete subject with an abstraction, such as "field." The style is heavy on nouns and light on verbs. All the verbs are *to be* verbs, inconsistent with a proposal writer's goal of projecting a person who can act and make things happen. Identifying the choices that inflate the paragraph prepares for editing by giving direction.

Edited Paragraph

The issue of job complexity in industry encompasses philosophy, economics, social theory, psychology, sociology, and technology. In rehabilitation, sheltered workshops for persons with mental retardation have used the job simplification model almost since their beginnings. As in industry, the reasons for adopting the job simplification model vary, but in the workshop the reasons are philosophical and technological rather than economic.

Although the editing began with some specific observations, the editing process itself revealed even more instances of an inflated style. The category word "field" in the first sentence is inaccurate, so "issue" was substituted. Maybe on first reading you slip by the "myriad of other disciplines," but having observed the writer's inflation, you may question whether any issue can encompass a "myriad" (innumerable, ten thousand) of disciplines beyond the five named. This phrase is a sophisticated "et cetera" but not a meaningful one. In the second sentence of the original, "their inception" could be interpreted to modify "retarded persons" rather than "workshops" because of its placement. In that sentence, some phrases were rearranged to put key words in the subject, verb, and object slots of the sentence, and "beginnings" was substituted for "inception." "Are varied" becomes "vary" to put the action of the sentence in the verb rather than in an adjective. Three of the *to be* verbs are now action verbs. The writer's self-reference is deleted.

What difference has editing made? The image of the writer is more subtle, but for the problem statement of a proposal it is better to be invisible than to get in the way of the meaning. The edited paragraph is also shorter than the original, but rather than being an end in itself, conciseness has been a happy consequence of editing that makes sentences reveal their meaning. Eliminating inflation, using precise terms, and placing key words in the subject, verb, and object have focused attention on the problem that needs to be solved.

Analysis of style helps an editor evaluate the content. Some clarification of meaning, especially the substitution of "issue" for "field" and the deletion of the "myriad" phrase, have resulted from an analysis of style. One content addition in the first sentence, "technology," results from comparison of the list in the first sentence with the descriptors in the final sentence. Modifying word choices and sentence structures is risky because it can change meaning. Careful reading has preceded all of these changes to prevent errors.

Guidelines for Editing for Style

Although editors cannot depend on rules about style, they can use guidelines for effective writing that have been established by experience and by research. Analysis for style may consider context, sentence structures, verbs, and other words.

Context	Make style serve readers and purpose; consider the possibility of an international audience and the possibility of discrimination.
Sentence structures	Use structure to reinforce meaning.
Verbs	Convey the action in the sentence accurately and forcefully.
Nouns, adjectives	Choose words that are accurate, concrete, and understandable.

These guidelines cannot be used as rules. Like all guidelines, they point but do not prescribe. Using them requires judgment.

Context: Make Style Serve Purpose and Readers

Editing decisions regarding style depend on situations outside the text as well as internal textual choices. Context refers to readers, document uses, physical conditions, other related documents, and more (see Chapters 2 and 14). Just as no one style in dress or music is suitable for all situations, neither does one style work for all technical documents. In the example on sheltered workshops, the style is formal because research proposals are serious, academic documents. One edits a scientific paper and article for a popular magazine according to different criteria. Editing work that will be translated into other languages requires knowledge of the language and culture (see Chapter 20).

Sentence Structures: Use Structure to Reinforce Meaning

The architecture of a sentence guides readers in comprehending the content, and the writer and editor aim to match structures with meaning. Important structures to consider are the main clause (containing the subject and verb), patterns of coordination and subordination, and parallelism. These structures indicate the importance and relationships of ideas.

Sentence structures are hierarchical. The independent clause, which makes a sentence grammatically complete, is the strongest part of the sentence. It is the structural core of the sentence. The structural core is the logical place for the main idea of a sentence. (See Chapter 11 for a review of these terms and identification of basic sentence patterns.)

Other structural signals about meaning are the patterns of coordination and subordination. An idea expressed in a structurally subordinate part, such as a dependent clause, phrase, or modifier, seems inherently less important than an idea in an independent clause. If two ideas are related and equal in importance, coordination and parallel structure may reinforce that relationship. Burying the main idea in a dependent clause or phrase gives readers misleading structural signals about meaning.

The relationship between structure and meaning is the basis for this principle of editing for style: *make structure reinforce meaning*. Several guidelines apply this principle.

Place the Main Idea of the Sentence in the Structural Core

The main idea should appear in the structural core of the sentence, the subject and verb of the independent clause. Consider the following sentence:

The course of the twentieth century produced a cancer death rate that rose parallel to the advances in technology.

The core sentence is underlined, "course . . . produced," but the meaning appears elsewhere. The sentence is not about the course of the twentieth century and what it produced at all but rather about how the rising cancer death rate paralleled advances in technology. "Course" and "produced" are the least important words in the sentence, but they take the strongest structural part. The topic of the sentence differs from the grammatical subject. The meaning does not appear in the core sentence, and the structure does not reinforce meaning. Style interferes with comprehension. Two options for revision are these:

In the twentieth century, the rise in the cancer death rate paralleled the advances in technology.

In the twentieth century, the death rate from cancer rose parallel to the advances in technology.

Both revisions guide readers to the main idea using structural signals.

A good editorial strategy is to look for the grammatical subject of the sentence to see if it is identical with the topic of the sentence (that is, whether the structure matches what the sentence is about). Then look for the verb. If extra words get in the way, you may need to prune or rearrange to reveal the subject and verb. The following sentence may seem "wordy," but the problem is redundancies in both the subject and the verb.

The department's policies related to standards of behavior must be firmly maintained and affirmed.

One wordy phrase is the subject, "policies related to standards." What is the topic, "policies" or "standards"? "Policies" is the grammatical subject, but the real issue is standards, not the policies that define them. The department wants good behavior. Standards define behavior, and policies define standards. The policies are one level of abstraction further from the desired behavior. The end of the sentence is also redundant, especially with the adverb. (To "firmly affirm" seems like an overstatement that paradoxically weakens the insistent tone of the sentence.)

The Department's standards of behavior must be maintained.

Sentences that open with the words "there are" or "it is" often waste the core on a simple declaration. As a result, the main idea of the sentence may be buried in a dependent clause. Furthermore, such openers delay the significant part of the sentence.

It is often the case that a herniated disc ruptures under stress.

Often, a herniated disc ruptures under stress.

It is possible to apply for the scholarship by completing either of two forms.

You may apply for the scholarship by completing either of two forms.

There are two expenses to be justified.

Two expenses must be justified.

In each of these pairs, the core sentence in the revised version gives more information. The sentences are more efficient not just because they are shorter but also because the key idea resides in the core.

"There are" and "it is" may be preferable to alternatives, especially when they introduce a list. "Exist" rarely substitutes effectively for the *to be* verb.

> There are three reasons for this problem. [effective use of the "There are" opener because it prepares readers to expect a discussion of the three reasons]
>
> Three reasons exist for this problem. [substitution without a gain in effectiveness]

The purpose of the sentence is to prepare readers to hear the three reasons; the first version does just that. Before you delete "there are" in order to use the sentence core efficiently, consider the whole situation.

Use Subordinate Structures for Subordinate Ideas

When the sentence pattern is complex (a main clause plus a subordinate clause), the pattern itself communicates the relationship of main and subordinate ideas. If structure and meaning conflict—that is, if the structure affirms one relationship but the words affirm another—comprehension will be more difficult. Or the words may lead the reader to one interpretation while the aim was another.

A letter of application for a job as bank teller included the following sentence:

> I was elected treasurer of a social fraternity, which enabled me to collect and account for dues.

The situation helps an editor determine which facts are subordinate. "Was elected" affirms leadership ability and the confidence of peers. Yet a person hiring a bank teller may be interested more in whether the applicant can handle money accurately. For this reader, then, the information in the subordinate clause is more important than the information in the main clause. An appropriate stylistic revision emphasizes the skills that matter in this context:

> As elected treasurer of a social fraternity, I collected and accounted for dues.

The following sentences might shape a reader's desire to purchase an old home in need of repair:

> Although the beams show signs of dry rot, the house seems structurally sound.
>
> Although the house seems structurally sound, the beams show signs of dry rot.

The second version of the sentence leads a reader to conclude that the purchase is risky. The idea of structural soundness is subordinate to the idea of damage.

Use Parallel Structure for Parallel Items

Parallel structure can reinforce relationships between items in a series or a compound sentence. All the items should be the same part of speech (nouns, verbs, participles) or same type of phrase. In a compound sentence, both clauses should use the active voice (or the passive voice); a mixture requires a shift in the way the reader processes the information.

> We can help to keep costs down by learning more about the health care system, how to use it properly, and by developing self-care skills.

The faulty parallelism in this sentence makes it difficult to determine whether there are three ways to keep costs down (as the punctuation suggests) or two (as the parallel phrases "by learning" and "by developing" suggest). Two revisions are possible:

> We can help to keep costs down by learning more about the health care system, by using it properly, and by developing self-care skills.

> We can help to keep costs down by learning more about the health care system and how to use it properly and by developing self-care skills.

Parallel structure, with appropriate punctuation, encourages the correct interpretation. Which is the better sentence of the two revisions? The answer depends on the middle phrase, regarding use. Reasoning suggests that learning how to use the health care system in itself will not keep costs down, but using the system correctly will. This argument favors the first version, in which "using" is a separate strategy rather than something to learn. Furthermore, the first version is easier to read because the series of three is a simpler structure than the double compound in the second version (the two parallel phrases plus the compound object of the preposition, "learning").

The repetition of the pronoun *by* is optional, but it can signal relationships, and you should be consistent in whether or not you repeat it.

Sentence Arrangement

English sentence structure allows for different arrangements of parts, but for clarity and cohesion (the connection of sentences), some patterns of arrangement work better than others. The beginnings and endings of sentences matter the most.

Place the Subject and Verb Near the Beginning of the Sentence

A writer can anticipate that readers, even those who skim, will pay attention to the beginning of the sentence. An introductory modifying phrase may provide a transition to the previous sentence and establish a context for the information in the subject and predicate. Structured as a phrase, it signals the reader to expect the subject when the clause begins. However, a delay of the subject and verb keeps readers wondering what the sentence is about. Likewise, separating subject and verb makes it hard for readers to keep them together in their minds. The core of the sentence belongs near the beginning, with subject and verb close together.

Stylistic problems tend to cluster, and the example that follows reveals problems (including a dangling modifier) in addition to the delay of subject and verb, but the goal of placing subject and verb near the beginning of the sentence leads to other improvements as well.

> In discussions with staff members in other divisions, <u>observations</u> in support of the budget assessment by the manager <u>were made</u>. [The subject is delayed until word 9, and the verb is delayed until words 19 and 20.]

> <u>Staff members</u> in other divisions <u>concurred</u> with the manager's budget assessment.

Arrange Sentences for End Focus and Cohesion

Sentence structure not only clarifies the meaning of individual sentences but also connects one sentence to the next, so that the larger text structures (paragraphs, sections) make sense. These connections between sentences develop from the relationship between the end of one sentence (predicate) and the beginning of the one that follows (subject). In the subject and predicate, sentences follow a pattern of known to new information. The subject states familiar information. The predicate, in commenting on the subject, adds new information. This information, now known or familiar, becomes the subject of the next sentence. This pattern of known to new explains cohesive ties between sentences.

The paragraph you have just read illustrates the known-new pattern and the cohesive ties that result from arranging sentences to place known information in the subject and new information in the predicate.

[1]*Sentence structure* is a familiar or known topic because it is the subject of the chapter. A new word, *connects*, appears in the predicate.

[2]*Connect* from the predicate of sentence 1, now a known term, becomes the subject, *connections*, in sentence 2. Key words in the predicate of sentence 2 are *end* and *beginning*, *predicate* and *subject*. These words are echoed in the modifying phrase at the beginning of sentence 3. Sentences 4, 5, and 6 work together to elaborate on sentence 3. "Known to new" of sentence 3 appears explicitly in sentence 7.

[1]<u>Sentence structure</u> not only clarifies the meaning of individual sentences but also <u>connects</u> one sentence to the next, so that the larger text structures (paragraphs, sections) make sense. [2]These <u>connections</u> between sentences develop from the <u>relationship</u> between the <u>end</u> of one sentence (<u>predicate</u>) and the <u>beginning</u> of the one that follows (<u>subject</u>). [3]In the <u>subject and predicate</u>, sentences follow a <u>pattern of known to new</u> information. [4]The <u>subject</u> states familiar <u>information</u>. [5]The <u>predicate</u>, in commenting on the <u>subject</u>, adds new information. [6]This information, now known or familiar, becomes the subject of the next sentence. [7]This <u>pattern of known to new</u> explains <u>cohesive ties</u> between sentences.

Because readers expect the predicate to add a substantial comment about the subject, good writers (and their editors) structure sentences to place important information at the end. Sentences generally should not dwindle into modifying phrases or weak predicates, such as those that merely declare that something "is important." Instead, the predicate should add new information that the next sentence will comment on. Structuring sentences to emphasize the end is called *end focus*.

The sentence that follows wanders through a long beginning and then drops off in the predicate into an almost meaningless statement ("is apparent"). The sense of empty words is increased by the lack of an agent (who is to seek and develop?) and a weak verb (*is*). The first revision shifts the modifying phrase to the beginning, uses a human agent as subject, and ends with an important point. The second revision avoids the ambiguity of "in the future," which may seem to justify postponing action, and substitutes "have the opportunity," which may better encourage action right now.

> Room for aggressively seeking and maintaining new markets for existing products and developing new products in the future is apparent.

> In the future, we must aggressively seek and maintain new markets for existing products and develop new products.

> We have the opportunity to seek and maintain new markets for existing products and to develop new products.

Prefer S-V-O or S-V-C Word Order

The subject-verb-object (S-V-O) or subject-verb-complement (S-V-C) pattern is the most common one in English. Its familiarity makes it easy to understand. Inversions of the pattern require extra mental processing. The sentence you just read is an example of S-V-O order:

> Inversions of the pattern require extra mental processing.
>   S   V   O

A subject complement is either a substitute for a subject or a modifier, as in this sample of S-V-C order:

> A complement is a substitute.
>   C   V

Inversions of the order, because they use patterns that are less familiar to readers, call attention to themselves and possibly away from meaning.

> The jurors will no favors grant.
>   S   O   V

> Favored by committee members is the plan to renegotiate his contract.
>   C   V   S

Sentence Length and Energy

Sentence length influences readability and the sophistication of content. Length can make sentences seem to drag or to sound as though they are from a child's book. A reader's impression of length depends not just on the number of words but also on other choices that make sentences seem long, such as stacked phrases, impersonal subjects, separation of subject and verb, and negative constructions.

Adjust Sentence Length to Increase Readability

If sentences within a document all contain roughly the same number of words, the reading will become monotonous. The rhythm will lull the readers rather than keep them alert. Variety in sentence length may be desirable for its own sake, but variety also can be used to emphasize key points. A very short sentence surrounded by longer ones will draw attention and thus influence a reader's perception of the significance of the content.

Long and complex sentences are generally more difficult to understand than short and simple ones because they require a reader to sort and remember more information and more relationships. Yet they can be easier to understand than a series of short sentences just because they do establish relationships and help with interpretation. For example, a complex sentence (with a dependent and an independent clause) could show a cause-effect relationship. The first two sentences in this paragraph do just that. By contrast, the following series of simple sentences have a childlike quality:

> Long and complex sentences are hard to understand. Short and simple sentences are easy to understand. Long and complex sentences require a reader to sort information and relationships. They require a reader to remember more information than short sentences.

Length is relative to readers. Adult readers who know the subject appreciate the information they receive from complex sentences. But readers who have limited backgrounds on the subject will need to absorb the new information in smaller and simpler chunks. Also, length describes more than just a number of words. Sentences constructed from a series of phrases become too long more quickly than do sentences constructed with several clauses. If the sentence core gets lost in a series of modifying or prepositional phrases, the sentence is too long for easy comprehension.

Use People as Agents When Possible

Human agents engage readers. Keeping people out of sentences creates stuffy reading that puts readers to sleep. Sentences full of abstractions seem long compared to sentences in which human beings perform actions.

The following sentences from an employee policy statement aim to clarify the terms of vacation eligibility. Employees become eligible for vacation after they have worked a year, but these sentences raise the question of what constitutes a year if an employee misses work time because of an accident or illness. The long sentence and separation of subject and verb increase reading difficulty, but employees will also struggle to see themselves in the action because they are never named. Edited version B takes advantage of headings and bulleted lists to increase readability, but it also uses human agents, shorter sentences, and closer connections between subject and verb.

Original

> After the successful completion of the first 90 days of employment, <u>time lost</u> as a result of an accident suffered during the course of employment as <u>recognized by the</u>

Workers' Compensation Board, shall be added to time worked during the vacation year to qualify under the provisions of vacation leave. Also, after the 90-day period, time lost as a result of a non-occupational accident or illness not to exceed 520 hours shall be added to the time worked during the vacation year.

Edited Version A (sentence core near the beginning; human subjects)

After you complete 90 days of employment, the work time you lose because of accident or sickness may count as time worked for purposes of calculating vacation eligibility. If the time lost results from a work-related accident that the Workers' Compensation Board recognizes, the time you lose will be added to the time you worked for counting when you have worked a year. If the time lost results from a non-occupational accident or illness, up to 520 hours of the time you lose will be added to the time you worked.

Edited Version B (editing for style plus format)

After you complete 90 days of employment, the work time you lose because of accident or sickness may count as time worked for purposes of calculating vacation eligibility.

Work-related accident. If the Workers' Compensation Board recognizes that your accident occurred as you were completing work duties, the time you lose for the accident counts toward time worked.

Non-occupational accident or illness. Up to 520 hours that you lose because of an accident or illness that is not related to work counts toward time worked.

Prefer Positive Constructions

Negative constructions will make sentences seem longer than they are because they increase reading difficulty. A double negative requires extra steps of interpretation. These steps slow down reading and increase the chance of error in comprehension.

It is not uncommon for employers to require writing samples from applicants.

Readers have to use "not" to cancel "un" before they understand that writing samples are common in job applications. They would comprehend more quickly if the sentence began, "It is common . . ." or "Employers commonly. . . ."

How long does it take you to figure out the meaning of the following sentence?

The elimination of disease doesn't guarantee that we won't die according to genetic timetables.

The three negatives ("elimination," "doesn't," "won't") are especially troublesome because the concept of "genetic timetables" is difficult. Though the following versions alter the emphasis, they are easier to understand.

Even if we eliminate disease, people may still die at the same age they do now because of genetic timetables.

Elimination of disease doesn't guarantee a longer life. Genetic timetables, rather than disease, may establish the lifespan.

Simple negative constructions in two clauses also load extra interpretation responsibilities on readers:

It is not possible to reduce inflationary pressures when the federal government does not reduce its spending.

These edited versions are easier to understand:

Inflation will continue if the federal government keeps on spending at the same rate.

Inflation will decrease only if the federal government reduces its spending.

In this next version, the negative construction remains in one clause, for emphasis. Even elimination of one negative construction, however, makes the sentence easier to understand.

Inflation will not decrease unless the government reduces its spending.

Positive language has a psychological benefit, as well as being easier to understand. In the next example, the company that resolves a complaint may negate its efforts to help if its letter ends with a reference to the problem:

If you ever have any more problems with our company, do not hesitate to call.

The edited version anticipates a positive future relationship:

If our company can serve you in the future, please feel free to call.

Using Your Knowledge

As you work with sentence structures, aim for these goals:

1. Use structure to signal meaning. Put the main idea in the main clause and subordinate ideas in subordinate clauses or phrases. Use parallel structure for parallel ideas.
2. Arrange content in sentences so that the main clause comes early, without long delays. To develop paragraph cohesion, end sentences with words that connect one sentence to the next.
3. Use people as agents when you can, and prefer positive constructions.

Always read carefully for meaning, and confer with the writer when you have any reason to question your understanding.

Further Reading

Jones, Dan. 1998. *Technical Writing Style.* Longman.

Lanham, Richard A. 2000. *Revising Prose.* 4th ed. Longman. Also see *Revising Business Prose.*

Smith, Edward L., and Stephen A. Bernhardt. 1997. *Writing at Work: Professional Writing Skills for People on the Job.* NTC.

Williams, Joseph M. 2002. *Style: Ten Lessons in Clarity and Grace.* 7th ed. Longman.

Discussion and Application

1. *Sentence core.* Edit the following sentences to use the sentence core effectively and to place subject and verb together near the beginning of the sentence.

   a. The focus of the test for Experiment II was tangential to and not a direct approach to leadership.

   b. The major framework of her essay involves presenting a discussion of health care funding.

   c. Some studies have revealed that there has been a small increase in mastitis cases involved with BST-supplemented cows.

   d. The expected results of the use of BST supplements is an increase in the profitability of the dairy producer operations.

   e. The reason why video vignettes are less used than lecture in teaching ethics is because of the higher cost and less available resources.

2. *Parallelism and subordination.* Edit the following sentences to reinforce meaning by use of parallelism and subordination.

   a. The student health center offers a variety of services such as physician appointments, mental health, health education programs, and lab and X-ray work.

   b. The report considers factors important to students in choosing a medical facility, ratings of services at the student health center, reasons why students do not use the center, and offers suggestions for increase in student usage.

   c. To become a mutual fund shareholder, an investor places an order with a local securities dealer or by contacting the fund sales staff directly.

   d. Even if callers refuse to state their name, messages can still provide helpful information. For example, some people call to report a change of a meeting, to file a complaint, or any other kind of message.

   e. There are two main elements that a tennis racket should deliver—power and control.

3. *Subordination and interpretation.* Describe the different interpretations these two versions of the same sentence might invite.

   a. Although the region is engaged in extensive tourist promotion, the historical site is accessible only by a secondary route.

b. Although the historical site is accessible only by a secondary route, the region is engaged in extensive tourist promotion.

4. *Persona and readers.* Discuss the suitability of the persona and style of the following paragraph assuming that the passage appears in a manual for volunteers assisting a probation officer. Then discuss the suitability of the persona and style assuming the passage appears in a law textbook. How do context and readers influence your description of the style? What particular stylistic features make the passage appropriate or not for the two contexts?

> Probation is a method of disposition of a sentence imposed on a person found guilty of a crime. It is a court-ordered sentence in lieu of incarceration. The offender remains in the community under the supervision of a probation officer for a predetermined period of time. If compliance with the terms and conditions of probation set by the court is made by the offender, he or she is discharged from the court's jurisdiction and the debt to society is considered paid. If compliance is not made, another method of sentence may be imposed which may include incarceration.

5. *Editing for style.* Sentences from the previous paragraph are reproduced here. The beginnings of revisions follow the sentences. With one or two classmates, continue editing the sentences with these aims: use a human agent (person, offender, judge, probation officer) where you can; use the subject and verb for important words; place the subject and verb close together, near the beginning of the clause.

 You may use two sentences in place of one or combine sentences so long as you don't change the meaning.

 a. Probation is a method of disposition of a sentence imposed on a person found guilty of a crime. It is a court-ordered sentence in lieu of incarceration.

 When a person is found guilty_____ , the court may _____.

 b. If compliance with the terms and conditions of probation set by the court is made by the offender, he or she is discharged from the court's jurisdiction and the debt to society is considered paid.
 If the offender _____ , the court _____.

 c. If compliance is not made, another method of sentence may be imposed which may include incarceration.

 If the person _____ , the court _____.

6. *Consequences of editing.* Discuss the consequences of your editing of the previous sentences on clarity, length, and tone. What has editing achieved? What has been sacrificed? What readers would prefer the edited sentences? What readers might prefer the first version?

7. *Editing process.* Discuss your process of analysis and editing in application 5. What issues did the group members need to resolve before agreeing on the words and arrangement of sentences? What were the uses and limitations of the guidelines for sentence structure?

8. *Paragraph cohesion.* Rearrange and restructure sentences in the following paragraph to reinforce the connection of ideas from one sentence to the next.

> Obesity results from an energy imbalance over time. It is easier to gain weight than to lose it because the body accommodates overeating more readily than it ignores hunger. Eating more calories than are exerted through physical activity creates energy imbalance. The increase in calories in diets has resulted from several economic factors. Food is proportionately cheaper than it was a few decades ago, meaning that people can afford more food. People are also eating more high-calorie restaurant meals. At the same time that people are eating more, they are exercising less and thus using fewer calories. Over time, this imbalance means weight gain.

9. *Style manuals.* Consult at least one professional journal and the most commonly used style manual in the field in which you anticipate writing and editing. (See Chapter 8 for some titles of style manuals.) Do they offer guidelines to authors for style? If so, summarize the guidelines for sexist language, passive voice, and other issues that they emphasize. In class, share these summaries to determine where the disciplines agree and disagree.

16 Style: Verbs and Other Words

Once you understand the effect of sentence structure on style and comprehension (see Chapter 15), you will have a good basis for editing words. Good structures depend on good choice of subjects and verbs in particular. In order to convey meaning, these parts of the sentence must be accurate and specific. Good choices of words also add energy to writing and make it more readable.

This chapter continues the discussion of style begun in Chapter 15 with an emphasis on verbs and concrete nouns. It also suggests ways to choose words to avoid unintentional discrimination.

Key Concepts

Verbs communicate the most important information about the subject, and good verbs enliven writing. To improve style at the word level, focus first on verbs. Effective writing also requires specific nouns and modifiers. Because changing words can change meaning, editors make meaning a priority over a style preference.

Verbs: Convey the Action in the Sentence Accurately

Sentences often falter stylistically because of their verbs. Writers can easily identify subjects and use nouns; it is harder conceptually to figure out what the nouns *do*. Evaluating verbs is a necessary step in determining whether a sentence conveys its meaning. When you edit for style, consider verbs after you evaluate context and sentence structure.

Build Sentences around Action Verbs

Readers approach sentences wondering "Who did what?" or "What happened?" The answer to the question should appear in the verb. When the action that should be conveyed by the verb is lost in imprecise substitutes or in other parts of speech, including nouns and adjectives, sentences fail to communicate effectively. Editors can help writers find their verbs.

In the following sentence, the verb does not tell what the subject does:

A one-year <u>warranty</u> <u>was placed</u> on the digital camera through March 2006, guaranteeing that all parts and labor would be covered during this time.

The fact that the warranty "was placed" is less important than how long it lasts and what it covers. "Placed" is a weak, imprecise verb. A reader has to insert mentally the answer to the question of what the subject (the warranty) did. Putting the action in the verb, the structure where readers expect to find the answer to their question, would clarify the meaning.

This digital camera's one-year warranty <u>extends</u> through March 2006 and <u>covers</u> all parts and labor.

The new verbs, "extends" and "covers," are stronger than the original verb, "was placed," because they tell more specifically what the warranty does. The action in the original sentence is buried in a past participle ("would be covered") in the case of the second verb and is only implied in the case of the first ("through March 2006"). The revision increases reader comprehension and reduces possible misinterpretation by eliminating the word "guaranteeing." A "warranty" and a "guarantee" mean different things in a legal sense.

This sentence also hides the action by using an imprecise verb.

The <u>stream of air</u> that escapes the larynx <u>experiences</u> a drop in pressure below the vocal folds.

An editor examining the sentence core might ask: Can air "experience"? Does this verb tell what the subject does? The verb works with the subject grammatically but not logically. An editor can spot the hidden verb in the noun "drop."

The <u>stream of air</u> that escapes the larynx <u>drops</u> in pressure below the vocal folds.

If you can help writers find their verbs, the main action will appear in the sentence core, and structure will reinforce meaning.

Choose Strong Verbs

Many writers draw on a limited repertoire of verbs. Comparatively weak verbs describe everything that happens in their writing:

is (or other variations of *to be*: are, was, were, will be)			
have (has, had)	make	involve	provide
add	give	concern	become
deal with	do	reflect	use

Nothing much happens in these verbs. These all-purpose verbs work in many contexts because they describe many possible actions. As an editor, you won't eliminate all uses of these verbs, for occasionally they may be the best choice. But if such verbs predominate, you will have to hunt for the real meaning and substitute the appropriate verb. Consider the following example:

The report <u>will deal with</u> the third phase of the project.

Although this sentence tells the reader the general subject of the report, the reader might have learned more had the verb been more specific. Will the report describe, state results, evaluate, or give instructions for the project's third phase?

The paragraph in the left column below uses weak verbs (*hold, are, have*). The second paragraph uses stronger verbs (including infinitive forms of verbs) to convey more action and decisiveness. You would probably prefer to read a report with verbs such as those in the paragraph on the right. Some verbs in the left column were implied in the nouns—called *nominalizations* (*meeting, discussion, expansion, agreement, investigation*).

The City Council will <u>hold</u> a meeting for discussion of the possible expansion of the basketball arena. If Council members <u>are</u> in agreement about the need, they will <u>have</u> a consulting firm <u>proceed</u> with investigation of the feasibility.
Verbs: hold, are, have, proceed

The City Council will <u>meet</u> to <u>discuss</u> whether to <u>expand</u> the basketball arena. If Council members <u>agree</u> about the need, they will <u>hire</u> a consulting firm to <u>investigate</u> the feasibility.

Verbs: meet, discuss, expand, agree, hire, investigate

Avoid Nominalizations

Burying verbs in nouns or adjectives weakens the verbs. A verb turned into a noun is a nominalization ("nominal" refers to names or nouns). Suffixes, such as *-tion*, *-al*, and *-ment*, convert verbs into nouns, as the following list illustrates:

Verb	Suffix	Nominalization
admire	-tion	admiration
agree	-ment	agreement
transmit	-al	transmittal
depend	-ence	dependence
rely	-ance	reliance
solder	-ing	soldering

Nominalizations do serve valid purposes in writing. They become problems only when they obscure what is happening, as when the action a sentence seeks to communicate is disguised as a thing. The form conflicts with meaning and therefore makes comprehension more difficult. Furthermore, because sentences full of nominalizations are wordy and lifeless, they discourage reading.

Verbs can become adjectives, too, in their participle form (with the addition of the suffixes *-ing* and *-ed*).

Some <u>magazines</u> <u>are</u> more specialized than others by dealing with just one topic, such as science or art.

The verb is the weak *to be* verb, "are." The writer threw in "dealing with" as well, sensing the need for a more specific action. The real action, however, appears in the past participle, "specialized."

Some <u>magazines</u> <u>specialize</u> in just one topic, such as science or art.

Well-chosen verbs will clarify meaning and enliven prose with action.

Prefer the Active Voice

Voice refers to the relationship of subject and verb. In the active voice, the subject performs the action represented by the verb. The subject is the agent of action.

The board <u>reached</u> a decision.

      subject   verb

The subject of the sentence, "board," performs the action identified by the verb. In the passive voice, the subject receives the action identified by the verb. The subject is the passive object of action, or the recipient of action.

A decision <u>was reached</u> by the board.

The subject of the sentence, "decision," does not do the reaching. Rather, it receives the reaching. Don't confuse passive voice with past tense or with weak verbs. A sentence in passive voice always has these components:

a *to be* verb

a past participle

Some statements seem passive because nothing much happens in them, but they are not in passive voice unless they include a *to be* verb and a past participle.

Reasons to Prefer the Active Voice

The use of the verb "prefer" in this heading is intentional. Active voice is preferable in many situations, but editors should not arbitrarily convert passive voice sentences to active voice. It is usually preferable, but sometimes it is not.

- **Adds energy to writing.** Active voice conveys directly that people do things or that things happen. Passive voice emphasizes the result, the thing rather than the action.

- **Establishes responsibility.** Active voice is also preferable in many situations for ethical reasons. Sometimes writers use passive voice because they do not want to reveal who the agent of the action was.

A decision <u>was reached</u> that you should be fired.

The writer of that sentence may not want the reader to know who made the decision. The passive voice protects the agent from identification.

Passive voice can mask responsibility for future actions as well as for past ones. Consider the group that plans to place a microwave oven in the company lunchroom. One person asks about the effect on workers who wear pacemakers. "A sign will be posted," says one. By whom? Who will take responsibility for posting the sign? A sentence that does not declare responsibility is less likely to shape subsequent action than one that tells who will do the task. "The supervisor will post a sign" or "I will post a sign" gives a person a task and thereby increases the chance that the task will be performed. If responsibility is not assigned, the members of the group may all assume that someone else will do the job.

Reasons to Prefer the Passive Voice

The arguments for the active voice are persuasive, so the reasons for choosing the passive voice—in some situations—must be strong.

- **The agent is insignificant or understood.** Sometimes it doesn't matter who has done or will do something. The proper emphasis, then, is on the recipient of the action rather than on the agent.

  > Computer <u>chips</u> <u>are made</u> of silicon. [passive voice: emphasizes the action of making]

  The purpose of this sentence is to identify the material that forms computer chips, not who makes them. The agent is irrelevant (unless context tells us otherwise).

  > <u>Manufacturers</u> <u>make</u> computer chips of silicon. [active voice: introduces an irrelevant agent]

  The context may make the agent of action clear and negate the need to name the agent in the sentences. A policy statement directed to supervisors, for example, may establish in an opening paragraph or heading that the statement identifies policies for supervisors. To write in active voice throughout the policy statement would require the repetition of "supervisors" as the subject in many of the sentences. This repetition would therefore subordinate the policy, and the action that results from the policy, to the agent.

- **Readers expect the passive voice.** A publication or an organization may establish the passive voice as the preferred style. Some of the sciences, for example, maintain the convention of passive voice sentences with the writer's voice, or at least the "I," invisible. The reason for minimizing the "I" is to convey a sense of the writer's objectivity—the facts rather than the interpreter predominate. Thus, a science writer may write

  > It was determined that . . .

  rather than

  > We determined that . . .

  Though a number of studies have challenged the presumed objectivity of the passive voice, you should respect an organization's expectations in your editorial decisions. Readers will be distracted by variations from the norm and may discredit the findings of a writer who does not use the language of the community and therefore seems not to belong to it.

- **Creates cohesion and focus.** Because of the known-to-new pattern of sentences, the subject of one sentence generally restates an important idea in the predicate of the preceding sentence (see Chapter 15). Passive voice gives writers the option of using as the subject of a sentence a recipient rather than an agent of action. Stronger cohesive ties and focus may result. In the following pair of sentences, the first sentence focuses on readers; the second one focuses on style. If the point of the sentences is to suggest the effect of style on readers, the first revision, using passive voice to place "readers" in the subject position of the sentence, does a better job of achieving that focus.

Style affects a reader's comprehension. Readers may be distracted from meaning by a style that is inconsistent with their expectations. [passive voice to create a cohesive tie between the subject of the second sentence and the predicate of the previous sentence]

Style affects a reader's comprehension. A style that is inconsistent with their expectations may distract readers from meaning. [active voice to emphasize the topic of style]

Use Concrete, Accurate Nouns

Readers depend on accurate words for the details as well as for forming concepts. If the words are inaccurate or difficult to understand, the reader must either apply extra mental effort to substitute the correct word or "learn" incorrectly. Writing that features abstract nouns, phrases, or pairs rather than single words, and complex rather than simple words, may be hard to understand. Yet, since changing a word changes meaning, editing of words must be cautious.

Concrete words evoke the senses—sight, sound, taste, touch, odor, motion. Words that help us "sense" the meaning are easier to understand than are abstractions. Technical writing is often more concrete than philosophical writing because the subject matter is objects and actions (though it can be philosophical and abstract as well).

All-purpose nouns are the close cousin of all-purpose, weak verbs. They often fail to convey a precise meaning, as these examples illustrate:

areas	aspects	considerations
factors	matters	

Try to imagine what the following sentence means:

These aspects are important considerations for this area.

Abstractions can be functional when they introduce a list. The concreteness appears in the list that follows the category.

Some abstract nouns appear in the next section in phrases and redundant categories. Their translations offer more concrete nouns.

Prefer Single Words to Phrases or Pairs and Simple to Complex Words

When a writer unnecessarily complicates information, readers work harder to comprehend.

Phrases versus Single Words

Some writers try to sound sophisticated and knowledgeable by using phrases in place of single words and multisyllabic words when simpler words are more accurate. A restroom may be referred to as a "guest relations facility" and a hammer as a "manually powered fastener-driving impact device." Such a style may please a writer, but it rarely pleases a reader. In the novel *1984*, George Orwell called such circumlocutions "doublespeak." The National Council of Teachers of English (NCTE) gives out doublespeak awards for the "best" examples each year. The military and government have won with these creations:

Phrase	Translation
unlawful or arbitrary deprivation of life	killing
permanent pre-hostility	peace
violence processing	combat
collateral damage	civilian casualties in war
frame-supported tension structure	tent

Military and government writers are not the only inventors of phrases when a word would do. A hospital described "death" as "negative patient care outcome." In finance, a "negative investment increment" means "loss."

These examples illustrate that a negative subject may motivate inflation of language: the phrases are euphemisms for unfortunate or tragic outcomes. But even without this motivation, writers often wander around a specific subject without identifying it. A style that inflates and abstracts loses clarity. If the subject matter involves potentially dangerous mechanisms or chemicals, such violations of clarity may even cause harm to the readers.

Multisyllabic Words

Multisyllabic words are more difficult to understand than their one-syllable synonyms; they also take up more room on the page and take longer to read. They may create the appearance of pretense rather than of sophistication. Readers appreciate the simpler version.

Multisyllabic Word	Synonym
utilize	use
effectuate	do
terminate	end

Won't readers respect the writer who is able to use the multisyllabic word more than the writer who uses the single-syllable synonym? This is a question you may be asked by a writer who objects to editing for simplification of words. The answer is not easy. This guideline does *not* advise you as editor to substitute imprecise generic terms for specific technical ones. It does *not* insist that you eliminate jargon. It does *not* advise you to edit all documents to the same simple reading level. However, it does advise you that writers rarely gain respect on the basis of an ability to use inflated words. They are respected because they have gathered and interpreted data in a credible way, or because they have solved a problem, or because they have helped a reader perform a task accurately. If writers rely on multisyllabic words because their substance is weak, readers will not be impressed. Readers appreciate being able to move through a document without artificial barriers of extra syllables.

Redundant Pairs

If writers can't focus on the exact subject, verb, or modifier, they may insert two or more, hoping to cover all the possibilities. A writer may announce the "aims and goals" of a meeting, for example, not to distinguish between aims and goals but to avoid choosing just one of the words. A computer manual may instruct a user to "choose or select" a command without really meaning that the user has an alternative. The result is wordy, unfocused writing that may confuse readers because *and* and *or* signal multiple possibilities. In your initial analysis of a writer's style, look for pairs joined with the connectors *and* or *or*. The pairs may state legitimate alternatives, or they may indicate definitions in apposition. However, if they are merely synonyms, they are redundant.

Redundant Categories

Redundant categories are abstract restatements of concrete words. The abstraction puts the more concrete word into a category. A weather announcer, for example, may predict "thunderstorm activity." The person planning whether or not to carry an umbrella to work that day imagines thunderstorms, not "activity," which is not something from which one needs the protection of an umbrella. Yet, because of the redundant category, the thunderstorm in the sentence is merely a modifier. The weather announcer could more directly predict thunderstorms. The category is redundant if the specific term establishes the class.

> Joe expects to set up a business in the Los Angeles community. [Los Angeles, by definition, is a "community," so the word is redundant.]
>
> Joe expects to set up a business in Los Angeles.

Redundant categories demote key terms from noun to modifier, as well as adding extra words.

Redundant	**Edited**
a career in the medical profession	a career in medicine
hospital facility	hospital
time period	time
red in color	red
upright position	upright
money resources	money; resources
field of industry	field; industry

Application: Editing for Style

In Chapters 15 and 16 we have been looking at sentences and words in order to define the components of style. The brief proposal that appears in Figure 16.1 illustrates the cumulative effect of stylistic choices at the word and sentence levels. It also illustrates the effect these choices have on the reader's comprehension and attitude. Like most documents, its needs for editing are not limited to style. You should spot at least one grammar error, and the insertion of headings could aid both comprehension and access. However, the analysis will focus primarily on style.

[1]This request to the United States Geological Survey is in reference to the $15,000 allocated by the Office of Coal Management to the Ames District for use in hydrologic assistance. [2]At this time and stage of access and interpretation of existing data on record in the form of computer storage and publications in print that we may not be aware of is our main concern. [3]Due to the U.S. Geological Survey having vast storage of and access to this data we would like to suggest the available funds be used in the following two areas if possible.

[4]The first area may be handled by the Geological Survey district office in Wilson due to accessibility and central locale to all literature and data sources. [5]By compiling this data a comprehensive interpretation of surface water, i.e., quantity, quality, salinity, etc., for site specific coal leases or areas immediately adjacent those leases can be provided to the Ames District hydrologist. [6]Thus, due to time constraints, time may be spent on analyzing these interpretations and conducting onsite calculations.

[7]Secondly, another area we foresee as a positive and very useful endeavor is the expertise that can be provided by the Water Resource Division of the Geological Survey in Mountainview. [8]Because the subdistrict office and the White River Resource Area office are both located in Mountainview, we may obtain their help in the form of infrequent consultations, informal review of tract analysis and field reconnaissance on a one time basis of any lease area lacking available hydrologic data.

[9]The foregoing should provide adequate justification for requesting the U.S. Geological Survey's assistance.

Figure 16.1 Original USGS Proposal

Analysis

Analysis should reveal specific editing objectives so that the editing can be purposeful and systematic. It will begin with general responses based on awareness of the context and work toward specific objectives.

- **Context and content.** Readers of proposals (the people who can determine whether to grant the funds or not) are likely to ask: How will the money be spent? Is this expense worthwhile? Are the proposers capable of doing what they propose? Readers need to be persuaded that the proposed expense represents the best use for their funds. They would prefer not to read the proposal twice or more to find out what it is about. Yet this proposal is confusing on first reading. The final sentence compliments the writer and document rather than anticipating what a reader will need to know or do at this point.

- **Sentence structure.** The sentences are long, and it is difficult to find the sentence core. Verbs are weak ("is" and "would like" in the first paragraph). The "due to" construction (sentences 3 and 4) substitutes a prepositional phrase for a clause.
- **Verbs.** In addition to overreliance on weak verbs, the writer uses passive voice frequently (sentences 4, 5, and 6), making it difficult to determine who is to do what, as well as creating a dangling modifier in sentence 5.
- **Words.** The word "area" in sentences 4 and 7 is ambiguous. Is it a geographic area or a subject area? The mention of specific sites ("Wilson," "Mountainview") suggests a geographic area, but "area . . . is . . . expertise" in sentence 7 suggests a subject area. The writer also creates some ambiguity with redundant pairs and categories ("time and stage of access and interpretation of existing data on record in the form of computer storage and publications in print" in sentence 2, "positive and very useful endeavor" in sentence 7). Some modifiers are excessive ("immediately adjacent" in sentence 5, "available . . . data" in sentence 8).

Based on the analysis, an editor may establish the following editing objectives:

1. Shorten sentences and emphasize the sentence core by placing it earlier in the sentence.
2. Use action verbs. Prefer active voice; clarify responsibility if passive voice remains.
3. Make terms concrete. Delete unnecessary repetitions.
4. Leave the reader with a good impression.

Figure 16.2 shows the edited copy.

The editor revised electronically, tracking changes. Figure 16.2 shows the changes (underlines track additions, and balloons track deletions).

Evaluation and Review

The editing of the proposal began with analysis of the proposal as it will be used by readers. The analysis follows the process described in Chapter 14. It is a necessary first step in editing for style because the style guidelines do not prescribe changes and must be used with awareness of meaning and purpose. Because some of the sentences were so ambiguous, an editor should consult with the writer before sending the proposal forward.

Some content and visual design questions arise as a result of stylistic editing. For example, proposals usually include specific budgets. The reader may wonder how much of the $15,000 will go for each purpose and how much each data search and consultation will cost. Headings or numbers could help to identify the

> This request to the United States Geological Survey (USGS) refers to the $15,000 allocated by the Office of Coal Management to the Ames District for use in hydrologic assistance. We are concerned about our lack of access to existing data in publications and computer databases. Because the USGS has access to this data, we suggest that the available funds be used in the following two ways.
>
> The USGS office in Wilson has access to all literature and data sources. It could compile and interpret data on surface water (quantity, quality, salinity, etc.) for site specific coal leases or areas adjacent to those leases. The Ames District hydrologist could then spend his or her time analyzing these interpretations and conducting on-site calculations.
>
> Second, the Water Resource Division of the USGS in Mountainview could provide expertise both through the subdistrict office and the White River Resource Area. We could obtain their help through infrequent consultations, informal review of tract analysis, and initial field survey of any lease area lacking hydrologic data.
>
> Please consider our request carefully and contact us at extension 388 if you need further information

Figure 16.2   Edited USGS Proposal

two proposed uses of the money. The editor may raise these questions with the writer and suggest additional data. But this level of editing exceeds editing for style and should be approved before it is done. Even if this proposal is edited only for style, it will be easier to understand and therefore more persuasive.

One consequence of editing for style has been to shorten the proposal. The revised version contains 185 words compared with 265 in the original. But the marginal explanations for editorial emendations in Figure 16.3 do not point to conciseness as an end in itself. Often the best way to achieve conciseness is to use strong subjects and verbs early in the sentence.

The Language of Discrimination

Word choices, as they relate to human subjects, may imply bias against groups on the basis of their gender, age, race, religion, culture, politics, sexual orientation, or physical or mental disability. Rarely do writers intend discrimination. They may find incredulous the idea that the phrase *the disabled* is more negative than *people with disabilities*, or that the term *handicap* or the generic *he* may offend some readers. How the writer and editor feel though, is less important than how the readers will react and how word choices will affect their attitudes and comprehension.

[1]A strong verb replaces "is."

[2]Subject and verb come early in the sentence.

Publications are listed before computer databases to clarify that "computer" does not modify "publications."

[3]A phrase is converted to a clause for easier comprehension.

[4, 5]The core sentence comes early in both of these sentences for easier comprehension. Active voice replaces passive voice. Verbs are stronger ("compile," "interpret").

[6]Active voice replaces passive voice to establish who will do what.

[7, 8]Core sentences come early. Unnecessary modifiers are deleted. Abstractions ("endeavor," "area") are minimized.

[9]Instead of complimenting themselves, the writers offer the reader further assistance.

---

[1]This request to the United States Geological Survey (USGS) refers to the $15,000 allocated by the Office of Coal Management to the Ames District for use in hydrologic assistance. [2]We are concerned about our lack of access to existing data in publications and computer databases. [3]Because the USGS has access to this data, we suggest that the available funds be used in the following two ways.

[4]The USGS office in Wilson has access to all literature and data sources. [5]It could compile and interpret data on surface water (quantity, quality, salinity, etc.) for site specific coal leases or areas adjacent to those leases. [6]The Ames District hydrologist could then spend his or her time analyzing these interpretations and conducting on-site calculations.

[7]Second, the Water Resource Division of the USGS in Mountainview could provide expertise both through the subdistrict office and the White River Resource Area. [8]We could obtain their help through infrequent consultations, informal review of tract analysis, and initial field survey of any lease area lacking hydrologic data.

[9]Please consider our request carefully and contact us at extension 388 if you need further information.

Deleted: is in reference

Deleted: At this time and stage of access and interpretation of existing data on record in the form of computer storage and publications in print that

Deleted: may not be aware of is our main concern. Due to the U.S. Geological Survey having vast storage of and access to this data we would like to

Deleted: areas if possible.

Deleted: first area may be handled by the Geological Survey district

Deleted: due to accessibility and central locale

Deleted: By compiling this data a comprehensive interpretation of

Deleted: , i.e.,

Deleted: immediately

Deleted: can be provided to the

Deleted: . Thus, due to time constraints, time may be spent on

Deleted: ly, another area we foresee as a positive and very useful endeavor is the expertise that can be provided by

Deleted: Geological Survey

Deleted: . Because

Deleted: office are both located in Mountainview, we may

Deleted: in the form of

Deleted: and

Deleted: reconnaissance on a one time basis

Deleted: available

Deleted: The foregoing should provide adequate justification for requesting the U.S. Geological Survey's assistance.

Figure 16.3   Changes Tracked in the Edited Copy, USGS Proposal

Whatever your personal feelings about particular phrases or pronouns, you edit for neutral, unbiased language for two good reasons:

- Language that appears discriminatory to readers creates such significant noise in the document that it may block comprehension. Readers will never hear the intended meaning if discriminatory language interferes.
- Most professional associations and journals have policies to discourage discriminatory language, particularly as the language relates to gender. You edit the document to conform to these policies, just as you edit reference lists and punctuation to conform to the accepted form.

These reasons supersede your own feelings about the significance of word choices. They also reflect a professional respect for the people who will read the words that you edit.

Unnecessary demographic information can make language discriminatory. For example, the sentence "Dr. Alice Jones, paralyzed from the waist down, was named dean of the College of Human Sciences" introduces information that is irrelevant to her appointment and to her ability to do the job. That information about her paralysis would be appropriate only in a human interest story. Information that indicates race, age, gender, religion, politics, sexual orientation, or physical attributes is rarely appropriate in professional situations because it does not relate to professional credentials. It discriminates by implying that a given age, for example, may affect competence or by conveying surprise that the achievement is inconsistent with the demographic characteristic.

Application: Discriminatory Language

The paragraphs in Figure 16.4 appear in an architect's program for a visitor's park design. The program is a report with recommendations for design based on analysis of the facility function, use, and setting. It guides the designer by establishing a concept of the facility as well as specific goals and criteria for design. In these paragraphs, the writer addresses the issue of visitors who have disabilities and how the park should be designed to accommodate them.

The message of the paragraphs is positive: the architect is concerned not just with physical accommodations, such as wheelchair ramps, but also with the visitors as whole persons. However, the language contradicts the message by making "the handicapped" an abstraction, and by separating these visitors from others. The language can shape the designer's response to the message by subtly suggesting that "the disabled" are significant primarily in terms of their handicaps. Though the message discourages differentiation of visitors with handicaps and other visitors, the language permits it.

An analysis of the style will reveal nominalizations and weak verbs as well as some discriminatory language. Passive voice is appropriate to the extent that the designer is understood throughout the program to be the agent of implementing the concepts. Yet overuse of passive voice may reduce readers' access to and acceptance of the ideas. Figure 16.5 shows how this report might be edited for style.

Handicapped people have expressed that they do not need or desire segregated outdoor activities. They prefer not to be singled out, but instead appreciate efforts made to accommodate their special needs. A sensitive approach without differentiation is preferred, and demonstrations of sympathy should be avoided.

The disabled possess different learning styles which tend to focus on using the sensory perceptions to the greatest extent possible. Depending on their particular disability, they utilize their hands, eyes, and ears to perform as informational transmitters. To enhance the overall experience for the disabled visitor, and encourage his participation, all kinds of sensory experiences should be incorporated into exhibit and facility design.

Activities should be created which will enable participation by the physically disabled. Easy access to areas in the way of ramps, minimal inclines, and railings should be provided. Developing designs and activities that accommodate the handicapped population can provide them with a more enjoyable and secure experience.

Figure 16.4 Original Visitor Park Report

Editing for a Nonsexist Style

The generic *he* is now shunned by most professional associations. Research also shows that readers associate male images with the male pronoun far more often than they associate female images. One study required subjects to complete sentence fragments such as "Before a pedestrian crosses the street, . . ." and "When a lawyer presents opening arguments in a court case, . . ." Subjects also had to describe their images of the people in the sentences and give them names. Subjects who used *he* in completing the sentences gave the person a male name five times more often than a female name, and they imagined the person as a man four times more often than as a woman. By contrast, subjects who used *they* or *he or she* selected a male name only twice as often as a female name and imagined the person as a woman as often as a man.

The male bias in imagery caused by the generic *he* is especially inappropriate in evaluative situations. The following guidelines for employee evaluation encourage in a subtle (and probably unintentional) way the identification of outstanding employees as male. Supervisors are asked to classify employees into one of three rankings: needs improvement, good, and outstanding. The definitions of

with disabilities
~~Handicapped~~ people have expressed that they do not need or desire segregated outdoor activities. They prefer not to be singled out, but instead appreciate efforts made to accommodate their special needs. A sensitive approach without differentiation is preferred, and demonstrations of sympathy should be avoided.
People with disabilities use their
~~The disabled possess different learning styles which tend to focus on~~ ~~using the~~ sensory perceptions to ~~the~~ *a* greatest extent ~~possible.~~ *in learning* Depending on their particular disability, they ~~utilize~~ *use* their hands, eyes, and ears to *transmit* ~~perform as informational transmitters.~~ To enhance the overall experience for the disabled visitor, and encourage ~~his~~ *their* participation, all kinds of *to* sensory experiences should be incorporated into exhibit and facility design.
Physical disabilities require
~~Activities should be created which will enable participation by the physically disabled.~~ Easy access to areas in the way of ramps, minimal inclines, and railings ~~should be provided. Developing~~ designs and activities that accommodate ~~the handicapped population~~ *visitors with disabilities* can provide them with a more enjoyable and secure experience.

Figure 16.5 Edited Visitor Park Report

performance levels for the first two descriptions contain no gendered pronouns, but the description of the outstanding employee contains four references to males.

> The employee is clearly superior in meeting work requirements, and he consistently demonstrates an exceptional desire and ability to achieve a superior level of performance. His own high standards have either increased the effectiveness of his unit or set an example for other employees to follow. This rating characterizes an excellent employee who consistently does far more than is expected of him.

Professional associations have developed guidelines for avoiding sexist language. The National Council of Teachers of English (NCTE) suggests these alternatives for terms that include "man":

Sexist Language	Alternative
mankind	humanity, human beings, people
man's achievements	human achievements
the best man for the job	the best person for the job, the best man or woman for the job
man-made	synthetic, manufactured, crafted, machine-made

the common man	the average person, ordinary people
chairman	coordinator, presiding officer, head, chair
businessman	business executive or manager
fireman, mailman	firefighter, mail carrier
steward and stewardess	flight attendant
policeman, policewoman	police officer

The NCTE advises these alternatives for masculine pronouns:

1. Recast into the plural.
2. Reword to eliminate unnecessary gender problems.
3. Replace the masculine pronoun with *one, you,* or *he or she,* as appropriate.

You can alternate *she or he* with *he or she* and *she/he* with *he/she* so as not to privilege male or female by the order in which you use the pronouns.

Using Your Knowledge

1. Begin editing for words by analyzing patterns in the writing. Does the writer favor weak verbs? Passive voice? Abstract nouns? Redundancies? Any of these characteristics will point to specific editing objectives, such as strong verbs, active voice, and concrete nouns. Also look for discriminatory language that may offend readers.
2. Edit with respect for the power of word and sentence choices to change meaning. Compare the sense of the edited version with the original. Preserving meaning is always a more important objective than applying a style guideline.

Further Reading

Jones, Dan. 1998. *Technical Writing Style.* Longman.

Lanham, Richard A. 2000. *Revising Prose.* 4th ed. Longman. Also see *Revising Business Prose.*

Smith, Edward L., and Stephen A. Bernhardt. 1997. *Writing at Work: Professional Writing Skills for People on the Job.* NTC.

Williams, Joseph M. 2002. *Style: Ten Lessons in Clarity and Grace.* 7th ed. Longman.

Discussion and Application

1. *Using verbs effectively.* The following sentences do not use verbs effectively. Analyze each sentence to determine the source(s) of the problem (action not in the verb, weak verb, passive voice, nominalization), and then edit to use verbs more effectively. Maintain the writer's meaning; prepare a query to the writer if you cannot determine the meaning.

   a. Prolonged use of the battery can cause it to become drained of its energy.

   b. The crimper on the alfalfa mower breaks the stem every inch to allow the fluids in the stem to be released.

   c. (*from instructions for playing tennis*) Place your legs in a bent position with your toes pointing outward at an angle of 45 degrees.

   d. Because there is a trend toward fewer and larger farms, it will cause an increase in the demand for machinery, decreasing the demand for farm laborers.

   e. The report shows a recommendation toward simple, cost-effective advertising with the aid of either an advertising agency or an account executive from a media service.

   f. Further research needs to be entailed into the project.

2. *Passive voice, weak verbs, past tense.* Distinguish passive voice sentences from those with weak verbs or past tense. (Look for the *to be* verb and past participle to identify the passive voice sentences. Verify your identification by determining whether the subject performs or receives action.) Convert the passive voice sentences to active voice. Note that "voice" pertains only to transitive verbs, not to linking and *to be* verbs.

   a. The report was written collaboratively by three engineers.

   b. The report was informative but too long.

   c. The engineers have sent the report to the editor.

   d. The report has been shortened by three pages.

   e. This method of writing and editing is effective for us.

3. *Sentence structure and meaning.* Describe the sentences below in terms of core sentence, use of verbs, and voice. Do they all mean the same thing? Which version is "better" stylistically? What are the bases for deciding the answer to that question?

   a. There must be thorough preparation of the specimens by laboratory personnel.

   b. Laboratory personnel must prepare the specimens thoroughly.

   c. The specimens must be prepared thoroughly by the laboratory personnel.

   d. Preparation of the specimens by laboratory personnel must be thorough.

4. *The risk of changing meaning.* Compare the following sentence pairs to determine how the revision for style has also inappropriately changed meaning. Define the change in meaning. Suggest a way to edit the first sentence in the pair to improve style without changing meaning.

   a. Technical writers are now finding themselves in roles in product design and managing the production as well.

      Technical writers now find roles in product design and production management.

   b. The problem involves derivation of objective methods for evaluating the effect of adriamycin on the heart.

      The problem derives objective methods for evaluating the effect of adriamycin on the heart.

5. *Doublespeak.* Locate examples of "doublespeak," and conduct a doublespeak awards contest in your class.

6. *Analysis: technical document.* Select a passage from a technical document, such as a computer manual, or from a textbook on a technical subject. Analyze the style, using the terms and principles discussed in this chapter and in Chapter 15. Then, taking note of the presumed readers and purpose for this document, evaluate the style. What are its strengths? What, if any, editing goals would be appropriate?

7. *Style for online documents.* Writers for websites and online help aim for brevity so that users will not have to scroll through multiple screens. What stylistic devices help them achieve brevity? Look for a site with information comparable to that in the print document you examined for application 6. For example, compare a topic in a program's online help with the information in a print manual. What are the differences? Are these differences stylistic, or has the content been shortened?

8. *The language of discrimination.* Edit the definition of an outstanding employee that appeared in the "Language of Discrimination" section to remove references to gender.

9. *Your own style.* Analyze your own writing style. Look for placement of subject and verb in sentences and for use of verbs, active and passive voice, and concreteness of nouns.

10. *Vocabulary.* If you are not confident about explaining the meaning of the following terms, check the glossary or review the chapter: *nominalization, passive voice, past tense.*

17 Organization: The Architecture of Information

When architects design buildings, they are interested, among other things, in uses and usability of the building. Users need to find its functional spaces, like offices, meeting rooms, and restrooms. In doing business, they need to get from place to place efficiently. Signs, office numbers, and even colors can help to reveal the way the building is organized, but arranging the spaces to begin with influences a user's sense of the building and the ability to use it.

Like buildings, information needs structure if users are to make sense of it, find what they need, and do what they came to do. In documents, organization refers to categories of information and their arrangement. It affects a user's performance and a reader's comprehension. Organization by task can help a reader complete a task. A document structure that matches the content structure helps a reader learn. A document organized with the expectation that it may be reused can be tagged so that the relevant parts can be retrieved when they are needed.

Organization is a powerful aid (or hindrance) to performance, comprehension, and reuse. Because organization is so closely linked to meaning, editing for organization also may reveal content gaps or inconsistencies.

This chapter defines two broad uses of technical documents for readers: performance and learning. These purposes roughly correspond to user-based organization and content-based organization. The chapter emphasizes the sequence of

Key Concepts

A well-organized document makes information easy to learn and easy to use. Organization is the architecture of information. When a reader uses information to perform a task, organization by task reflects what readers need to do. When a reader uses information to learn, organization that matches the inherent structure of the content helps a reader recognize patterns for making sense of information. Some documents are structured according to templates that are common in the field or organization. Standard organizational patterns include chronological, spatial, comparison-contrast, cause-effect, and topical. Because structure reinforces meaning, these patterns help readers interpret meaning by showing how the parts connect to form a whole. Grouping of related ideas also helps readers comprehend. Documents may also be structured for reuse.

information as it influences performance and comprehension. It also considers paragraph organization, and it introduces the topic of organizing information at the company level to enable retrieval and reuse of information. Chapter 18, "Visual Design," shows how to provide selective access to information and how to reveal the structure visually.

Organization for Performance: Task-Based Order

Many problems in software documentation and other instructions occur because writers organize according to program or product function rather than according to the tasks the users need to perform. For example, the programmer for software to create a database for a bail bond company wanted to organize the manual according to the screens, each one of which records different types of information. However, the user normally begins with the eighth screen in the program, "entering personal data," and thus would have been frustrated by wading through seven categories of instructions before finding the ones needed to get started. The editor advised organizing according to tasks the user would perform in the probable sequence of use. The user of the program does not care as much about program functions as about getting the job done. The sequence of screens in the program is invisible and irrelevant to the user. Likewise, developers of websites, especially of organizations, sometimes see the purpose of the website as representation of that organization, defining "who we are and what we do" rather than imagining tasks that users come to the website to complete. A site's FAQs may be a list in the order received rather than by related topics. Organizing by user tasks can create a happier experience for those users.

The editor's role in organizing according to task should take place at the planning stage, because reorganization after the manual is drafted is a huge project that somewhat repeats the efforts of organizing to begin with. The task-based order begins with a task analysis: identification of tasks that users need to do. For complicated procedures, the task analysis includes major tasks plus individual tasks. These major tasks define the divisions of a long manual or complex website.

In addition to helping plan the task-based order, the editor will check it during development and when a draft is complete. The editor can look for grouping of related tasks, sequence, and completeness according to the task analysis. The editor can check for whether there are too many or too few steps within a section (which could become confusing), in addition to editing for style, visual design, and illustrations.

By trying a product while reviewing instructions, an editor provides an initial usability test. Formal usability testing, however, can provide additional information about possible gaps in the instructions or problems with the product itself. The most reliable usability test examines representative users in their work settings using the product to perform their work tasks.

Online help arranged by topics offers the advantage of pointing users to the particular tasks that interest them without the distraction of irrelevant information.

But a tradeoff is that users have to find the right information, and they may not know the terminology of the program. Topical information may lack contextual cues to understanding: tasks are isolated from others. To compensate for these limitations, writers and editors must anticipate terms, even the wrong ones, that users might try in locating information. Each topic might also have a "related topics" link to help users both locate what they are seeking and connect pieces of information in related tasks. These strategies help users impose their own organization on the information by pointing to relationships.

Organization for Comprehension: Content-Based Order

All technical communication is informative, and readers aim to learn. Readers of user documentation seek explicit instructions. Readers of policy statements, reports, proposals, and product descriptions seek information in order to make decisions. Thus, technical editors work to enable readers to understand accurately and easily. The organization of the whole document or whole sections will affect comprehension.

A fundamental principle of learning is that people organize bits of information into structures and patterns in order to make sense of it. You understand "building," for example, not as a set of rooms in random order but as a structure whose parts are organized in a predictable way. Your experience of one office building or one classroom building shapes the way you first learn about another building of similar purpose. For example, you might expect the receptionist's office for a company on one of the floors of a high rise building to be near the elevator. You would use familiar information to learn new information.

Forming and remembering patterns of information requires both understanding of the whole and of the parts. You recognize "building" as the whole, but you can also identify the parts that form the whole (doors, stairs, offices). Typical organization of documents reflects this whole-part relationship. The introduction identifies the whole to orient the readers, the body identifies the parts, and the conclusion "puts the parts back together" into a whole that has been modified by new information. In a proposal, for example, the introduction defines the problem, the sections identify the procedures for solving the problem, and the conclusion points to the solution to the problem.

People use knowledge of patterns to understand abstract concepts as well as objects. Usable information has order. Parts relate to each other. The parts combine to form a whole that has meaning. We have already seen, in discussing sentences, how their structure can reinforce meaning. Whole-document organization also can clarify or confuse, depending on whether the structure matches the concept and probable use. The parts of a document can be linked in predictable ways, including chronological, spatial, cause and effect, and topical. The best choice depends on the nature of the information.

Principles of Content Organization

Editing for organization requires your best analytical skills. You have to understand the content, and you may have to get past the existing order to imagine a more appropriate order.

This section of the chapter presents some principles that will help you improve a document's clarity and effectiveness by reorganizing it. The principles suggest guidelines for editing for organization. The order in which the principles are presented also suggests a process for applying them; that is, if you consider the principles in the order that they are presented, you will work efficiently. Briefly, when editing for organization, you should do the following:

1. Follow pre-established document structures.
2. Anticipate reader questions and needs.
3. Arrange information from general to specific and from familiar to new.
4. Apply conventional patterns of organization: match structure to meaning.
5. Group related material.
6. Use parallel structure for parallel sections (chapters, paragraphs).

Let us examine each guideline in more detail.

1. Follow Pre-Established Document Structures

Someone other than the writer may establish the structure of a document. For example, an organization that publishes an RFP (request for proposals, inviting proposals for providing a service or solving a problem) may specify what components should appear on what pages. The RFP for the proposal in Figure 17.3 (p. 295) specified that an abstract and a work schedule appear on the title page of the proposal. A manual in a series of manuals follows a structure established for the series (document set).

Certain organizational patterns are widely accepted for documents in different disciplines. Readers who know these conventions expect documents to follow them. For example, a research report in a scientific journal typically follows the pattern of problem statement, literature review, methodology, results, and discussion. These terms are likely to be section headings. Many business executives, especially those who manage according to management by objectives (MBO), expect that business plans will address goals (broad aims), objectives (specific aims), strategies (means of achieving the goals and objectives), and evaluation procedures (means of measuring whether the goals and objectives are achieved). A company may have templates for all of its manuals that define parts and their sequence. A company or university website also includes standard parts for each of its sections. For example, each department in a university website will probably include contact information, program and curriculum information, and a faculty list. There may be other standard parts, such as information on how to apply for admission. An editor checks for completeness of content by matching the existing structure against the template.

Because of the structure established by an organization's specifications or a discipline's conventions, readers expect that they will find certain kinds of information in certain places. When readers encounter variations from established structure, they may become frustrated and also lose some confidence in the writer.

Your first task in editing for organization is to compare the structure of the document to be edited with any external patterns to which it must conform. This comparison will help you spot missing parts as well as ineffective structure. For example, if you are editing a management plan that lacks evaluation procedures, you will advise the writer to complete that section.

2. Anticipate Reader Questions and Needs

Editing, like writing, requires imagining what readers know and need. Readers ask predictable questions from a document, including "What is it?" "What is this about?" "Why is it important?" "Who is affected by it?" "How do I get started?" These are versions of the investigative journalist's questions: who, what, when, where, why, how, and so what? As editor you can use these questions to review whether probable reader questions have been covered in the order in which readers are likely to ask them. You can be almost certain that if the document defines or describes an unfamiliar concept or object, the first question will be "What is it?" The predictability of this question explains why introductions often include definitions.

Anticipating user needs will also help in organizing a website. Top-level menu items reflect a user's reasons for visiting a website. These reasons identify the types of information to include, such as branding information (who the organization is and what it does), contact information, program or product descriptions, and policies.

3. Arrange from General to Specific and from Familiar to New

Concepts and familiar information help readers learn new information. The familiar information may be the problem that led to the feasibility study or proposal, prior research on a subject, the purpose of a task, a destination, or an update on an ongoing project. This familiar information places the new information to follow in a context and links the new with the old.

As editor you evaluate the beginning of a document for information that orients readers. Writers may inadvertently omit information that is familiar to them but that readers need. They may not think to include identifying information or may tell what to do without telling how to do it. You can look more objectively at the introduction from a reader's point of view. If you know more than readers probably will, you can still anticipate basic reader questions and evaluate opening paragraphs to see whether the questions are answered. These questions to orient readers are typical investigative questions: What is it? How? Who? Where?

The amount of information to precede details will depend on the document's purpose as well as the complexity and familiarity of the information. Instructions do not generally need to begin with detailed conceptual information, but information that provides metaphors, vocabulary, and principles may help readers follow steps and work intuitively. Readers are eager to get on with the task. The general information in instructions may be as minimal as a definition of the scope of the task (what a reader will learn to do), expected competencies, and equipment and materials needed. The introduction may also include a brief description or explanation of relevant equipment and processes if these will help the reader complete the task accurately, efficiently, and safely. When the reader's task is comprehension rather than performance, the general information needs to orient the reader not just to the topic but to its background and significance and to relevant principles. Beginners are likely to need more explanation and background than experts need; and people who will use the information casually need less orientation than do people who will use the information in research.

In some circumstances, the introduction should anticipate future readers. In an organization, where ongoing tasks are to solve problems and to develop policies, future readers who face related problems may want to know the history of the problem. A common gap in documents for people who share common organizational knowledge is the problem details of who-what-when-where-why. Meeting minutes include standard parts (date, time, people present) to prompt inclusion of this type of information. Some companies will define expectations for other types of organizational documents as well, but as editor you can evaluate the inclusion of this information in the introduction, thinking of the document being referenced over time.

After you review overall document structure according to pre-established structures and probable reader questions, you review the introductory material for the whole and for each major section. Does it orient the reader to the subject and purpose of the whole document or of the section? Does it begin with information that is familiar to the intended reader?

4. Use Conventional Patterns of Organization: Match Structure to Meaning

Readers have learned patterns of organizing information, such as chronological, spatial, comparison-contrast, and cause-effect. These patterns reflect relationships of information rather than externally imposed patterns such as the scientific report and MBO patterns described previously. The pattern of the document should match the inherent structure of the information.

A chronological structure is appropriate for narratives, instructions, and process descriptions. A spatial order (for example, top to bottom, left to right, in to out) is useful when the text describes a two- or three-dimensional object or space. Comparison and contrast helps readers weigh options. It can follow point-by-point or item-by-item patterns depending on what is to be emphasized. A document organized to illustrate causes and effects should consistently move either from causes to effects or from effects to causes. All of these patterns help readers perceive meaning because they match the document structure with content structure and show how parts relate.

In a report to aid decision making, such as a feasibility study, the middle section addresses each of the issues that affect the decision, such as cost, legal restrictions, available workforce, and competition. The topics in such a report are arranged by importance and by relationship to one another. All financial topics, for example, would be grouped together. As you consider the organization of these topics, you can also assess whether they are complete—that is, are there other topics that will bear on the decision that are not discussed?

The order of importance pattern is more abstract than the others. Who is to decide what is more important than something else? The problem—the reason for the document to begin with—should give cues. If the problem originated in meeting environmental standards, for example, the investigation on that impact is more important than the research on costs or equipment. As an editor, you need to be particularly cautious about rearranging material to emphasize one point at the expense of another. Your perception about what is important may not be the same as the writer's. The writer, usually the subject matter expert, should have final say about rearrangement to suggest degree of importance.

All of these patterns—chronological, spatial, comparison-contrast, cause-effect, topical, order of importance—are subordinate to the overall pattern of general to specific. Many documents use more than one pattern. For example, a chronological narrative may also use comparison-contrast within the sections on various time periods. One pattern, however, should predominate.

5. Group Related Material

Grouping is a familiar concept for writers and editors. An editor tries to keep paragraphs focused on a topic rather than wandering, to avoid mixing unrelated information, and to develop lists meaningfully rather than randomly.

The list in Figure 17.1 illustrates a list in need of sorting. The list appeared in the first draft of an employee handbook for a video rental store; its purpose was to delineate duties of the manager. The writer's free association is evident. For example, does item 15, "filing on people," refer to filing in a file cabinet (as in item 16) or to filing in courts to pursue writers of bad checks (as in item 14)? More organization in the list would help managers to develop a concept of their job and to identify and recall specific duties.

One task in editing this list, then, is to group various tasks into categories, such as finances (budget, accounting, and bookkeeping), personnel, marketing, customer relations, maintenance, and inventory/supplies. Figure 17.2 shows an initial grouping based on these categories.

For your information, "filing on people" (item 15) does refer to filing charges against customers who fail to return the movies they have rented. "P.O.P." (item 21) is an abbreviation for "Point of Purchase" and refers to displays supplied by the wholesaler. "Coop" (item 33) refers to a cooperative advertising arrangement between the movie distributor and the retail store. "Commtron" (item 40) is the

Duties: Manager

1. waiting on customers
2. selling memberships
3. inventory of tapes, control cards, etc.
4. inventory of store
5. employee meetings
6. keeping employee morale up
7. accounts payable and receivable
8. computer input
9. computer reports—late, end of week, end of month, etc.
10. daily deposits
11. totaling time cards
12. payroll—total system done in store
13. checkbook balanced
14. hot check system
15. filing on people
16. filing system
17. ordering/merchandising
18. budget for store
19. prebooking
20. customer orders and order system
21. P. O. P. ordered and regulated
22. receiving
23. mail outs (free movie postcards)
24. static customer mail-outs
25. tape repair
26. cleaning VCRs
27. having anything that is broken fixed right away—the longer it sits the more money tied up
28. brown book—bookkeeping record
29. monthly report book
30. payroll book
31. schedule for employees
32. keep store appearance bright, eyecatching, clean, fun; movies playing by customer counter and children's movies in children's section at all times
33. keep coop checked on and updated
34. taking care of customer complaints or problems
35. taking care of employee problems
36. ordering computer supplies
37. ordering office supplies
38. keep plenty of paper supplies on hand
39. reshrinkwrap tape boxes that are getting worn out
40. Commtron bill
41. video log (introducing and explaining it to customers)
42. machine inventory every Wednesday—better control over what you have in the store; just in case one is stolen the time span isn't very long
43. promotions and marketing
44. know what's hot and new and learn the titles and actors
45. know your stock and what you have on hand—better control of movies and VCRs

Figure 17.1   Unedited List of Duties of Video Store Manager

major supplier of movies to this store. This information was not obvious from the document; queries to the writer were necessary.

Figure 17.2 shows how the list might look after one editorial pass. The editing is not yet complete, but the grouping will help the editor when reviewing the list with the writer.

The grouping will probably reveal some missing duties. For example, will the manager train new employees? It may raise questions about whether the duties are really managerial duties. Might a subordinate handle orders? Grouping may also encourage clarification of items. For example, is "promotions and marketing" (item 43) a category term, or does it refer to a specific duty? Thus, editing for organization, using the guideline about grouping, should help the editor and writer make the content more complete and accurate as well as easier to comprehend.

After the content is complete and meaningfully organized, the editor can work on consistency (for example, "mail-outs," "mail outs," or "mailouts"?) and parallelism of phrases. (The example in Figure 17.4, pages 298–299, will illustrate the importance of grouping related ideas in paragraphs.)

Duties: Manager

1. **Finances**
   accounts payable and receivable
   daily deposits
   checkbook balanced
   hot check system
   budget for store
   brown book—bookkeeping record
   Commtron bill
   filing system
   computer reports: late, end of week, end of
        month, etc.
   monthly report book
   filing on people

2. **Personnel**
   employee meetings
   keeping employee morale up
   payroll—total system done in store
   schedule for employees
   taking care of employee problems

3. **Inventory/Supplies**
   inventory of tapes, control cards, etc.
   inventory of store
   ordering computer supplies
   ordering office supplies
   keep plenty of paper supplies on hand
   machine inventory every Wednesday
   computer input

4. **Marketing**
   selling memberships
   P. O. P. ordered and regulated
   ordering/merchandising
   mail outs (free movie postcards)
   static customer mailouts
   keep store appearance bright . . .
   keep coop checked on and updated
   know what's hot and new and learn the titles
        and actors
   know your stock and what you have on
        hand—better control of movies and VCRs
   prebooking
   promotions/marketing

5. **Customer Relations**
   waiting on customers
   customer orders and order system
   taking care of customer complaints and
        problems
   video log—introducing and explaining it to
        customers

6. **Maintenance**
   tape repair
   cleaning VCRs
   having anything that is broken fixed . . .
   reshrinkwrap tape boxes that are getting
        worn out

Figure 17.2   Manager's Duties Grouped after One Editorial Pass

As a comparison of Figures 17.1 and 17.2 illustrates, it will be easier to learn and remember 45 duties sorted into six categories than a long, unsorted list. Some research has suggested that seven items, plus or minus two, are the maximum number that the human mind can handle. The number would depend on the type of information and readers, but a few categories will be easier to remember than a long list.

The principle of restricting the number of chunks of information relates to prose as well as to instructions and lists such as the one in Figure 17.1. Chunks may be the major sections in a report, proposal, or chapter as revealed by level-one headings. If such a document has many sections, readers may lose a sense of how the individual sections relate to one another and to the whole. Excessive efforts to reveal the structure of the information with divisions of text may actually diminish the coherence.

The principles of grouping and chunking also explain the effective organization of online documents. Top-level categories reveal the main tasks or main organizational divisions to be featured. Secondary items are then grouped within these main categories, with the grouping revealing how the parts are related to each other and to the whole. The depth of the structure—number of levels of menu choices or categories of information—depends on the complexity of the information but also on how much a reader can grasp in a chunk. Some grouping will be preferable to undifferentiated sections.

6. Use Parallel Structure for Parallel Sections

The principle of parallelism at the sentence level is important for larger structures, too, such as paragraphs, sections of chapters, and chapters. The structure can clarify the content relationships of sections. For example, if you are editing a progress report that is ordered by tasks, the discussion of each task will proceed from work completed to work remaining. The editor of this book looked for a pattern in each chapter of introduction, principles and theory, and application or guidelines.

Paragraph Organization

Paragraphs, like whole documents, make more sense if they are well organized. The subjective descriptor for good organization in a paragraph is *flow*. A paragraph that flows enables a reader to move from sentence to sentence without halting or rereading. In more objective terms, what creates flow (or cohesion) is the linking of sentences and repetition or variation of key words.

Linking Sentences

Sentences must connect to each other to make sense in a paragraph. Abrupt shifts in topic will puzzle readers. A workable pattern for those connections is to tie the beginning of one sentence to the end of the previous sentence. The subject offers the familiar information, whereas the predicate offers the new information. The new information in the predicate of one sentence becomes familiar and can be the subject of the next sentence. That linking helps to create cohesion. (See the discussion of the known-to-new structure in Chapter 15.) Sentences don't have to rigidly follow this pattern, but a sentence should echo a key point of the one that precedes it.

Repetitions and Variations

In addition to links, repetitions or variations of key words keep a paragraph focused on a topic. Variations can be restatements in different terms, examples, opposites, and pronouns. They can also be effects or causes. In the preceding paragraph on linking sentences, with the heading announcing "linking" as the topic, there are two repetitions of "linking," but there are also these variations of the concept:

Variation of "Linking"	Type of Variation
connect, connections	restatement in different terms
abrupt shifts	opposite
tie	restatement
subject/familiar–predicate/new	example
cohesion	effect

These terms keep readers focused on the topic of linking. A subtopic is the idea of making sense, announced in the first sentence. The second sentence varies this concept with the opposite, "puzzle."

Application: The Problem Statement for a Research Proposal

This section applies the guidelines for editing for organization to increase comprehension. Figure 17.3 shows the problem statement from a proposal for research on the bulb onion. The proposal was written by a graduate student competing for research funds distributed by the graduate school of a university. The graduate school distributed an RFP and promised to fund some—but not all—of the proposed projects.

The writer's goal is to persuade the proposal evaluators that his project is feasible and worthy of being funded. To demonstrate feasibility, he has to show that he has the facilities and skills to complete the work and that he can finish within the given time. To demonstrate worth, he has to show that the research will yield significant information. (As a technical editor, you know these things because you are familiar with typical technical documents, such as proposals. If you do not know the purposes and methods of proposals, you consult a technical writing handbook or textbook. You also consult the RFP to determine criteria for these grants.) The problem statement reproduced here is mostly concerned with demonstrating worth.

The proposal evaluators are not experts on plant genetics. They are professors from various departments in the university. They are intelligent and well educated but probably uninformed about onion research. Furthermore, they will read quickly through a stack of proposals and will not have the luxury of rereading and mulling over meanings. (As a technical editor, you discover this information through queries to the writer or possibly to the contact person specified on the RFP.) As editor, you can probably act as a good stand-in for the intended readers; that is, unless you have a good background in the study of plant genetics, the material will be new to you, as it will be for some evaluators. Your responses, therefore, will be useful in assessing likely responses from the evaluators. Your analysis of the context is complete enough at this point for you to assess the document itself.

Guideline 1 (see p. 287) for the editing for organization will not apply here because you are looking at just part of the proposal. But guideline 3—arranging information from general concept and familiar information to specific details and new information—will be very important.

Read the problem statement through. Note where you must reread sentences or look back in the text to verify assumptions. Also note where the meaning seems especially clear. These notes will guide you in determining editing objectives and suggesting emendations. After you have noted both the confusing and the clear places, you can set some editing objectives. You can determine what you

Isozyme Variability of Allium cepa Accessions

Statement of the Research Problem

1   Genetic variability is essential in the improvement of crop plants. Until recent years most of the single gene markers used in higher plant genetics were those affecting morphological characters, i.e., dwarfism, chlorophyll deficiency, or leaf characteristics. Molecular markers offer new possibilities of identifying variations useful in basic and applied research.

2   Proteins are an easily utilized type of molecular marker. Protein markers code for proteins that can be separated by electrophoresis to determine the presence or absence of specific alleles. The most widely used protein markers in plant breeding and genetics are isozymes.

3   Electrophoretic studies of isozyme variation within a plant population provide information on the genetic structure (Sibinsky et al. 1984) without depending upon morphological characters, which are easily influenced by the environment. Genetic studies of isozymes have been conducted on more than 30 crop species (Tanksley and Orton 1983). Considerable variation can exist among plant populations as well as among individuals within a given population. Domestic and wild barley accessions were assayed and substantial differences within and between accessions were found (Kahler and Allard 1981). Significant variation in allelic frequency and polymorphism has been detected in lentil collections (Sibinsky et al. 1984) as well as in *Zea mays* (Stuber and Goodman 1983).

4   Researchers in the proposed project are particularly interested in analyzing isozyme variability within the bulb onion (*Allium cepa L.*). The bulb onion is a major horticultural crop in Texas with a cash value of $75 million in 1982 (*Tx. Veg. Stats.* 1983). As onions are a major crop, techniques are continually being explored to facilitate the breeding of superior varieties. However, a major factor limiting advances in breeding is the identification of selection criteria, i.e., genetic variation.

5   Electrophoresis as discussed above provides a tool for selection. Electrophoretic techniques for analyzing isozyme variability in onions have been established (Hadacova et al. 1983; Peffley et al. 1985). Isozyme markers have many applications including introgression of genes from wild species, identification of breeding stocks, measurement of genetic variability, determination of genetic purity of hybrid seed lots, and varietal planting and protection (Tanksley 1983).

6   This research intends to explore isozyme variability of *Allium cepa* accessions with the intent of identifying molecular markers useful in onion breeding and genetics.

Figure 17.3   Unedited Problem Statement for a Research Proposal

would need to do as editor to clarify the problem and its significance for intelligent but uninformed readers.

Unless you are a plant genetics expert, you probably got lost early in the problem statement. The first sentence is easy enough because most educated readers know that crop scientists improve crops through genetic manipulation. You may have been relieved when you got to paragraph 4, where you can relate to the idea that onions are a major cash crop (even if you are surprised by this information).

If you stumbled over other parts, you may be thinking, as many editors do, that defining terms, either parenthetically or in a glossary, will solve the problem. Indeed, the numerous technical terms can slow down non-expert readers. If you follow the processes of comprehensive editing described in Chapter 14, though, you may rightly choose to think first about organization and completeness of information. Just as readers comprehend a text top down, you will edit top down. Your first task should be to define the concept of the research problem. Definitions may be an option later. If you are not convinced, look up a few of the terms that are repeated frequently: "electrophoresis," "isozyme variation," and "accessions." Chances are that the definitions do not clarify the significance of the problem statement. One reason is that the predictable reader question, "So what?" (or "What is the significance?"), is not really answered except indirectly in the information about cash value of the crop.

Reorganizing may make the concept and terms clearer. Guideline 3 suggests that you place general or conceptual information before specific. Thus, the concept of the research (the need and purpose) should appear in the first or second paragraph. While considering organization, you may also have noted that the discussion of electrophoresis in paragraphs 3 and 5 is interrupted by paragraph 4 ("as discussed above" is a clue). Thus, an editing objective may be to group the sentences about electrophoresis together (guideline 5). Additional objectives may include the definition of terms, correction of grammar and punctuation, and other sentence- and word-level emendations.

To prepare for a thoughtful and systematic reorganization, begin with guideline 3: place general information before specific. To identify the general information (concept), look back through the problem statement to see if you can answer the basic question: What will the proposed research do that previous genetic research has failed to do?

The first paragraph contrasts two kinds of gene markers: those affecting morphological characters (the kind formerly used in research) and molecular markers (the kind to be used in the proposed research project). Why are molecular markers better? To answer this question, you might need to ask why morphological markers are limited. The answer to that second question is buried in a dependent clause in paragraph 3: "which are easily influenced by the environment." If the morphological characters are easily influenced by the environment, the data generated by using them may be unreliable. These markers may measure environmental effects rather than genetic variation. This statement is implicit in the first paragraph, but it is not stated. Thus, this key information becomes available to uninformed readers only upon rereading (and mental reordering of the information).

This kind of questioning and probing for the central ideas that you do as an editor parallels a reader's questioning. Your editing task is to make the concepts so clear that a reader won't have to probe. Once you can answer the basic questions, you should have the conceptual information that should appear early in the document. This process of asking questions about the concept is central to top-down editing. If the questions cannot be answered from the existing text, you query the writer.

A simple paraphrase of this problem statement might be this:

> Researchers have been limited in their attempts to breed superior bulb onions by an inferior method of marking the genes. The old method used markers for morphological characters. Because these characters could be influenced by the environment, the data they yielded were unreliable. A new method, electrophoresis, allows the identification of molecular markers (isozymes). In the proposed research, electrophoresis will be used to analyze isozyme variability in onions.

Figure 17.4 shows the problem statement reorganized to present conceptual information early and explicitly and to form paragraphs by topic. The sentences in boldface have been moved, and a phrase in boldface italics (paragraph 2) has been added to clarify information. No other editing has been done. The terms remain undefined, and the writing style is the same. But the document is now more comprehensible and persuasive for readers who are not genetics experts. The terms become meaningful once the concept is clear. Given the hasty way in which the proposal will be read, it will be sufficient for readers to know that electrophoresis is a method and that an isozyme is a molecular marker. Depending on how much time is available, the editor may work on sentences for clarity and persuasiveness. But even if he or she must stop now, the document will be more effective than the original.

The version in Figure 17.4 is not necessarily the "right" one. Because we are not applying rules, we cannot assert with certainty that one way is right and another is wrong. However, each emendation here can be explained in terms of how readers learn from a text. Nothing has been done arbitrarily. Furthermore, the editing is based on an analysis of the context of use, including the purpose of the document, the intended readers, and the conditions of reading. Thus, decisions have been made to accommodate those readers and purpose. Had the problem statement appeared in the research report for publication, the organization would have been different. The review of literature in paragraph 5, for example, would have moved forward because of the established pattern in research reports of setting the research context with a literature review.

You may be interested to know that the edited proposal succeeded in winning funding!

Isozyme Variability of *Allium cepa* Accessions

Statement of the Research Problem

[1] Genetic variability is essential in the improvement of crop plants. **Re-** ◄──
searchers in this project are particularly interested in analyzing isozyme variability within the bulb onion (*Allium cepa L.*). The bulb onion is a major horticultural crop in Texas with a cash value of $75 million in 1982 (*Tx. Veg. Stats.*
1983). As onions are a major crop, techniques are continually being explored to
facilitate the breeding of superior varieties. However, a major factor limiting ◄──
advances in breeding is the identification of selection criteria, i.e., genetic
variation.**

[2] Until recent years most of the single gene markers used in higher plant genetics were those affecting morphological characters, i.e., dwarfism, chlorophyll
deficiency, or leaf characteristics. **However, these characters are easily influenced by the environment,** *so the data are unreliable.* Molecular markers offer
new possibilities of identifying variations useful in basic and applied research.

[3] Proteins are an easily utilized type of molecular marker. Protein markers
code for proteins that can be separated by electrophoresis to determine the presence or absence of specific alleles. The most widely used protein markers in plant
breeding and genetics are isozymes.

[4] **Electrophoresis** <*as discussed above*> **provides a tool for selection of ge-** ◄──
netic characteristics. Electrophoretic studies of isozyme variation within a plant
population provide information on the genetic structure (Sibinsky et al. 1984)
without depending on morphological characters. **Electrophoretic techniques for
analyzing isozyme variability in onions have been established (Hadacova et al.
1983; Peffley et al. 1985).**

[5] **Isozyme markers have many applications including introgression of genes** ◄──
**from wild species, identification of breeding stocks, measurement of genetic
variability, determination of genetic purity of hybrid seed lots, and varietal
planting and protection (Tanksley 1983).** Genetic studies of isozymes have been
conducted on more than 30 crop species (Tanksley and Orton 1983). Considerable variation can exist among plant populations as well as among individuals
within a given population. Domestic and wild barley accessions were assayed
and substantial differences within and between accessions were found (Kahler
and Allard 1981). Significant variation in allelic frequency and polymorphism
has been detected in lentil collections (Sibinsky et al. 1984) as well as in *Zea mays*
(Stuber and Goodman 1983).

[6] This research intends to explore isozyme variability of *Allium cepa* accessions ◄──
with the intent of identifying molecular markers in onion breeding and
genetics.

Figure 17.4 The Problem Statement in Figure 17.3 Reorganized

Code: Boldface indicates rearranged material; boldface italic indicates inserted information;
< > indicates material that should be deleted because of rearrangement.

The original first sentence remains to establish the concept of the research. The familiar "genetic variability" also prepares for "isozyme variability." (See guideline 3.) This easily understood information announces the significance in familiar and persuasive terms. The paragraph also establishes the context for the proposed research—the ongoing research at the university. (See guideline 3.)

The last sentence introduces the rest of the paper by stating the limitations of the old method of research. The rearranged and new information states the specific limitation of the old method of research. By knowing the limits of the old, reviewers will see the importance of the proposed method.

This paragraph and the next one are rearranged to create separate paragraphs on electrophoresis and on isozyme markers, rather than mixing the subjects as the original paragraphs 3 and 5 do. (See guideline 5.)

The subject of isozymes could precede the subject of electrophoresis. However, the material on isozymes is mostly background material. This material on electrophoresis focuses attention on the proposed project. (See guideline 3: place important information early in the sequence.)

This background information can be skimmed; it basically demonstrates the proposer's knowledge rather than giving information on the proposed research.

The explicit purpose statement leads nicely to the next section of the proposal, a statement of specific objectives.

Organizing for Reuse

In an effort to avoid miscommunication in different parts of a company and to keep all company information consistent and up-to-date, some companies are using *single sourcing* of information. This means, for example, that the technical documentation division and the marketing division would use the same product description. Probably the information would be stored in a database and coded with XML (see Chapter 5). Some gains in accuracy, consistency, and efficiency can result from single sourcing. If you work in a company that single sources, you will almost certainly work from document templates that identify sections of documents. These templates are analogous to any other externally imposed document structure, such as the structure for scientific reports. You can use the templates to check for completeness and for structural consistency. You may also be asked to code the sections with XML as you edit.

Even if you use these templates, you will still have to evaluate the structure of the information in the body of the document and impose a pattern that is consistent with the inherent structure of the information or that enables users to complete tasks.

The management of information by single sourcing and databases is sometimes called "information architecture." This architecture organizes the information of the company as a whole with the goal of keeping information consistent and up-to-date across divisions and of saving the time of recreating the same information within different company divisions. Thus, as this chapter has emphasized, the term and objectives are broader than the objectives of organizing information within a single document.

Using Your Knowledge

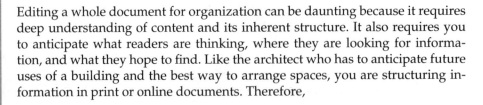

Editing a whole document for organization can be daunting because it requires deep understanding of content and its inherent structure. It also requires you to anticipate what readers are thinking, where they are looking for information, and what they hope to find. Like the architect who has to anticipate future uses of a building and the best way to arrange spaces, you are structuring information in print or online documents. Therefore,

1. use task-based order for procedures.
2. to encourage comprehension, match the structure of the document with an externally imposed structure or according to one of the conventional patterns of organizing information.

Further Reading

Colomb, Gregory G., and Joseph M. Williams. 1985. Perceiving structure in professional prose: A multiply determined experience. *Writing in Nonacademic Settings.* Ed. Lee Odell and Dixie Goswami. Guilford. 87–128.

Huckin, Thomas N. 1983. A cognitive approach to readability. *New Essays in Technical and Scientific Communication: Research, Theory, Practice.* Ed. P. V. Anderson, R. J. Brockmann, and C. R. Miller. Baywood. 90–108.

Discussion and Application

1. *Website structure.* Check the websites for two or more departments in your college or university. Are the parts consistent from department to department? Are the parts arranged in the same order? Are there good reasons for any differences you see? Does the structure of the site reflect the organizational structure ("who we are") or the reasons why users come to the site (for example, to find courses required for the major)?

2. *Patterns of organization in research journals.* Determine conventional patterns of organization in your subject field by consulting a research journal. In a class discussion, compare the patterns of organizing research articles in different subjects, such as psychology and chemistry.

3. Compare several print or online user manuals. Are there any consistent patterns in the manuals? What organizational strategies seem particularly effective? If there are inconsistent patterns, have the document designers had a good rationale for modifying the patterns? How do the structural patterns of the print and online versions compare?

4. *Match of text and visuals, parallelism.* Find the assignment for Chapter 11 ("Use of the Internet by Individuals") at the textbook website (www .ablongman/rude). Compare the structure of information in the text and in the table. How might you reorganize one or the other or both, and why would using the same structure for the visual and verbal presentations be helpful to readers?

5. *Parallelism in paragraphs and sections.* The following paragraphs and tables are from the time and cost analysis section of a proposal for landscaping an office park. (The primary structure is chronological: the two phases.) Analyze the order of ideas and grouping in the paragraphs and tables. For example, consider whether the Phase 1 and Phase 2 sections present information in parallel order. Also consider how the tables group and sequence information. What would be some options for spatial patterns for Table 1? Edit for organization according to the guidelines presented in this chapter.

   [1]The estimated cost of this project is based on a cost of $6 per square foot for the patio and $4 per square foot for the sidewalks. [2]The first phase includes the upper patio, lower walks, planters, and stairway. [3]The cost for the patio is higher because it is made of bricks. [4]There are 80 linear feet of planters at $10 per square foot and 4,600 square feet of lower sidewalk at $4 per square foot. (See Table 1.)

   [5]The estimated time of completion of phase one is four to six weeks depending on the weather.

TABLE 1 Phase One Cost Analysis

	Sq. Feet	Cost
Upper level	6,400	$38,000
Planters	80	$800
Lower walk	4,600	$18,400
Stairway	300	$3,000
Total	11,380	$60,200

TABLE 2 Phase Two Cost Analysis

	Sq. Feet	Cost
Sidewalk 1	1,926	$7,680
Sidewalk 2	2,400	$9,600
Sidewalk 3	1,280	$5,120
Sidewalk 4	1,280	$5,120
Sidewalk 5	1,760	$7,040
Sidewalk 6	640	$2,560
Sidewalk 7	2,000	$20,000
Total	11,286	$57,120

[6]The second phase of construction will consist of the installation of six concrete sidewalks and the reconstruction of a blacktop sidewalk into concrete.

[7]The estimated time of completion of phase two is two to three weeks depending on the weather.

[8]The total minimum cost for the proposed changes is $117,720.

6. *Grouping by user needs.* A six-page guide for users of a major city library is organized alphabetically. Topics are listed on the next page. Each topic is followed by one to four sentences explaining location or use. Discuss the merits and limitations of alphabetical organization for this document if the readers are first-time users of the library. How well will the alphabetical organization work if the readers are experienced users? What other patterns of organization might work for this document?

   Arrange the topics in another order based on the needs and interests of first-time users. What are their questions? What information may they need to reference more than once? (If they have to refer to certain facts, one way to make that information easy to find is to place it early.)

   Does the reorganization suggest any revisions in content? In other words, should any topics be added or deleted? For your information, the comment for the "personal property" topic advises patrons not to leave their property unattended.

acquisitions	gifts/exchange	online catalog workstations
arrangement	government	oversized books
of the library	documents	periodicals
bindery	hours of the main library	personal property
book returns	information	personnel office
borrowing books	interlibrary loan	rare books
call numbers	Internet access	reserve materials
cataloging	library cards	restrooms
change machine	loan periods	sorting areas
circulation	lost and found	stacks
computers	lounge areas	stairs
copy center	magazines (see	study areas
current periodicals	periodicals)	technical processing
director's office	materials processing	telephone directories
elevators	meeting rooms	telephones
fines	microforms	water fountains
food and drink	newspapers	

7. *Grouping, FAQs.* Find a website with a list of twenty or more FAQs. Software and hardware companies, suppliers of services (such as downloading books or songs), and universities are among those that try to solve user problems with FAQs. How are the FAQs organized? How do users find what they want? How do the FAQs work with search functions? Share your observations with class members. What principles about organizing FAQs might you define on the basis of the sites the class has examined?

8. *Cohesion: your own writing.* Examine several paragraphs from your own writing for paragraph cohesion. Mark the verbal links from sentence to sentence. Take special note of how important words in predicates become subjects of following sentences. Underline repetitions or variations of key terms. How could you edit your own writing to increase cohesion?

18 Visual Design

In Chapter 17, we compared the work of an editor organizing information to the work of an architect organizing spaces in a building. Both professionals aim to help users comprehend and use information. The architect's work is similar to the editor's in another way: the architect employs visual signs about the building and its structure. These signs may be explicit identifiers, like building directories or numbers on office doors. But the building's visual features, such as exterior materials (brick, wood, glass), overall shape (tower, one-story), and landscaping, are also signs of the building's uses. These features help people to identify the building (a factory does not look like an office building). The visual design affects recognition and usability.

Editors can signal structure and give a document a visual identity with type, headings, columns, color, and a table of contents or on-screen menu. *Visual design* refers to the visual and physical features of the document. Visual design, like organization and style, is functional, not just aesthetic: it points readers to information and helps them interpret the information.

This chapter defines terms used in discussing visual design—*document design, graphic design, layout, format, genre,* and *template*—and identifies visual design options. It then discusses functions of visual design as bases for editorial choices. It includes a section on headings that pertains to several design functions. A case study illustrates the way visual design influences comprehension. The chapter concludes with guidelines for design choices.

Key Concepts

Visual design aids comprehension and usability and motivates readers. Design is always subordinate to meaning and use; that is, editors do not use design to decorate a page but to enable a reader to understand and use the document. Desktop and online publishing offer multiple design options but also require responsible and informed choice of options. More is not necessarily better, and simpler is usually safer. Editors must make judgments about design based on their knowledge of the document as it will be used.

Definitions of Terms Related to Visual Design

Visual design is a component of *document design,* a broad concept that refers to all the choices a designer makes about a document, including content, organization, and style. Like a designer in any other field, a document designer (who may be the editor) considers uses of the item being designed and makes choices to enable those uses. For technical documents these uses include reading to get information, to understand, to learn, to complete tasks, or to make decisions. Design choices are good if they enable these functions. Visual design is an important component of document design because it indicates the structure of the information, provides guidelines for locating information, and invites or discourages reading.

Like document design, *information design* is broader than visual design. Information design addresses complexity and volume of information. Visual design is one way of helping readers sort information, but information design also implies storage and retrieval of content in databases and content management systems.

Document design, information design, and visual design are related to *graphic design,* but whereas visual design is concerned with how the presentation enhances function, graphic design is concerned with specifics of production and with aesthetics. The document designer considering visual presentation decides to use running heads to help readers find particular sections in the document (a function), whereas the graphic designer determines the typeface, type size, and space from the top margin for the running head, based in part on aesthetics (what looks good). The document designer plans two columns of text, but the graphic designer decides how long the lines of type are and how much space is between columns. With desktop and online publishing, the document designer or editor may also do the graphic design.

Layout is the graphic designer's plan for arrangement of text and visuals on a page and includes specifications for type and spacing. The layout may include a grid to show line length, placement of headings and visuals, and justification. The grid includes a sample for any format choices, such as a list, that the writer or editor has made. It organizes space and creates visual consistency. Figure 18.1 shows a designer's grid for the layout of a page in this book.

Visual design is similar to *format,* a term which emphasizes the result of the choices about placement of information on the page or screen rather than the process of design. Because format is conventional for some types of documents (genres), such as letters and memos, format can visually identify the genre, and the terms are sometimes confused. But some formats apply across genres: a report may look like a proposal or a memo—the genres may share a format, though they have different functions. The term *visual design* emphasizes the process and avoids the multiple meanings of *format.*

A *template* is a collection of styles in word processing or page layout software that determine what the type will look like on the page. (A template may also be a list of parts that defines a type of document.) A template is the skeleton of the document, whereas genre and format both imply some content identity as well.

Figure 18.1   Designer's Grid for This Book

The editor participates in the visual design and evaluates whether the choices as implemented work well. This chapter reviews principles of design so that editors can make informed choices.

Visual Design Options

Writers and editors have some choices about design, just as they have choices about writing style. The following list introduces some bases for choosing among the various options and indicates some of the ways in which the intended use of a document affects design choices.

Page Layout

- **Page orientation.** "Portrait" orientation is standard for print. Books that are taller than they are wide fit on most bookshelves. However, the wide, or "landscape," orientation may better suit the text or its storage requirements. Pages with the landscape orientation have multiple columns or wide margins so that the lines will not be too long for easy reading. Shifts in orientation, such as printing tables landscape in a portrait text, distract from reading or performance because they require readers to flip the document around.

Many monitors are wider than they are tall. Designers of online documents section the space to enable quick reading.

- **Line length and number of columns.** Lines that are too long or too short slow down reading. The maximum line length for easy reading in print is 2 ½ alphabets or two times the point size. This means a limit of 65 characters per line across the page, whatever the type size. It also means that a line 20 picas long is about right for 10-point type, whereas a line of 24 picas is appropriate for 12-point type. ("Point" and "pica" are printers' measures, defined in more detail in Chapter 23.) A large page or wide screen may require large margins or multiple columns to keep the line length right. Newspapers and magazines use multiple narrow columns to facilitate quick reading down the page rather than line by line across the page. Many good websites section their pages into columns of equal or unequal width because lines that span the monitor width are too long to read easily.

- **Indention.** Indention signals a new paragraph in print. A hanging indent, where a heading protrudes into the left margin, can distinguish a major heading from a minor heading. Indention of a long quotation (block style) visually signals a quotation, just as quotation marks signal a shorter quotation within a paragraph.

- **Alignment.** Text may be aligned on the left margin (left justified), on the right margin (right justified), on both margins, or centered. Left alignment or justification provides a common point for the eye to return to in left-to-right reading. Centering and right justification can be used for headings but not for large blocks of text.

- **Binding, folding, size.** Printed documents may be bound or folded, depending on use, storage, and economics. For example, a brochure that will be mailed may need to fit into an envelope. A spiral binding lets a manual rest open on a desk while it is used, but it lacks durability, and it does not permit a printed title on the spine.

Type

- **Font.** Different fonts have different personalities and different degrees of legibility. Amateur designers of print should probably stick with commonly used fonts, such as Times Roman and Helvetica, in standard sizes (10- or 12-point). Verdana has been designed specifically for the Web.

- **Variation in type style.** A change in type invites attention. Shifts from roman to italic type, from lowercase to capital letters, from the primary typeface to boldface, or from a serif to sans serif typeface all attract notice that can be useful to point to safety issues or key points. Too many shifts create a busy look that readers shun.

- **Underlining.** Like changes in type, underlining emphasizes the material, but it reduces legibility. The line itself interferes with perception of the descenders on letters (the tails below the line), information a reader uses in recognizing the word. Boldface, italics, and variations in type size are better options. Underlining in online documents not only reduces legibility but may also be confused with hypertext links.

Display of Information

- **Number and length of paragraphs.** Short paragraphs are quicker to read, but longer paragraphs permit the development of complex thoughts.
- **Prose or visual representations.** Paragraphs are best for explanation and narrative. When the subject is highly complex or when the reading is selective, a graphic display of text may aid in comprehension and selective reading. For example, flowcharts visually show the sequence of steps, and outlines show the hierarchy of ideas. Tables are appropriate when the readers must locate specific items of information.

  For an international audience or for subject matter about two- or three-dimensional objects, line drawings or photographs may be preferable to words. Visuals lessen the need for translation, and they show spatial relationships readily.
- **Lists.** When there are parallel ideas or a clear sequence of items, a list will reveal the structure more quickly than will embedding the items within paragraphs. But too many lists can create the sense of reading an outline rather than the finished text. Grouping the items within the list by related topics helps readers interpret the meaning of the information.
- **Numbers, letters, bullets.** Numbers, letters, and bullets distinguish items in a list. Numbers and letters may be used in lists embedded in paragraphs or in offset lists. They imply that the order of items is significant. Numbers can show different levels in the structure, as in the decimal system of numbering where "2.3.3" would mean that the part is the third item within main part 2, subpart 3. Numbers and letters are both useful as cross-references, as when a step in instructions refers the reader back to "step 2." Bullets (dark or open circles or squares or other shapes preceding an item) are used with offset lists in which the sequence of information does not matter. If itemization alone is the goal, bullets are sufficient and do not suggest meaning they don't have.
- **Boxes, white space, shading, pointers, symbols.** A box around a warning, white space surrounding type, and shading all draw the reader's eye. Pointers are arrows or hands with pointed fingers that direct a reader's attention to specific information. Certain symbols, such as a skull and crossbones, have developed universal meanings as signals of danger and can call attention to warnings or substitute for verbal warnings.
- **Color.** High contrast, as achieved with black type on a white or yellow background, makes type easy to read. Reversing the pattern (light type on a dark background) can attract attention but at some sacrifice of readability. Colors have specific cultural meanings. Nancy Hoft (1995) reports, for example, that green symbolizes "environmentally sound" in many countries

but "disease" in countries with dense jungles (p. 267). The international warning symbols use color in consistent ways with red meaning danger to life, orange warning of risk of serious injury, and yellow urging caution. Color adds to the cost of print documents, but gains in comprehension, usability, and appeal may warrant the cost.

Structural Signals and Navigation

- **Headings.** Headings identify key points, serve as transitions, show the overall structure of the document, and identify specific sections for selective reading.
- **Menu bars.** In online presentation, the arrangement of the categories of information in a menu bar enables readers to find the information they want within the document and shows the structure of the document.

Functions of Visual Design

Like design for buildings, interiors, or cars, design for technical documents is functional. Design enables people to use spaces or equipment. Document designers make choices with type, page layout, paper, and information so that people can use and understand a document. The three main functions of visual design—comprehension, usability, and motivation—establish principles for editing.

Comprehension

As you learned in Chapter 17, comprehension depends in part on perception of the structure of the information. An organized document is easier to understand than a disorganized one, and the structure of the document must match the structure of the information. Page or screen design and headings aid comprehension because they reveal visually the structure of the document and, in turn, the information. Headings offer an outline of the information—the hierarchical, sequential, and other relationships that readers use to interpret what they read. Paragraphs signal new topics. These visual signals complement verbal ones, such as forecasting statements in introductions, that explain how a chapter or section will develop.

Using design for emphasis indicates what ideas are most important. If you highlight information with boldface type, color, shading, or listing, you alert readers to pay attention. Emphasis should be thoughtful: not every series of three in a sentence should be offset in a list because the information may not be important enough to warrant the attention. Only the most significant warnings should be boxed.

Usability

As Chapter 2 notes, readers are likely to read technical documents selectively. They *use* documents more than they *read* them. They may look for specific points of information, or they may look away from the text to complete a task and then find the place where they were. Good design enhances usability because it points readers to the information they need.

Selective Reading: Finding Points of Interest

The table of contents and index point to topics, but visual signals on the page also help readers find information. Headings within the text and running heads (section or chapter titles—sometimes abbreviated—at the top or bottom of each page), page numbers, and chapter numbers and titles identify topics and their location. Also, boldface or italic type or shading direct readers to material of particular importance.

A usable online document may provide access to specific parts with a menu bar, site map, search utility, index of terms, and links to related material. Like print documents, online documents may use space, color, and variations of type to draw a reader's attention to what is most important.

To ensure usability, you imagine the probable methods a reader will use to search a text and assess the adequacy of access devices for those methods. You also check the accuracy and frequency of the headings and the wording in the table of contents against the headings in the text. In an online document you check the links for accuracy and completeness.

Page and Screen Design: Finding Information in Predictable Places

Consistent placement on the page aids readers. For example, North American business letters include the return address and date in the heading at the top of the letter, enabling readers to respond. Writers do have the option of aligning all the parts of the letter on the left margin (full-block format) or aligning the heading, closing, and signature on a second margin just right of center (semi-block format). But both options are consistent in the placement of the major parts. The heading will always be at the top, and the signature will always be at the bottom. Readers can depend on finding the information they need to respond. Business letter format varies in different countries, and good editors, always concerned about readers, learn the formats that will be familiar to their readers.

Conventions of screen design are still in flux. Menus are sometimes at the top, sometimes on the left, and sometimes on the right. Illustrations may be almost anywhere on the screen. Users find that a search box at the top of the screen is convenient. The design should be consistent from screen to screen so that users can "learn" the structure of the information at the site. If a website is related to print documents, some visual consistency between the two media will provide product identity and a familiar interface for users.

Frequent changes of design place the visual identity of a document at risk, can disorient readers, and are expensive in terms of decision time and redefinition of publication templates. It is rarely appropriate to change a design of a technical document simply for variety. But if design is inadequate for comprehension, usability, motivation, or use, it should change. Design changes may be legitimate just to freshen the look of the document. Even when function overrides aesthetics, the document can still be attractive.

Ongoing Use

If a print document is to be used more than once, editors make decisions about size, shape, material, and binding by imagining the document in use or storage: fitting into an envelope, hanging on a wall, stored in a file folder, staying open on a desk, carried in a pocket, standing on a bookshelf, or attached to a piece of equipment, as with a safety warning on a portable electric tool. If the document will be used outside or carried around, lamination can protect it from weather and from torn edges. If it will be used in low light, large type size and white space compensate for the low visibility. Spiral binding will be a good choice for a manual that remains open but unsatisfactory for a book stored on a shelf and identified by its spine.

An explicit title in the HTML or XML code of an online document will identify the document in a user's list of bookmarks.

Motivation

Visual design creates the first impression of a document and invites or discourages reading. But what is motivating in design depends in part on readers. Some readers are highly motivated and will read no matter what the design (assuming that the content meets their expectations). Professional journals use limited formatting and small print because the readers must keep up with developments in their profession, and content is more important than design. (These journals also have small budgets because advertising revenues and circulation are limited.) The austere design also identifies the professional nature of the journal. Popular magazines, by contrast, compete with other magazines for a reader's discretionary time. Their elaborate and colorful designs attract attention and interest. Likewise, a website that aims to sell a product is probably flashier than a website that offers health information.

A document's design may discourage readers. Users of equipment who can't find the instructions they need may abandon the manual and simply experiment with their machines or procedures. A user who gets impatient waiting for the large file of a graphic-intensive website to download may go elsewhere. Managers may skim introductions and conclusions without reading the whole report. Such experimentation may be satisfactory and even desirable; an editor should not feel a moral obligation to make readers absorb every word. However, unmotivated readers may overlook necessary information. Users may discover just a small part of a machine's capacity or may fail at some procedure. And if they fail or get stuck, they may need a lot of costly time and effort from the customer support department, or they may take their business elsewhere.

Design can direct a reader's attention to vital information about health and safety. Highlighting with boldface type, boxes, space, or icons should encourage readers to read at least those sections. Design of a report can draw the attention of a manager to a warning and recommendation for a modification in equipment.

Headings

Headings can aid comprehension and usability. The headings must be effectively worded, the levels must accurately represent the hierarchy and relationship of ideas, and the frequency must match the readers' needs for signals.

Wording

Informative headings provide more information than structural headings. For example, the headings "Equipment Costs," "Installation Costs," and "Purchase Costs" tell more than do "Part 1," "Part 2," and "Part 3." Exceptions are "Introduction" and "Conclusion," which indicate both content and structure. One-word headings generally tell less than do headings containing several words; for example, "Costs" tells less than "Installation Costs," and "Discussion" tells less than "Analysis of Environmental Impact."

Headings phrased as questions ("How Do I Apply?") may appeal more to readers of brochures than headings phrased as statements ("Application Procedures"). Either heading, however, identifies the two key concepts, "application" and "procedures."

Parallel structure in headings provides consistency. Groups of headings within a section may all be phrases, questions, or complete sentences. Steps in a task may be identified in descriptive terms or as commands but not as a mixture of the two.

Heading Levels

Different levels of headings reveal structural levels of document parts—main parts and subordinate parts. A short document, such as a brochure, will probably have only one heading level, but a longer, more complex document, such as this book, may have as many as four levels. Theoretically, it could have even more levels, but a complex structure can get confusing. A report of four to eight pages is not structurally complex enough to support more than one or two levels.

If the headings are to reveal the structure, readers must visually distinguish first-level (main) headings from second- and third-level headings. Centering or hanging indent, space before and after the heading, boldface type, type size, capitalization, color, rules, and shading all create the visual power associated with main headings. Italics are less visually powerful than roman type and therefore used more for minor headings, such as paragraph headings, than for main headings. Paragraph headings run into a paragraph rather than being displayed on a separate line.

Boldface type is more readable than either capitalized or underlined type. Variations in type size can show differences in levels of headings, but readers probably won't recognize a difference of less than four points. (A point is a printer's measure; 72 points equal approximately one inch.) For example, if the second-level heading is set in 10-point type, the main heading will have to be at least 14-point type unless (or even if) there are other distinguishing features such as spacing.

Graphic designers advise left-justified headings rather than centered ones because of the left-to-right reading pattern. The left margin is a common point of orientation for readers from Western countries whether the text is display copy (headings and titles) or body copy. Centering may not be noticed on the headings for the narrow columns of a brochure. At least two lines of text should follow a heading on a page; an unusually large bottom margin is preferable to allowing a heading to appear alone at the bottom of the page.

The rules for outlining generally apply to headings. Just as a roman numeral I implies that at least a II will follow and an A implies at least a B, one heading at any level implies at least a second heading at the same level.

An editor verifies that the headings accurately reflect the structure of the document and that the different levels, if any, can be distinguished. An editor may also advise on the style and placement of headings.

Heading Frequency

If headings are good, are more headings better? Not necessarily. Headings are signals. They should not replace the text, nor should they interfere with the continuity of reading. Too many headings can distract from the content.

There are no rules for the amount of text per heading. Some types of documents, such as brochures, have proportionately more headings than do textbooks or reports. Each chunk of information may be identified by a heading in a brochure. If each paragraph of a book had a heading, however, it would be difficult to read long sections—the book would seem to hiccup.

To judge an appropriate frequency of headings, think about reading patterns and the structural complexity of the document at hand. For a brochure, with small chunks of information meant to be absorbed quickly, headings will predominate on the page. Long reports and textbooks need longer sections in order to develop complex ideas.

Application: Radar Target Classification Program

To illustrate the effect of visual design on comprehension, usability, and motivation, various designs for a single document are shown in Figures 18.2–18.5. The document was part of a program plan for research on the use of radar systems on aircraft or missiles to identify stationary military targets. Such a radar system has to distinguish targets from background clutter (that is, everything else in the scene—trees, grass, buildings).

Target identification begins with a detection algorithm, a computer program that uses a mathematical procedure to search the radar image for bright objects. Then a classification algorithm (the subject of this example) would examine each detected bright object and classify it as a tank, a truck, or neither. The classification algorithm would evaluate the object's size and characteristic features (for example, trucks have wheels, whereas tanks have treads and gun barrels).

An algorithm is robust if it works well under a variety of conditions (wet, dry, snowy). Development of classification algorithms involves devising different approaches, testing how well they work, and then using the test results to plan improvements. A "seeker" is a missile programmed to search for a particular type of target. "MMW" means millimeter-wave, a particular band of radar wavelengths. "SAR" stands for synthetic aperture radar, a type of radar that produces detailed imagery. A "standoff sensor" is a radar or other system on an aircraft designed to view a battlefield from a long, relatively safe distance.

The purpose of the program plan is to persuade a manager that the research is well conceived and will lead to the development of effective algorithms. The manager is expected to have a good technical background but to want to read quickly.

The four versions of the document in Figures 18.2 to 18.5 illustrate how different format choices influence both the speed of reading and the accuracy of comprehension. Figure 18.2, the original version, uses minimal formatting. It is one long prose paragraph, with indention to identify it as a paragraph, plus a heading with numeric label. The sentences in the original were not numbered, but numbers are included here for reference.

The writer indicates document structure verbally. In sentence 4, the transition "For example" invites the reader to look for examples of scenarios. In sentence 5, the transition "Another scenario" signals that the list of scenarios continues. The words "intermediate complexity" recall "simple" in the previous sentence, thus signaling a simple to complex structure in the list. ("Simple" would have made more sense when readers first encountered it if a forecasting statement had been included to establish the order from simple to complex.) Sentence 6 affirms the structure of both the list and the order from simple to complex: "Third" signals that this is the third example, and "more complex" defines this scenario in relation to the first two examples. These verbal signals increase the chances that the reader will share the writer's concept of the scenarios in terms of number and complexity.

In the last sentence, the word "Finally" is somewhat misleading. Because of the strong verbal signals about the three scenarios, a reader might first assume that the word signals the fourth item on the list. However, because the sentence does not describe a fourth scenario, the reader can conclude either that "finally" merely signals the end of the paragraph (which we can see visually) or that it identifies the final goal or task in the research project. The misdirection of the

4.3 CLASSIFICATION ALGORITHM PERFORMANCE AND
   UNDERSTANDING

[1]This section of the program plan describes the goals and objectives of
the stationary target classification effort and outlines the proposed utilization of radar data to achieve these goals and objectives. [2]The most fundamental goal of the target classification effort is to develop and understand
target classifiers which work reliably in a highly variable background clutter environment. [3]Different levels of algorithm complexity would be based
upon various target classification scenarios. [4]For example, one simple scenario is the MMW seeker searching a small acquisition region in search of a
priority (e.g., tank) target. [5]Another scenario of intermediate complexity
might be the MMW seeker having the ability to gather multiple looks at the
target or even a reduced resolution SAR. [6]A third, more complex scenario is
the standoff sensor application which would have the capability of utilizing
high range and cross-range SAR data with a 2-D classification algorithm.
[7]Algorithms for these three typical applications might be considerably different in complexity and performance, but the goal is to develop algorithms,
understand the underlying signal processing, and to characterize their robustness in highly variable clutter environments. [8]Finally, the promise of
new algorithms to be developed in the future, both by universities and industry, will be evaluated as part of the effort in search of a promising solution to this difficult problem.

Figure 18.2 Original Version of a Section from a Program Plan

signal might require readers to reread at least the final sentence, if not others, to
determine how the sentence relates to preceding ones.

Figure 18.3 breaks the one paragraph into three, displays the embedded list,
and provides subheadings for the list items. The visual signals now predominate
over the verbal signals. The space between paragraphs indicates when a new
paragraph begins, making indention redundant.

Because of the boldface subheadings, a reader who is skimming will quickly
see that the concept of complexity is important. The subheads thus elevate the
concept of complexity to a high level in the document structure and make readers
more aware of this concept. To justify such an emphasis, the concept must be important—a matter of judgment based on knowledge of the subject.

The version of the document in Figure 18.4 tries to solve the problem of the
troublesome "Finally" in the last sentence. It uses a list format, but the two levels
of lists show a hierarchy of ideas. Subheadings and boldface are omitted from the
"scenarios" list because they increase the clutter of the page. Verbal information
added to the list introduction establishes the simple-to-complex organization.

4.3 CLASSIFICATION ALGORITHM PERFORMANCE AND
UNDERSTANDING

This section of the program plan describes the goals and objectives of the stationary target classification effort and outlines the proposed utilization of radar data to achieve these goals and objectives.

The most fundamental goal of the target classification effort is to develop and understand target classifiers which work reliably in a highly variable background clutter environment. Different levels of algorithm complexity would be based upon target classification scenarios:

1. **Simple.** The MMW seeker searches a small acquisition region in search of a priority (e.g., tank) target.
2. **Intermediate complexity.** The MMW seeker has the ability to gather multiple looks at the target or even a reduced resolution SAR.
3. **Most complex.** The standoff sensor application would have the capability of utilizing high range and cross-range SAR data with a 2-D classification algorithm.

Algorithms for these three typical applications might be considerably different in complexity and performance, but the goal is to develop algorithms, understand the underlying signal processing, and to characterize their robustness in highly variable clutter environments. Finally, the promise of new algorithms to be developed in the future, both by universities and industry, will be evaluated as part of the effort in search of a promising solution to this difficult problem.

Figure 18.3 List Format with Subheadings

This design elevates the importance of project objectives above the three scenarios. This version of the document will produce a different concept by the reader than the version in Figure 18.3 even though both use lists and contain the same information. The headings give readers an initial concept of the information; that is, they help readers form a concept and pattern of the data into which the verbal details will fit. If this emphasis is wrong—if the main structure is the three scenarios rather than the three goals—the format will mislead readers. Changing their concept later will be difficult because the concept outlined by the visual display will be established. Format that does not match content contributes to misunderstanding.

Is it more important for readers of the radar document to understand the three scenarios or to understand the three goals? The answer will tell you whether the version in Figure 18.4 is preferable to the version in Figure 18.3. The section of the document you have before you is too small a part of the whole for you to answer the question with confidence, though the introductory statement suggests that the section is about goals, and the "Finally" in the last sentence suggests that

4.3   CLASSIFICATION ALGORITHM PERFORMANCE AND
      UNDERSTANDING

This section of the program plan describes the goals and objectives of the stationary target classification effort and outlines the proposed utilization of radar data to achieve these goals and objectives. There are three main goals:

1. **To develop target classifiers which work reliably in a highly variable background clutter environment.** This is the most fundamental goal of the target classification effort. Different levels of algorithm complexity would be based upon three target classification scenarios of increasing complexity:

   a. The MMW seeker searches a small acquisition region in search of a priority (e.g., tank) target.
   b. The MMW seeker has the ability to gather multiple looks at the target or even a reduced resolution SAR.
   c. The standoff sensor application would have the capability of utilizing high range and cross-range SAR data with a 2-D classification algorithm.

2. **To develop and understand algorithms based on target classifications.** Algorithms for these three typical applications might be considerably different in complexity and performance. Subgoals for the understanding of algorithms will be:

   a. To understand the underlying signal processing.
   b. To characterize their robustness in highly variable clutter environments.

3. **To evaluate the promise of new algorithms.** These new algorithms would be developed in the future, both by universities and industry, as part of the effort in search of a promising solution to this difficult problem.

Figure 18.4   List Format with Two Levels; Emphasis on the Goals Rather Than on the Scenarios

the dominant structure consists of the goals. On the other hand, there is so little elaboration of the goals after the first one that an editor may rightly assume that the three scenarios are dominant or that there is one goal with subgoals. Reading closely is the first task of the editor.

In addition to close reading, you have the option of consulting other similar documents. Is the pattern of comparable sections in other program plans to list goals and objectives? Such a pattern would reinforce the decision to list them. Even more important, you can query the writer about the appropriate emphasis, perhaps illustrating two or more possibilities to clarify your query.

4.3   CLASSIFICATION ALGORITHM DEVELOPMENT

Introduction This section describes classification algorithm development as well as the proposed utilization of radar data in this development effort.

The basic purpose of the radar target classification effort is to develop target classification algorithms that work reliably in highly variable clutter environments, and to understand how they work. Different levels of algorithm complexity will be based on different target classification scenarios:

A. One relatively simple scenario would be an MMW seeker searching a small acquisition region for a priority target (e.g., a tank).

B. A scenario of intermediate complexity would involve an MMW seeker that could gather multiple looks at the target and then process them by (1) noncoherently integrating them to reduce radar scintillation or (2) coherently processing them to construct a reduced-cross-range-resolution SAR image.

C. A relatively complex scenario would involve a standoff sensor capable of utilizing high-resolution range and cross-range SAR data in a 2-D classification algorithm.

Algorithms for these three typical applications might be considerably different in complexity and performance; in each case, however, the goal would be to develop the algorithm, understand the underlying signal processing, and characterize the algorithm's robustness in highly variable clutter environments. An additional part of classification algorithm development will be evaluation of new algorithms that will be developed in the future by both universities and industry.

Figure 18.5   Paragraphs Edited for Format and for Style, Organization, and Completeness of Information

Your efforts at increasing comprehensibility through visual design may reveal gaps in the information. In the sample passage, the writer may wish to add details to some of the goals. Alternatively, the writer may collapse two or more goals into one—maybe there aren't really three goals after all. Some questions about word choice also emerge from the close reading that reformatting requires. For example, in the last sentence of Figures 18.2 and 18.3, is it accurate to say that the "promise" of new algorithms will be evaluated? Is the title of the section clear enough?

In Figure 18.5, the document has been edited for style, organization, and completeness of information as well as for visual design. Some of the content changes were based on inquiries to the researcher who wrote the original paragraph. The visual design is simpler than the design in Figure 18.4.

Which format—paragraph or listing—is better? There is no easy answer. Lists are not always preferable to paragraphs or vice versa, nor is the two-level list necessarily preferable to the single-level list. Formats are not good or bad in any absolute sense. Design must enable readers to comprehend or act. Headings are good if they enable readers to comprehend the text accurately, and they are bad if they distract or mislead. A list may help with complex material because it reveals the structure of the information. But it would become tiresome and counterproductive if used to the exclusion of paragraphs in a long book meant to be read with concentration.

If you as editor have questions about the structure of the information—about the number of items or about their relative importance—the reader probably will too. The reader is less likely than you are to probe for the correct meaning and may "learn" the information incorrectly—unless you help by providing the correct visual and verbal signals.

Using Your Knowledge

Design is an inexact task, but guidelines will help you make good choices. Some specific suggestions for visual design follow:

1. **Know the conventions of design.** You can find genre conventions, such as letter format, in textbooks and in style manuals (see Chapter 8). You can check your company's style manual and examples of its previously published documents. If the document is a response to a request for manuscripts, such as a proposal in response to an RFP or a journal article in response to a call for papers, consult the instructions for preparing manuscripts. If you are editing for an international audience, learn the conventions for countries where the document will be read.

2. **Make design decisions early in document planning**. Writers can prepare their drafts according to design specifications, applying software templates (see Chapter 5). That means less work for you as editor. Make decisions knowing production options and the budget. Reserve expensive design options for the most important documents.

3. **Preview design choices in the medium in which the document will be published.** You may do most of your work at the computer, but if a document will be printed, you should examine the designed pages on paper. Some documents will be published in multiple forms. The same design may not work for both forms. If you are editing online documents, consider that different browsers and platforms may result in different displays. Check the display using several browsers and platforms. Keep the design simple to accommodate different browsers and the comparative difficulty of reading from the screen.

4. **Read for meaning before emending visual design.** If you give the wrong visual signals about the structure of information, it will be difficult, if not impossible, for the words to correct that impression. Don't change visual design until you understand content.

5. **Match the level of visual design to the demands of the text.** Paragraphs are fine for short and simple documents and for documents or sections that will likely be read straight through. Documents that are meant to be read quickly and selectively, such as résumés, are highly formatted. Instructions need to differentiate steps in a procedure, perhaps with numbers and step names in boldface. Complex technical information may need to be represented visually in tables or graphs or lists.

6. **Check headings.** Confirm that levels of headings match the information—that is, that level-one headings indicate the main points. Check for parallel structure. Evaluate their frequency: Are there enough that readers can identify main points? Are there so many that they interrupt reading?

7. **Use design to enhance content, not distract from it.** Visual design should be subtle. Readers are distracted from content by unconventional choices or too many signals. The best design is the simplest one that will achieve the goals of aiding comprehension, usability, and motivation. Force yourself to articulate the reason for any emendation. For example, you might insert a heading saying to yourself, "The reader needs to know here that the discussion is complete and the conclusion begins." If you have changed the display merely because "it looks good" or "just for variety" or because you recognized an embedded list in a paragraph, your reasons are fuzzy and your priorities in editing are skewed.

These guidelines only point to choices and reasons for them. As in everything else you do in comprehensive editing, you imagine the document in use by readers, and you balance design ideals against constraints of budgets and time. Always, the need for clarity and accuracy of meaning takes priority over appearance.

Further Reading

Albers, Michael J., and Beth Mazur. 2003. *Content and Complexity: Information Design in Technical Communication.* Erlbaum.

Carliner, Saul. November 2000. Physical, cognitive, and affective: A three-part framework for information design. *Technical Communication* 47.4: 561–576.

Hoft, Nancy L. 1995. *International Technical Communication.* Wiley.

Kostelnick, Charles, and David D. Roberts. 1998. *Designing Visual Language: Strategies for Professional Communicators.* Longman.

Schriver, Karen. 1997. *Dynamics in Document Design.* Wiley.

Williams, Robin. 2004. *The Non-Designer's Design Book: Design and Typographic Principles for the Visual Novice.* 2nd ed. Peachpit Press.

Discussion and Application

1. *Defining terms.* Place the following terms into the appropriate category as indicated by the table below. Two of the terms will fit into two categories.

   proposal, chronological, paragraph, letter, headings, instructions, boldface, bulleted list, cause-effect, grouping

Format	Organization	Genre

2. *Concepts*
   a. Explain these apparently contradictory statements by identifying a situation when each might be true:
      (1) Complex reading material may require long paragraphs.
      (2) Complex reading materials may require short paragraphs with headings and/or lists.
   b. In your own words, explain how visual design affects comprehension. Give some specific examples of how an editor can use visual design to influence what a reader learns from a document.

3. *Design and genre.* Find three different documents, such as a letter, an announcement, and a manual. Identify the different visual design options used for each. Explain what functions each of the options fulfills, considering comprehension, usability, and motivation as well as genre identity.

4. *Journal design.* In the library, locate a professional journal and a trade journal, preferably on the same subject. For example, consult the *Journal of Clinical Psychology* and *Psychology Today*; or *Mechanical Engineering* and *Popular Mechanics*; or the *Journal of Finance* and *Money*. Identify the visual features of both journals, including the document as a whole (size, shape, binding, and paper) and page design (number of columns and line length, use of headings, white space, illustrations, and color). Using the evidence of content, format, and journal preface (if any), describe the readers for the two journals, considering their interest in the subject, probable reading habits, and backgrounds. Then explain how the design decisions relate to the assumptions about readers. Compare and contrast the two journals, and evaluate the design for both on the criteria of comprehension, usability, and motivation. If you were editor of either journal, would you recommend changes? If so, what would be your reasons?

5. *Website design.* Find three different websites and evaluate the design options as you did for the print documents in application 4. As your instructor guides, choose three sites representing the same kind of purpose or three sites with contrasting purposes.

6. *Displaying lists.* The subject of the following two paragraphs is radar. Both paragraphs, like the document in Figure 18.2, have embedded lists. Format the two paragraphs to display the lists. In example 1, you will have to determine which list(s) to display and which to leave in sentence form. Edit for punctuation, spelling, and completeness of information as well as for format. Then, with other class members or in writing, discuss the bases for making a decision about whether the original paragraph format, the list format, or perhaps a third alternative is the best choice for these paragraphs. If you have questions about meaning, formulate queries for the writer.

Example 1

Test planning—Mission planning defines the number and type of missions together with the actual flight plans for the mission to efficiently gather the data. The needed data includes distributed clutter, discretes, and targets in clutter under varying conditions such as: clutter types: meadows, trees, tree lines, desert; target types: both civilian and military, in these clutter environments and in various target configurations; and environmental conditions: wet, dry, snow. The mission planning will reflect inputs solicited from the MMW government and industry communities to make the data base widely useful.

Example 2

Three difficulties exist with 2-D images from a data analysis point of view. 2-D images are typically processed over a narrow angle of rotation (1 degree is typical). 2-D images in radar coordinates are also difficult to compare at different aspect angles because the target orientation is different in each image. The third difficulty is the elevation ambiguity inherent in 2-D radar imagery.

7. *Vocabulary.* These terms and concepts should be familiar enough to you that you can use them to make editorial decisions about visual design: *hanging indent, left justification, informative heading, structural heading, offset list.*

19 Editing Illustrations

Today's world communicates visually. Writers use typography, space, and illustrations to help readers understand content. Some documents are wholly visual, with illustrations substituting for text. The World Wide Web, with its graphic interface, has increased the use and power of visual communication. Users identify options with icons, and they expect pictures of products and procedures to be presented in an appealing way.

Illustrations, just like text, require comprehensive editing. Readers seek information in illustrations and ease of access to the information. Visual information must be organized in a way that shows the overall concept as well as the relationship of parts to the whole. Visual information must be accurate and ethical. The presentation must be free of the distraction of messiness or undue clutter.

This chapter summarizes the uses of illustrations in technical communication, reviews types of illustrations, and discusses the work an editor performs in comprehensive editing. It also examines the process of preparing illustrations for print. An example shows how even handsomely drawn illustrations may be inaccurate and confusing. Chapter 7 covers copyediting of illustrations, and Chapter 12 discusses tables.

Key Concepts

Effective illustrations in technical communication help readers understand and use information. They clarify meaning and represent it accurately. Their form and the organization of parts aid with the interpretation of their content. Graphic elements—emphasis, scale, size, perspective, "data ink"—encourage clarity when used well. Illustrations are integrated with text according to conventions and reader expectations.

What Illustrations Do

In technical communication, illustrations help readers understand and use information. They may also motivate readers, communicate values, and brand the organization.

Help Readers Understand and Use Information

Illustrations help readers interpret information by showing relationships. A bar graph might compare a company's earnings, by quarter, over two years. Such a graph helps a reader form a concept of the company's performance. Line drawings and photographs show spatial relationships—where the parts are in relationship to each other, the whole, and perhaps the setting. Shaping the information visually may even help the writer to understand the information in new ways.

Illustrations may also help users find and use information. For example, you might use a table to find out how much you need to save each month in order to accumulate a certain amount for a down payment on a house purchase. Illustrations in procedures help readers relate the information in the document to the object in their hand or on a screen. If you have ever assembled furniture, you may have relied on the illustrations to understand how the parts fit together.

Illustrations may support verbal text, but they may also stand alone. If you have traveled by plane, you may have seen procedures for emergencies on a card in the seat pocket. They were probably completely or mostly visual, in part so that international travelers can understand them.

The use of illustrations to give information establishes your first editing question: Does the illustration present information accurately and in a way that is easy to understand?

Motivate Readers, Convey Values

Readers like illustrations, and they may pay more attention if illustrations break steady columns of text. A cartoon may relax a reader struggling with intimidating text.

Illustrations communicate more than information, especially when they represent people. A photograph of a person can rouse sympathies, and photographs can dramatize as well as record damage to property and equipment. Illustrations communicate values.

Examples:

- A brochure about services for senior citizens shows an older man in a wheelchair while his younger caregiver stands above him. The purpose of the information is to suggest the independence of the man who uses the services. But the dominant posture of the caregiver unintentionally exaggerates the man's neediness. An editor suggests a new photograph with the two subjects seated at a table on the same level, to suggest that the caregiver's role is consultation and support.
- An advertisement inviting parents to send their children to a preschool that illustrates only blond children suggests that others may be intentionally excluded. Illustrations brand the organization that uses them, conveying its values, whether positive or negative. An editor verifies that diversity is valued and then suggests more inclusive visual representations.

An editor reads illustrations critically and questions messages in them that conflict with stated goals and policies of the organization.

The power of illustrations to convey values establishes your second editing question: Does the illustration represent the organization's values accurately? Are those values consistent with the best values of society as a whole?

Types of Illustrations

Different types of illustrations serve different functions and convey different information. For example, a table provides access to specific data while graphs provide a concept of how pieces of quantitative data relate to one another. The form of a graph offers an instant interpretation of information, whether it emphasizes comparison, trends, or part-to-whole relationships. Table 19.1 summarizes types of illustrations and their uses. As editor, you will need to know what illustrations achieve what purposes in order to edit for form, organization, and content. One of your tasks will be to check the match of form and content. Reviewing the types of illustrations in this table will help you use the correct terms in referring to types of illustrations, a mark of your professional expertise.

Illustrations may be classified according to four broad categories: tabular, graphic, structural, and representational.

- **Tables** place verbal or quantitative items in rows and columns for easy access to particular pieces of information and for comparison.
- **Graphs** display quantitative information for purposes of comparison.
- **Structural illustrations**—flowcharts, schematic diagrams, and maps—emphasize the structure of the data rather than comparisons or representation.
- **Representational illustrations**—drawings and photographs—depict two- and three-dimensional items and show spatial relationships.

Editing Illustrations for Accuracy and Clarity: Content, Organization, and Style

Illustrations share with verbal text the features you know to consider in comprehensive editing: content for specific purposes and users, organization, and form. The right choices result in illustrations that present content clearly and accurately, without apparent distortion. The illustration must make sense. The parts must cohere, and readers must be able to identify and interpret what they see. Editing may result in restructuring of the illustrations or in additions or deletions. As with all comprehensive editing decisions, editors depend on guidelines, judgment, and the ability to imagine readers interacting with a document to improve and not just change the illustrations.

TABLE 19.1	**Types of Illustrations**

Tabular

table

INDUSTRY	1997	1999	2001
Food and kindred products	20	37.2	39.2
Tobacco products	21	1.8	2.5
Textile mill products	22	32.2	27.2

Displays information, quantitative or verbal, in rows and columns. Tables enable location of exact quantities or items.

Graphic

bar graph

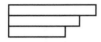

Compares quantitative information in horizontal bars. Bars are preferable to columns (vertical) when the information is linear, as when the comparison is of distance or speed. Identifying labels are easy to read on bars because they are horizontal.

column graph

Columns better than bars show trends over time.

multiple bar/column graph

Compares groups of items on a single point. Emphasizes the value of the individual items.

segmented bar/ column

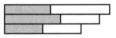

Compares groups of items on a single point. Emphasizes the whole formed by the individual items.

pictograph

Forms bars or columns from images of the objects being compared.

line graph

Plots change over time, temperature, or other independent variable.

multiple line graph

Plots multiple changes over time, temperature, or other independent variable.

pie or circle graph

Shows the relationship of parts to the whole.

TABLE 19.1	*(continued)*

Structural

schematic diagram		Shows the structure of an object but not in a representational way. Used frequently in wiring diagrams and science.
flowchart or organizational chart		Identifies steps in a process (flowchart) or the hierarchy of an organization.
schedule chart	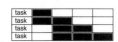	Shows when tasks in a project will be done and their relationship to each other when the tasks are not simply sequential.
map		Shows the spatial relationships of geographic places.

Representational

line drawing		Outlines an object without shading, background, and other realistic details.
exploded drawing		Separates the parts of an object to show how they are connected.
cutaway drawing		Shows the object as though its front had been peeled away lengthwise. Reveals the internal structure.
cross-section		Shows the object as though it had been cut in half crosswise. Reveals the internal structure.
photograph		Represents objects realistically; the least abstract of the visuals. Background clutter may distract from a part to be emphasized.

Content: Appropriateness and Number, Accuracy and Clarity

Editors consider whether and how illustrations may convey information or enable action better than text alone. Because readers tend to pay more attention to illustrations than to text, the inclusion of an illustration announces that this information is important. Determining when to include illustrations depends on the nature of the information as well as on document purposes and reader needs.

Some information is inherently visual and might best be represented visually. Spatial information is an example. Line drawings and maps show spatial relationships. Data is often easier to understand if it is graphed. These generalizations don't mean that every reference to a location requires a map or that every set of numbers requires a graph. You have to make judgments about the importance and complexity of the information.

Budgets and conventions of the document set may influence how many and what type of illustrations a document may include. If other manuals in a series or other newsletter issues are illustrated in one way, the set has established conventions that subsequent documents will respect. Policies about visuals should be established during planning. For example, a writer should find out before having thirty photographs made that the budget allows only for three line drawings.

Illustrations are easy to include in online documents, but the more data (including colors) they include, the bigger their size and the slower they are to download. Readers with slow browsers may not wish to wait for the download. Visually impaired users may be excluded unless alternative verbal text is provided.

Once you have determined that illustrations are appropriate, you need to evaluate whether they represent the content accurately and clearly. These qualities are related to other features of illustrations as discussed in the following pages.

Match of Form, Content, and Purpose

In Chapter 18, you saw that the form of the text, whether in long paragraphs or short lists or columns, could help a reader interpret the information. With illustrations, the form may have even greater power. The form of the illustration (for example, line graph, table, photograph) should reinforce the content. It should also enable readers to use the information in it. This principle relates to the selection of types of illustrations.

A graph shows almost instantly the shape of the data, so a reader's initial interpretation is very fast. The graph thus helps in concept formation. The graph in Figure 19.1, however, confuses readers by conveying a misleading message. Its purpose is to compare three people's scores on a test. However, the line graph conveys the message of movement over time rather than a comparison of values. It also suggests visually that subject A progresses toward subjects B and C.

If the purpose were to compare one person's score at three different periods, the line graph would work fine to show the change. A column graph would work

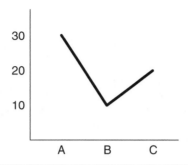

Figure 19.1 Mismatch of Form and Purpose

This line graph misrepresents the data by suggesting some progression from point A to point C. All three scores are one-time scores by different people. A column graph would better compare absolute numbers.

better than a bar graph if the intent is to compare the three scores of three separate subjects. Measuring scores implies "high" and "low" scores. The vertical presentation would suggest that vertical concept.

Organization: Sequential and Spatial

In Chapter 17, you learned that organization influences comprehension. Likewise, the organization of data within an illustration and the sequence of illustrations in a text influence what readers learn from the illustration. Illustrations must make sense in relationship to one another and to the text. The arrangement of parts within an illustration provides interpretive information.

In a sequence of illustrations, the whole should come before the part—a general-to-specific arrangement. Readers need to comprehend the whole before the parts make sense. Readers might not recognize a part of a mechanism if they have not seen the context in which the part exists.

Ordering of details within the illustration shows the relationship of parts and helps readers form concepts. Some standard arrangements are more important to less important, larger to smaller, and first to last. Thus, the bars in a bar graph might be ordered from largest to smallest unless some other information (such as chronology) suggests a different arrangement. You would ask whether a reader most needs to understand size or time.

Style: Discriminatory Language and Good Taste

The guidelines for avoiding discriminatory language and maintaining good taste in writing (see Chapter 16) apply to illustrations. Illustrations can discriminate whether they use words or not. For example, a document illustrating white males in professional positions and females in menial positions shows bias whether the accompanying text is biased or not. Representations of people in line drawings and pictographs ideally suggest respect for diversity of age, race, and gender.

Good taste involves respect for adult readers and the seriousness of document purpose. For example, comics can motivate and reinforce ideas, but lewdness insults. Corny cartoon characters are risky because they can equate readers with children or with incompetent adults. The formality of illustrations should match the formality of the style and page design to maintain a consistent persona and tone. While cartoons may work for casual brochures, they would be detrimental to a bank's annual report printed on heavy, cream-colored paper.

Editing for Graphic Elements

By expertly evaluating the content, form, organization, and style of illustrations, you will contribute substantially to their effectiveness. But you will add value to your editing expertise if you also understand some conventions of display. Graphic elements of illustrations include emphasis, scale, and perspective. Illustrations may be rescaled or resized depending on the desired emphasis. The clutter of extra ink in the form of lines as well as unnecessary information may need to be deleted. Decisions about graphic elements affect the clarity of the illustration.

Emphasis and Detail

Emphasis of details within the illustration helps readers determine what is important. Details may be highlighted in these ways:

- **Size**: Large attracts more attention than does small.
- **Color or shading**: Bright colors and dark shading attract the most attention.
- **Labels**: Important parts are labeled, others are not.
- **Foregrounding**: Important parts appear in the front.
- **White space**: Details surrounded by space attract more attention than do details crowded by others.
- **Shape**: In a matrix of circles, one notices the lone triangle; in a segmented circle, one notices the segment separated from the others.
- **Sequence**: First and last seem more important than middle.
- **Typographic cues:** Boxes and arrows focus attention.

The graph in Figure 19.2 appeared in the brochure of a charitable organization seeking donations. Many donors want to know that their gifts will be used for charitable purposes rather than for administration and base their decisions about giving in part on how efficient the organization is. Anticipating that some donors would question expenditures and being proud of its relatively low administrative expenses (13%), the organization prepared this pie graph—an appropriate form for illustrating how income is distributed.

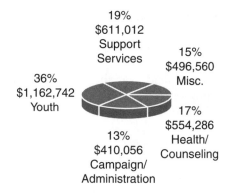

Figure 19.2   Inappropriate Emphasis in a Graph

Inadvertently, instead of emphasizing the organization's efficiency, the graph emphasizes its administrative costs through the placement of this segment at the front and through the three-dimensional display. These methods make the 13% seem visually a larger part of the expenses than the numbers suggest. In fact, the campaign and administration expenses segment seems almost as large as the segment reflecting the amount spent on youth, though the numbers verify that almost three times as much is spent on youth as on campaign and administration. Because the visual representation is more powerful than the numbers, most readers will not use the numbers to reinterpret what the graph has interpreted for them. The graph may interfere with the purpose of convincing potential donors that their dollars will be well spent in this organization.

The document's purposes and readers' needs determine which details should be emphasized. Readers must be able to identify in the illustration any details they will use, such as parts they need to assemble or numbers they need to interpret. Details must be accurate and complete. By contrast, details unrelated to the concept or task diminish the power of visuals. The form can be buried in the clutter, so that readers spend as much time plowing through excessive details as they would reading the information in paragraph form. Thus, as editor, you should simplify or use multiple visuals if the clutter hides the important information. Pie graphs with so many segments that it's hard to identify or compare the segments or multiple line graphs that look like spaghetti do not convey information effectively. Likewise, the detail of a photograph may make it difficult for readers to locate specific parts they need to connect, in which case several photographs, or perhaps a line drawing, should be used.

In evaluating the level of detail, ask what readers need to know and do from the illustration. If it presents more detail than necessary, delete what is unnecessary. If readers need to know all the information but the details intimidate or inhibit quick reading, consider using more than one illustration.

Perspective, Size, and Scale

Perspective, size, and scale give readers useful information and influence the interpretation of the information. Perspective shows depth: the place of an object in a three-dimensional field from the front of the illustration to the back. Whether an object is standing on end or on a horizontal plane is a matter of perspective. Because objects appear smaller in proportion to their distance from the viewer, objects that seem farther away also seem proportionately small. The inappropriate emphasis in Figure 19.2 results partly from choosing a three-dimensional graph, which places the front segment closer to the reader.

Size is a factor in identification. Illustrations should be big enough that they can be read, but rarely so large that they dominate the text and never so large that they seem to shout on the page. Size can also indicate importance. Illustrations should be sized proportionately within a document so that shifts in size indicate shifts in importance. Assuming comparable complexity and importance, one line graph will have essentially the same size boundaries as another line graph, but if one line graph is more important or complex than another, it could be larger. Parts of objects may be drawn proportionately larger than the whole to show detail.

The scale of graphs should reflect numeric values. Distortion results from expansion or condensation of the x or y axis or from improper measurement. If the graph gives the wrong first impression, it will be difficult to correct it later with words or other details because the schema will already be formed in short-term memory.

Bar and column graphs and pie graphs can be scaled quantitatively. Assuming that the bar or column begins at zero, each bar can be measured so that its length equates to its numeric value. Computer graphics programs will calculate the appropriate scales on graphs from the data you provide. However, a three-dimensional pie graph or bar graph distorts the information by introducing volume and perspective. Thus, when you choose a chart type, make accuracy a higher priority than a cool appearance.

Scale on line graphs is difficult to establish because two dimensions can be manipulated. The correct scale is a matter of judgment rather than measurement. The two line graphs in Figure 19.3 present the same data. However, the graph on the left gives the impression of a much greater incline than the graph on the right. The contraction of the x axis increases the incline. Contraction of the y axis can also steepen an incline.

Psychological factors and conventions will influence your judgment about whether a scale is right. For example, if a 20% increase in sales is an astonishing gain for a company, a fairly steep incline will be appropriate. But if a company gains 20% every year, the increase will be correctly represented in a flatter line.

In technical documents, even persuasive ones, the goal is always to provide accurate information. Purposeful distortion of data may be accepted in sales documents, whose readers expect to be sold, but it is not acceptable in technical documents, whose readers expect to be informed.

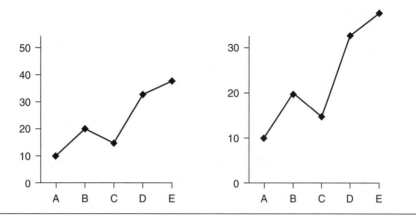

Figure 19.3   Scale in Line Graphs

Expansion of the vertical scale in the graph on the right exaggerates the incline.

Maximizing Data Ink

Edward Tufte (1983) theorizes that most of a graphic's ink should represent the interesting variations of data (see Further Reading). But many visuals are cluttered with "chartjunk." Excessive details and callouts, three-dimensional portrayals of lines and bars in graphs, unnecessary shadows and shading, grids that are too much foregrounded, elaborate legends typed in all caps, lines between rows and columns in tables when space suffices to separate them, and decorations all interfere with the reader's ability to interpret what is important. Illustrations should provide readers with meaning, and readers don't appreciate the distraction of sorting through material they do not need.

Tufte advises maximizing data ink by erasing non-data ink and redundant data ink. He also observes,

> Just as a good editor of prose ruthlessly prunes out unnecessary words, so a designer of statistical graphics should prune out ink that fails to present fresh data-information. Although nothing can replace a good graphical idea applied to an interesting set of numbers, editing and revision are as essential to sound graphical design work as they are to writing. (*The Visual Display*, p. 100)

Tufte focuses on printed visuals, but plenty of comparable junk accompanies graphics and graphic representation on the Web and in other online documents. For example, graphic backgrounds for screens can overpower the information that is on the screens, and animations or flashing banners for the sake only of drawing attention can become irritating if a reader has to stay on a page very long or has to return to it. Editors can advise during planning and revision that "more" does not usually mean "better."

Integrating Text and Illustrations

Illustrations convey information, help readers form concepts, and enable action. Often, though, text is needed for explanations and interpretation of significance. Even in documents that are mostly visual, text is the glue that unifies the document and creates coherence.

Words may identify parts and explain information. Integrating text and illustrations requires some mention of the illustration in the text. Words and visual information should complement each other. The text and illustration should use the same terms and convey the same emphasis. For example, if the text refers to a piece of data or a part of a mechanism, that information should be labeled on the illustration. Likewise, if the illustration names a part, the text should probably discuss it. Placement of the illustrations in relation to the text can also reinforce the parallels between verbal and visual information.

Placement on the Page or Screen

Usually the reference in the text precedes the insertion of the illustration, but other conventions of placement may be followed. For example, in this textbook, the illustrations are usually at the top or bottom of the page.

For instructions, keeping the related text and illustrations close together helps a reader follow the instructions. On a one-column printed page, the text introduces and precedes the illustration. In a two-column format on page or screen, the illustrations should consistently be either to the left or to the right of the instructions. The left-to-right reading pattern in Western countries and the general-to-specific learning pattern suggest that the left column should contain the conceptual information (whether visual or verbal) and the right column the details. The principle of consistency suggests that readers may be confused by a shifting pattern; that is, readers will come to expect all of the illustrations either on the left or on the right. Step numbers or headings may mark both the text and illustration to show that they belong together.

If the illustration depicts component parts, the arrangement of details should reflect the spatial relationships of the item represented. A computer's mouse, for example, would be inappropriately displayed above the monitor. Instead, it should appear to the right of the keyboard where it is used by right-handed people. This functional arrangement overrides aesthetic considerations, such as balance.

Should graphs and tables be placed in the text or in an appendix? The answer depends on whether most readers need the illustrations at the point where they are mentioned or whether the illustrations present supplementary information that only some readers may check. Supplementary illustrations belong in an appendix. If they provide concepts that all readers must know, place them in the text.

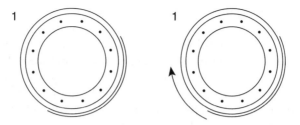

Figure 19.4   Visual and Verbal Information

The visual information in the dial on the left is insufficient to tell readers to mark the appropriate setting on the instruction sheet. The arrow in the dial on the right substitutes for a verb phrase: "turn the dial clockwise."

Nonverbal Instructions

Some equipment includes only visual instructions for installation and use so that the instructions can work for an international audience. Because visual information, like verbal, has limitations, the illustrations alone may not provide enough information. The measure of effectiveness for such documents is the same as the measure for verbal documents: Can readers interpret and use the information?

Some concepts and procedures are easier to represent visually than others. The goal of using nonverbal instructions only may not be possible for some topics. The illustration in Figure 19.4 appeared in the instructions for a machine that makes transparencies for overhead projectors. It is recognizable as a dial in context, and the arrow by the right-hand dial tells the user to turn the dial clockwise. However, the illustration by itself does not convey the entire instruction: a user has to determine, by trial and error, the best setting of the dial for a given purpose and mark the illustration so that subsequent users will know where to set the dial. Because this illustration is the first of four in a sequence, users might interpret that the first step is to set the dial, but then they will be frustrated trying to determine how and where. It is unlikely that they will use the illustration to mark the setting unless they have a verbal prompt.

Application: Cassette Instructions

Figure 19.5 shows a portion of the owner's manual for a cassette player. The manual as a whole is an 8 ½ x 11-inch booklet, with pages stapled down the center. Because readers look for shortcuts, they may try first to make sense of the instructions by looking at the pictures. Thus, the illustrations will have to be self-contained and self-explanatory. These particular instructions tell how to safeguard the cassette against accidental erasure. Editing follows the familiar procedure of analyzing the readers and communication situation, evaluating the illustration, and establishing editing goals.

Readers of these illustrations will perform an unfamiliar task using familiar objects, a screwdriver and cassette tape. Line drawings make sense because readers must recognize the objects and the location of parts on the objects in order to complete the task. These illustrations meet the criteria of appropriateness and match of form, content, and purpose.

Readers must be able to identify the objects and procedures in order to complete the task. The drawings and the type are large enough to read, and the white space around the parts both encourages attention to the illustration and establishes the boundaries of the separate details, the way that space between words establishes the boundaries of the words. The whole mechanism appears first, establishing the context for the specific parts that the cassette owner will manipulate. The cassette and screwdriver are easy to identify. Perspective is clear—the cassette does not stand on its end but rather lies on a horizontal plane, much the way a reader will hold it in order to complete the task. The device of circling indicates parts of the whole. Circling also clarifies that the scale for the illustration of the part is bigger than the scale for the whole. The callouts clearly attach to the part or process they identify.

Chances are your first impression is positive because of quality reproduction and professional use of type and space. However, the positive features of the illustration—appropriateness of the concept, easy recognition of items, use of space, overall attractiveness—may interfere with an evaluation of whether the readers will get accurate, usable information. Comprehensive evaluation of other features—the relationship of text and illustrations, the arrangement of information, and the emphasis and details—reveals some opportunities for improvement.

- **Relationship of text and illustrations.** The text and the illustration should convey the same message and use the same terms. Also, location on the page should indicate the relationship of verbal to visual information.

   Some terminology varies in the text and illustration. The text directs the owner to use a knife to break out the lugs, but the visual illustrates a screwdriver. The text refers to protecting "tracks," but the illustration refers to "sides." The device is called both a "cassette" and a "cartridge." The reference to "recorder" in the last line may misdirect readers. You cannot tell from the selection illustrated here, but the heading that follows the illustration is "Cassette Tape Player/Recorder." Thus, readers may assume that the next section provides the instructions for recording. In fact, the instructions appear three columns back, identified by a buried (indented) heading "Recorder." The potential for confusion could be eliminated by renaming the appropriate section "How to Record."

Safeguard Against Accidental Erasing

Every time a recording is made, the sound previously recorded is
erased. The cassette and the recorder are equipped with a special
device to safeguard valuable recordings from being erased accidentally. On the back of the cassette on either side are two lugs. If
you want to be sure that a recording cannot be accidentally
erased, break out these lugs with a knife. If only one track is to be
protected, break out the lug to the left when the tape is in position
for using that track. When the lug is broken out, the RECORD
switch cannot be depressed.

To record on a cartridge in which the lug has been broken, place a
piece of tape over that area and proceed as in the instructions for
Recorder.

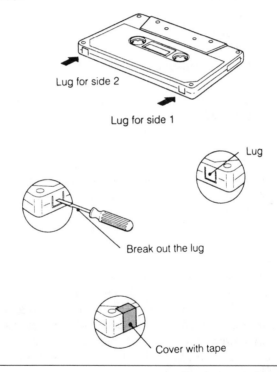

Figure 19.5   Part of an Owner's Manual for Using a Cassette Player

The form suits the content and purpose. Objects are easily identified, sequence clarifies the
whole-part relationship, and circling identifies parts and a shift in scale. Editorial analysis,
however, reveals some potential improvements.

The terms in the drawing must also parallel terms in the unwritten text, the terms readers use in referring to the illustrated items. "Lug" may be technically correct, but more readers will recognize "tab." Similarly, blank tapes typically identify the sides by letters rather than by numbers. Readers, relating the drawing to the objects in their hands, may more readily recognize sides A and B than sides 1 and 2. Terms should be edited to match in the text and the drawing.

■ **Arrangement.** The structure of the illustration is to move from definition of parts to steps for completing the process. The two instructions "Break out the lug" and "Cover with tape" suggest a two-part procedure. However, readers who follow the steps literally will defeat their purpose because the tape simply replaces the tab. Additional words are necessary in the callouts or as headings to indicate that the second step reverses the first rather than completing it.

> **To prevent erasure**
> Break out the tab
>
> **To record after the tab has been removed**
> Cover the opening with tape

Another possibility to show the relationships of steps would be to place the paragraph about recording on a cartridge in which the lug has been broken between the "break out" and "cover" steps. Figure 19.6 depicts this arrangement.

The check for arrangement continues with evaluation of the orientation of the illustrations in relationship to the way the readers will use the objects. A good feature of the illustration is the orientation for the "break out the lug" step. The screwdriver and cassette are positioned in the illustration to emulate the actual position if a right-handed owner performs this step. Thus, the orientation is accurate for about 85% of owners. The illustration of the cassette above, however, shows the opposite orientation. If readers begin literally or mentally with the orientation they see in the first illustration, they will have to reorient the cassette. Perhaps the orientation shift will not confuse readers for this simple procedure, but readers working with complex objects and unfamiliar procedures, such as assembling complex equipment, may be confused by arbitrary shifts. The cassette should be rotated in the first illustration to emulate a reader holding it in his or her left hand.

■ **Emphasis and details.** A check for emphasis and details reveals that the two procedures receive equal emphasis, yet the first procedure is the essential one. Separating these steps visually and with text, as suggested previously, will also solve the emphasis problem by clarifying the separate aims.

The illustration is confusing because owners who wish to safeguard just one side of the tape will have to work to interpret which lug to break out. If side 1 faces up in the illustration, the designation of side 1 and side 2 lugs is correct, but if side 2 faces up, they are backward. The words in the

Safeguard Against Accidental Erasing

Every time a recording is made, the sound previously recorded is erased. The cassette and the recorder are equipped with a special device to safeguard valuable recordings from being erased accidentally. On the back of the cassette on either side are two lugs. If you want to be sure that a recording cannot be accidentally erased, break out these lugs with a knife. If only one track is to be protected, break out the lug to the left when the tape is in position for using that track. When the lug is broken out, the RECORD switch cannot be depressed.

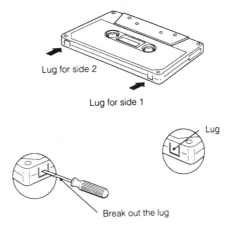

To record on a cartridge in which the lug has been broken, place a piece of tape over that area and proceed as in the instructions for Recorder.

Figure 19.6   Rearrangement of Text and Illustrations to Reflect the User Point of View

Separation of the two procedures with text helps to clarify that the procedures are not steps in sequence.

text help, but they may not be read: "Break out the lug to the left when the tape is in position for using that track." Not all users know when the tape is in position for using the track, especially for machines in which insertion is horizontal rather than vertical. Furthermore, the information about playing position is unnecessary (and therefore gives too much detail).

Finally, the sides of tapes are conventionally labeled A and B, not 1 and 2. Marking the cassette in the illustration "side A" will simplify the interpretation of which tab protects which side. The labeling will match that of the product in the user's hand. The instructions could read: "To protect side A, break out the tab to the left as side A faces you with the label at the top. Break out the opposite tab for side B."

In the illustration, the tabs are inaccurately attached to the cassette. The U shape indicates three open sides. On three major brands of cassette tapes, the tabs are attached on the opposite side from those in the illustration. Top and bottom may matter little in this situation, but they can matter quite a lot in other illustrations. Imagine an illustration showing the "on" and "off" positions for a switch in reverse position.

Figure 19.7 shows the edited illustration and text. The text has been simplified to follow principles of good style and to work better with the drawing. The reference to the "special device" is deleted because it is unnecessary and prompts readers to locate something that is never referred to again.

The original illustration has a number of positive features, and because it looks good, editors may assume that it is good. But an editor does not stop with first impressions. The discovery of the problems is a chainlike process that begins with imagining a reader using the illustration. When some obvious problems emerge, an editor looks further, guided by the goals of accuracy, clarity, consistency, and patterning instructions according to patterns of learning and use. The check of the mechanism itself results from the confusion over which tab to break out for a given side of the cassette.

Does the editing really matter? Owners can probably determine by trial and error which tab to break to protect a given side, and common sense may tell them that taping a hole just created is counterproductive. However, good instructions should prevent the need for trials by ensuring that owners do the job right on the first attempt. Otherwise, we'd just leave owners to their own devices to begin with and skip the manual. A few editorial emendations can save interpretation time and prevent errors. The gains from emendations such as those made here would be proportionately greater with a more complex procedure. When a procedure is complex, a reader does not so readily correct a simple misstep or make the adjustments in orientation required to do the job correctly.

Preparing Illustrations for Print or Online Display

The task of editing illustrations may include selecting photographs and noting instructions for print or online display as well as deciding about form, content, and labeling. You may also advise about size: reduction, enlargement, and cropping. For online displays, you also need to advise about file size, which affects download time.

Safeguard Against Accidental Erasing

Recording erases the sound previously recorded. To prevent accidental erasure by a new recording, you can break out the tabs on the back of the cassette with a screwdriver. To protect the recording on only one side of the tape, break out the tab to the left when the label for the side faces you and the label is at the top. When the tab is broken out, the RECORD switch cannot be depressed.

To record on a casette after the tab has been broken, place a piece of tape over that area and proceed as in the instructions in the section "How To Record."

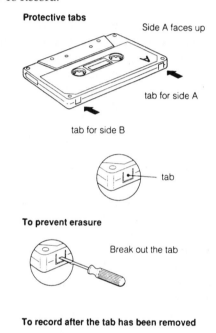

Figure 19.7   Instructions after Comprehensive Editing

Text and illustrations are parallel. The cassette is rotated to correspond to the mechanism in use and to subsequent illustrations. Words clarify which side of the cassette faces up so that the owner can select the correct tab. Headings identify the two procedures as separate steps.

You may recommend reductions and enlargements to fit the illustration in the available space, to indicate emphasis, and to ensure readability. Reduction may improve the quality of photographs and line drawings by de-emphasizing irregular or broken details in the original. If text is part of the image, it may need to be reconstructed to be big enough to read when the rest of the image is reduced. The amount of reduction or enlargement may be expressed as a percentage, but it is easier to communicate the desired dimensions by specifying one of the measurements; for example, "reduce to 18 picas wide." The reduction in height will be proportional. (A pica is a printer's measure, equivalent to about one-sixth of an inch.)

For online display, you might recommend a lower resolution, and thus a smaller file size, so that users will be able to open pages more quickly.

Cropping means cutting off part of the length or width of an illustration, removing extraneous borders or details so that the illustration focuses on the important information. A cropped illustration can subsequently be enlarged or reduced. Figure 19.8 illustrates reduction and cropping. Both procedures reduce the size, but they do so in different ways and with different objectives.

Photographs and shaded drawings, unlike drawings made with black lines, must be converted to halftones before printing in fullscale printing. That is because ink prints in one color rather than in shades. A halftone, made by lasers or by photographing the original through a screen, converts the shadings into dots of various sizes and density. (See an example in Chapter 23, Figure 23.1.) Large dots densely grouped create the image of a dark shade while small dots spaced further apart create a light shade. When the figure is printed, most viewers will not recognize the dot composition.

In printing, illustrations are handled separately from text. Thus, they should be separated from the typescript and marked with the instructions. Because some illustrations require special treatment, unless they are submitted in camera ready form, they should be handled early in the production process so that they will not hold up the text. (See Chapter 23 for more information on preparing illustrations for printing.)

Using Your Knowledge

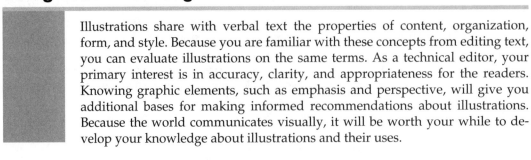

Illustrations share with verbal text the properties of content, organization, form, and style. Because you are familiar with these concepts from editing text, you can evaluate illustrations on the same terms. As a technical editor, your primary interest is in accuracy, clarity, and appropriateness for the readers. Knowing graphic elements, such as emphasis and perspective, will give you additional bases for making informed recommendations about illustrations. Because the world communicates visually, it will be worth your while to develop your knowledge about illustrations and their uses.

Figure 19.8   Cropping and Reducing Illustrations
Top, the original; *lower left*, cropped and enlarged; *lower right*, reduced.

Further Reading

Harris, Robert L. 2000. *Information Graphics: A Comprehensive Illustrated Reference.* Oxford.

Kostelnick, Charles, and David D. Roberts. 1999. *Designing Visual Language: Strategies for Professional Communicators.* Longman.

Tufte, Edward R. 1983. *The Visual Display of Quantitative Information.* Graphics Press.

Tufte, Edward R. 1997. *Visual Explanations: Images and Quantities, Evidence and Narrative.* Graphics Press.

Discussion and Application

1. The two columns of information in Figure 19.9 are part of a brochure for members of a professional society. It is a simple kind of annual report. Evaluate the text and graphs. Begin by analyzing the purpose of the information: What should readers know, and what should their attitude toward the organization be after reading this information? Evaluate the graphs according to the criteria identified in this chapter. Also consider the guidelines for copyediting illustrations in Chapter 7. Suggest editing goals.

2. The tax rate schedule in Figure 19.10 comes from the IRS website (www.irs.gov/formspubs/). Identify ways in which the table conveys information and enables action. Why is the tabular form appropriate? Can you imagine any ways to improve the table?

3. The three circle graphs in Figure 19.11 illustrate the proportion of women in the workforce who achieve managerial positions. The purpose of the graphs is to reinforce the text, which claims that women do not rise above the "glass ceiling" in proportion to their numbers in the workforce. They should enable readers to interpret the text or even to substitute for it. Evaluate the graphs according to match of form, content, and purpose. Recommend revisions.

4. *Vocabulary.* To test your understanding of key concepts:

   a. Distinguish these four types of illustrations and give an example of each: *tabular, graphic, structural,* and *representational.*

   b. Distinguish *figures* from *tables.*

   c. Distinguish *illustrations* from *graphics* as category words.

   d. Explain *sequential* and *spatial* arrangement.

   e. Define *emphasis, perspective, halftone,* and *cropping.*

OPERATIONS

The Society promotes a wide variety of activities under the auspices of some 67 committees. The work of these committees ranges from developing professional standards and Society goals to planning for future directions of the Society, conference planning, membership analysis and needs, and publications. Each of these activities requires financial support, some of which is donated by individual members or corporate sponsors. The bulk of the costs, however, is supported by the Society. The chart below shows how our funds are spent. Other expenses—particularly dues, rebates to chapters, and office operations—account for the balance. Note that our costs exceed the income received from membership dues; dues received pay for 87.4 percent of our expenses.

EXPENSES

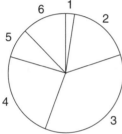

1	Miscellaneous	2.6%
2	Member Dues Refund to Chapters	17.6%
3	Office	35.7%
4	Publications	23.6%
5	Committees	8%
6	Awards & Grants	12.5%
	Total	100%

Income to the Society comes primarily from four sources: dues, conferences, sale of publications and interest from investments. Income varies somewhat from year to year depending on the number of members, the success of the publications sales program, and the financial success of our annual conference. The Society has been fortunate to have had two well-managed and financially successful conferences back to back—Pittsburgh and Boston. The success of these conferences has significantly improved the net worth of the Society.

INCOME

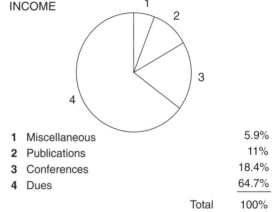

1	Miscellaneous	5.9%
2	Publications	11%
3	Conferences	18.4%
4	Dues	64.7%
	Total	100%

A widely-accepted standard within the professional society community calls for reserves (net worth) of about three years operating expenses. During the past five years our net worth has improved from $163,600 to $399,071, an increase of $235,471. Since 1980, the increase has been $183,171. The last two conferences, with other factors, have significantly contributed to our present position of 1.3 years of reserve for our more than $300,000 annual budget. We cannot rely on the continual financial success of the conferences. Therefore, the Board has taken under consideration several options to improve our financial base. Increased membership is one of these options as is an increased technical exhibit at the conference. We have made significant strides and are taking the necessary actions to assure our continued financial stability.

Figure 19.9   Part of a Brochure for Members of a Professional Society

2005 Tax Rate Schedules

Note: *These tax rate schedules are provided so that you can compute your estimated tax for 2005. To compute your actual income tax, please see the instructions for 2005 Form 1040, 1040A, or 1040EZ as appropriate.*

Schedule X — Single

If taxable income is over—	But not over—	The tax is:
$0	$7,300	10% of the amount over $0
$7,300	$29,700	$730 plus 15% of the amount over 7,300
$29,700	$71,950	$4,090.00 plus 25% of the amount over 29,700
$71,950	$150,150	$14,652.50 plus 28% of the amount over 71,950
$150,150	$326,450	$36,548.50 plus 33% of the amount over 150,150
$326,450	no limit	$94,727.50 plus 35% of the amount over 326,450

Figure 19.10   Tax Rate Schedule, Internal Revenue Service

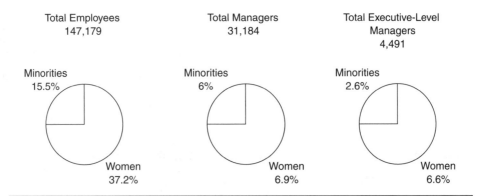

Figure 19.11   Graphs That Attempt to Compare Proportions of Women in Managerial Positions with Women in the Workforce

20 Editing for Global Contexts

By Bruce Maylath

With growth in world trade, editors quickly find that their documents must travel outside their region, country, and native language. This chapter considers the issues and tasks editors face in the globalization of the workplace, including the rhetorical expectations of different cultures, processes of globalization and localization, and issues in translation. A caveat is in order: although this chapter examines characteristics and traits of several cultures, editors must keep in mind that members of any culture may think and act in ways not typical of their culture. Sometimes their profession, education level, age, gender, socioeconomic background, or, most notably, time spent outside their culture may be a more important factor for them than their society's culture in how to read and understand a document.

Key Concepts

The meaning of written communication resides not just in words and their arrangement but also in the cultural and social experiences that readers bring to the text. Part of editing for global contexts, then, is considering the values and expectations of readers from different languages and cultures. Editors also can prepare documents for translation and localization by choosing words and structures to minimize ambiguity.

Preparing Documents for a Global Workplace

Businesses increasingly define themselves globally. Manufacturing and markets extend beyond the national boundaries of the company. Further, global mergers and acquisitions have made correspondence and documentation within single companies multilingual. Perhaps the best example comes from the automotive industry. In the late 1990s, Chrysler and Daimler-Benz merged to form Daimler-Chrysler. Although each company had long sold its vehicles abroad under such brand names as Dodge and Mercedes, suddenly all documents within the merged company's operations needed to be accessible to both its German-speaking and American English–speaking workforces.

This global expansion creates the need for communication through letters, email, faxes, and reports. In addition, manuals for products such as cars or computer hardware and software may require translation for international audiences. Both for business correspondence and translation of manuals, knowing the customs and the culture aids in communication. Furthermore, planning for translation before writing the original facilitates both the procedure and success of translation. Increasingly, translation and localization begin nearly as soon as a document is composed.

International Rhetorical Expectations

Varner and Beamer identify categories of differences in cultures that affect communication: ways of thinking and knowing (linear and logical or dualistic); attitudes toward achieving and activity; attitudes toward nature, time, and death; sense of the self; and social organization. They also distinguish "high context" from "low context" cultures, with a high context culture depending more on contextual information than on words to communicate meaning. These differences become apparent in various forms of communication, such as the business letter.

Whereas linear, activity-oriented Westerners favor a direct style of correspondence with the emphasis on the business transaction, people in Asian countries may be indirect about the transaction and try to establish a relationship with the correspondent. The first paragraph of a Western business letter conventionally states the business directly and purposefully, whereas a business letter from an Asian country may begin with a reference to the seasons and conclude with good wishes for the recipient's family. The Western style might be described as relatively hard and direct, whereas the Eastern style is more poetic. Miscommunication occurs when the Eastern reader is offended by the apparent indifference of the Western writer to the setting and the person or when the Western reader is frustrated by the apparent vagueness of the Eastern writer.

However, Western readers are sometimes offended by the even more direct style of such cultures as the Russian scientific and engineering community. For instance, a North American engineer writing a report on flaws evident in a project might first point out what has gone well. By contrast, a Russian engineer would more likely focus only on the flaws—and in a tone that would sound harsh to North American ears. (See the article by Artemeva in Further Reading.)

Various cultures have different expectations for formality, such as whether it is appropriate to address the recipient by first name. Languages place different meanings on words and colors. These cultural differences are displayed in the organization and format of the business letter as well as in the style.

If your work as editor requires you to review correspondence and reports for international readers, you will need to look beyond grammar and style handbooks to edit well. Learning about the customs of the culture will help you advise writers about communicating with sensitivity and with awareness of how readers in different settings may respond. The books on international communication listed at the end of this chapter provide a starting point.

Globalization versus Localization

An editor of an international or multilingual documentation project must decide whether to globalize or localize the documents. Globalization refers to developing products, including their documentation, for international users. If a product is usable in many cultures, it does not require as much adaptation (localization) for individual cultures. Localization refers to adapting material for the local culture. At the very least it means converting numbers and measurements, dates, and spellings to usage in the culture in which the document will be used. Often it means translating whole documents, adapting visuals and colors, or even writing a document afresh for the local culture. Globalization can be appropriate if different readers share at least a passing knowledge of the document's language and if that language is controlled to minimize misinterpretation. Localization is usually preferred in all other cases.

Globalization

Globalization, also known as internationalization (Cronin, p. 13), involves making a document or product usable around the globe. It is a one-size-fits-all approach. In terms of language, globalization means that a single language is used, to be understood by all—a *lingua franca*. Increasingly, that lingua franca is English, albeit with reservations on the part of many non-native English speakers (Phillipson). Although English does not have the most native speakers in the world—Chinese and Spanish have far more—it is the most widely spoken and most widely learned as a second or third language, thanks largely to the spread of the former British Empire and the later dominance of U.S. military and business power and cultural influence. The popularity of the World Wide Web, with its American roots, has made the English language particularly commonplace for people with education across the globe. Nevertheless, web pages in other languages are quickly catching up to and even surpassing the number in English. A document circulated among countries in the Arab League will be written in Standard Modern Arabic as the lingua franca. However, because Arabic dialects have grown different from each other, a document written just for, say, Morocco or Syria may appear in those nations' dialects, which can be incomprehensible to other Arabs.

Companies can sometimes use widespread knowledge of English or other lingua francas to their advantage, provided the message is simple or the terms used in the message are controlled. It is much faster than traditional translation, since accurate human translation is painstaking. Likewise, its accuracy is easier to verify. Machine translation, while quick, is often inaccurate. Unless the person inputting the original, or source language, text is also a translator of the so-called target language, one cannot tell with certainty where the machine has translated inaccurately. Keeping the text in the source language guarantees that the original authors and editors can check it and update it with relative ease. Finally, writing in a single language is often far cheaper than paying for multiple translations and localization for each language, nation, or region. However, companies that rely on globalization must instead devote some funds to teaching their employees worldwide a minimal competency in the lingua franca and to managing terms consistently. More and more, terminology management requires an investment in computer hardware and software and the training of employees to operate them.

Globalization begins at the planning and concept stage and is not something added on at the end. The development team should include people who know the target cultures. Planning for international users is audience analysis on a global scale. This analysis considers not just the tasks that users will perform or concepts they will learn but also their styles of learning, available equipment, and cultural values and customs. Because multiple cultures expand the number of variables, globalization aims to reduce the amount of culture-specific bias in products and documentation. It limits the need for translation and localization. Where globalization reaches its limits, translation and localization take over.

Terminology Management and Controlled Language

Terminology management means knowing which terms have been used before and using the same terms consistently. *Controlled language* refers to assigning a single definition to a term. Although synonyms are often seen by native speakers as beneficial, for the variety and liveliness they can add to a text, synonyms are a hazard when an editor aims to control the meanings of terms for readers with a limited knowledge of the language used. Synonyms can be powerful in argumentative essays read by an audience immersed in the language all their lives; however, they can be deadly in operating dangerous machinery when the operator does not recognize that the term never seen before refers to an essential lever or switch.

As an example, a simple device that most of us take for granted allows us to connect machinery to a cord and draw electricity. In English, it may be known as an electrical outlet, socket, or receptacle. Some speakers even refer errantly to the device as a plug, when a plug is instead the device that is inserted into the outlet or socket. An editor who leaves the term *outlet* in one section of a manual and *receptacle* in another can leave the reader of English as a second language utterly confused. The situation is made worse by the other meanings each word can carry. *Outlet*, for example, can refer to a place where water or traffic exits. *Receptacle* can refer to a trash bin.

The language tasks for an editor engaged in globalization are to

- create a glossary in which terms are defined,
- catalog where these terms have been used before, and
- permit only one definition per term.

Firms such as Boeing, Caterpillar, Medtronic, Philips, and Xerox have earned a reputation for greatly refining their levels of terminology management. By adopting controlled language as well, they have succeeded—for example, in Boeing's case—at providing frequently updated jet engine manuals that airplane mechanics around the world can read. By teaching the mechanics the essential terms, such as *turbine*, and by allowing only a single meaning for that term, Boeing ensures that jet engines can be maintained and the public can be kept flying worldwide.

Managing terminology and controlling language are continual and sometimes arduous tasks. Fortunately, the work is made much easier now by terminology management software that keeps improving. Though still expensive, such software can be well worth a company's investment and save an editor much grief and time.

International English

An editor who selects English as a global lingua franca must keep in mind that English comes in quite a number of varieties around the world. Although the renowned British linguist David Crystal has demonstrated by examining English-language newspapers around the globe that for the first time in world history there really is a global lingua franca—educated English, in everyday use—the native English can vary considerably in Australia, Canada, England, India, Ireland, New Zealand, Scotland, South Africa, the United States, and Wales. Countries in which English is a dominant second language, such as the former British colonies and protectorates of Nigeria and Malaysia, can vary even more, with their frequent borrowing of vocabulary and grammatical constructions from native languages. U.S. editors should note that where English is taught as a second or third language, it is rarely American English. The exceptions are Mexico and a few other Central and South American nations, because of their proximity to the United States, and the Philippines, a former U.S. colony. Nearly everyone else learns British English. Canadian English bears many similarities to U.S. English, but its users frequently use British spellings and some British vocabulary. Additionally, some terms are peculiarly Canadian. For example, both American and British editors write that autos *merge* when entering a highway, but editors in much of Canada are likely to write *squeeze (left)*. Similarly, what Britons and Americans call a *sofa* or *couch,* many Canadians refer to as a *chesterfield.*

An editor seeking a global standard of English will often opt for British English. This means more than altering such spellings as *-or* for *-our* (color/colour) or *-er* for *-re* (theater/theatre). It also means adopting British grammar and vocabulary.

Surprisingly, the spellings that separate American and British English have nothing to do with pronunciation. Such spellings as *plough* and *plow* are still pronounced the same when read by the same speaker. Where British and American spelling do differ, it is usually because of the influence of America's first dictionary writer, Noah Webster, who wanted to Americanize the newly independent nation's written dialect. Interestingly, his recommendations that the American public eventually accepted are ones that could easily be accepted throughout the English-speaking world (*-ize* instead of *-ise*, for example [apologize/apologise], which some writers in Canada and Britain use also). Although a few spelling differences stand out as unusual—British *kerb* instead of American *curb*, for example—most differences are common and follow recognizable patterns that are easy to learn.

Grammar can differ in ways that many Americans find startling. For example, Britons and Americans follow different rules concerning subject-verb agreement. The British view collective nouns like *family* and *team* as plural; Americans, as singular. Britons say and write "A family have every right to a home" or, referring to the national football (soccer) team, "England win whenever they play Wales." Americans almost invariably would change the verb forms in these sentences to "has" and "wins." However, if one is employing British English in international documents, these third-person-singular *-s* forms are not optional; when used with collective nouns, they are ungrammatical.

Perhaps most challenging to American editors, however, are the vast number of differences in terms. Although many American editors are aware that in British English an elevator is called a *lift* and a semi-trailer truck is called a *lorry*, other terms are likely to take an American editor by surprise. The list in Table 20.1 is a long but by no means complete list of American/British pairs.

Especially troublesome are words that look the same but have different meanings or a word that has the same meaning in some contexts but a different meaning in others. Linguists call the first situation *tautonymy*; the second, *heteronymy*. An example of tautonymy is the word *suspenders*. It is used in both British and American English. In the latter, it means a device for holding up trousers—the British call this over-the-shoulders device *braces*—but in British English *suspenders* is indeed a viable word: it refers to the device for holding up stockings—what the Americans call *garters*. An example of heteronymy is the word *truck*. In both dialects the word can refer to what Americans categorize as a pick-up truck. However, when the vehicle is bigger, Britons categorize it as a *lorry*. Thus, the American meaning behind *truck* is more expansive. A similar example is *pavement*. For Britons, pavement means only what Americans call a *sidewalk*. The American meaning is again more expansive, covering the hard material used to build roads as well as sidewalks. Thus, a technical document for a product, say a toy, that says "intended only for use on pavement" will be interpreted quite differently—and quite hazardously—according to which version of English a reader has learned.

Some editors have suggested using "Mid-Atlantic" English, a hybrid of British and American English. However, no one has established a widely accepted convention for just which terms and spellings and grammar rules should be standard in Mid-Atlantic. A company may decide on its own for its own internal use, but no company yet has had the influence to press such a standard on employees in other companies. However, with the continuing acceleration in global communication and commerce, this situation may change. Intecom, an umbrella organization for technical communication organizations around the world (which automatically includes members of constituent organizations such as the Society for Technical Communication) took a small step in this direction in 2003, when it published *Guidelines for Writing English-Language Technical Documentation for an International Audience* (available at www.intecom.org/guidelines.html under "The INTECOM Language Project").

TABLE 20.1	American/British Pairs of Words
American English	**British English**
(to) act upon (an issue)	table
advice columnist	agony aunt
antenna	aerial
bandage	plaster
billion	milliard or thousand million
bobby pin	hair grip
bouncer	chucker-out
camping trailer	caravan
construction ahead	works ahead
costume/masquerade	fancy dress
counterclockwise	anticlockwise
day-care provider	child minder
dead end	close
diaper	nappy
elevator	lift
escalator	moving stairs
exit	way out
expiration date	expiry date
first floor	ground floor
flashlight	torch
14 pounds (of weight)	1 stone
gasoline	petrol
gelatin dessert	jelly
grade (school evaluation)	mark
grade (year in school)	form
ground beef	mince
hardware store	ironmonger
hood (of car)	bonnet
landslide	landslip
(to) lease	let
outlet (electrical)	socket
parking lot	car park
pedestrian underpass	subway
pennies	pence
photo	snap
pitcher	jug
popsicle	ice lolly
public housing project	council estate
quadrillion	thousand billion

(continued)

TABLE 20.1	*(continued)*
American English	**British English**
quintillion	trillion
(to) rent	hire
rest area, wayside	lay-by
roller coaster	switchback
sausage	banger
scalper	ticket-tout
second floor	first floor
semi-trailer truck	lorry
7-Up (or other lemon-lime drink)	lemonade
shopping cart	trolley
shoulder (of road)	verge (for the parking portion; the portion that falls away into the ditch is the shoulder)
sidewalk	pavement
(to) slate	cancel
snap (on clothing)	gripper
(to) splurge	lash out
stove	cooker
subway (train)	underground or tube
suspenders	braces
sweater	jumper
(to) table (an issue)	set aside
toilet paper	bumph (can also refer to needless documents)
tongs	nips
traffic circle	circus or roundabout
trash	rubbish
trillion	billion
whine	whinge
wrench	spanner
(to) yield (on road)	give way

Using Visual Instructions

One way to globalize a document—and to save translation costs—is to avoid language altogether or to minimize its use in favor of visuals. When the instructions are simple and the languages of potential readers are numerous, pictures may substitute for words. For example, emergency procedures for air travelers are visual rather than verbal. Pictures, however, may be as ambiguous as words, and, like words, they reflect culture. Because some reading patterns are right to left or top to bottom rather than the Western left to right, marking of the sequence of frames by number may be necessary. Colors, symbols, and dress may have different meanings in different cultures. Moreover, pictures do not allow for explanations or alternative courses of action. If visual instructions will work, however, a company may be able to bypass translation. Editors should be called on to edit visual as well as verbal instructions, and a usability test should be mandatory (with users from different cultures) if safety is involved.

Localization

Localization refers to adapting material for the local culture. Even among English-speaking countries, usage varies. In Great Britain, *localization* would usually be spelled *localisation*, with the -*s*- taking the place of the -*z*-, and currency would be expressed in terms of pounds (or euros in Ireland), not dollars. Because Canadian and U.S. dollars have different values, numbers would either have to be converted or the amount qualified according to whether it is U.S. or Canadian. Differences expand when the language changes from English. Simple translation of words does not completely localize the material, although it is a first step.

More extensive localization could require modification of examples and illustrations to reflect customs of dress, interpersonal interaction, attitudes toward gender and age, corporate structure, and ways of thinking.

In Table 20.2, Nancy Hoft offers numerous categories for "cultural editing."

Anticipating these needs to localize, you may avoid as many references as possible to these cultural differences when developing the document.

In Table 20.3, Fred Klein has created a list of keywords as a starting checklist of features to consider when localizing a document.

Translation

Instructions for products distributed internationally must be understood by people using different languages. Readers much prefer reading in their own language to being forced to read a language they have learned in school. Today, nearly all packages sold in Canada, Mexico, and the United States are labeled and include instructions in English, French, and Spanish. By adhering to the language clauses in the North American Free Trade Agreement (NAFTA), companies can easily ship the packages across the borders separating Burlington, Vermont, and Montréal, Québec; or San Diego, California, and Tijuana, Baja California. Sometimes the packages show one language on a side (consecutive translation) rather than all languages on the same side (simultaneous translation). When multilingual packages first appeared on U.S. store shelves, I witnessed the consternation of some customers, who picked up packages set down by other customers with the French or Spanish side turned out. They had not yet realized that the language they preferred was on another side. Customers would sometimes set these packages aside in favor of a competing product nearby with the package clearly labeled in their own language—in this case, English. Imagine

TABLE 20.2	**Categories for Cultural Editing**

- Dates and date formats
- Currency and currency formats
- Number formatting
- Accounting practices
- List separators
- Sorting and collating orders
- Time, time zones, and time formats
- Units of measurement
- Symbols (in English, some symbols are / and &)
- Telephone numbers
- Addresses and address formats
- Historic events
- Acronyms and abbreviations
- Forms of address and titles
- Geographic references
- Technology (electrical outlets, computer keyboards, printer page size capability)
- Legal information (warranties, copyrights, patents, trademarks, health- and safety-related information)
- Page sizes
- Binding methods
- Illustrations of people
- Many everyday items (refrigerators, trash cans, post office boxes)
- Hand gestures
- Clothing
- Architecture
- The relationship of men and women in the workplace
- The role of women in the workplace
- Popular culture
- Management practices
- Languages
- Text directionality
- Humor
- Color
- Communication styles
- Learning styles (the relationship of the instructor and the students)

Source: Nancy L. Hoft. 1995. *International Technical Communication: How to Export Information About High Technology.* Wiley. 129–130. Reprinted with permission of John Wiley & Sons, Inc.

buying a self-assembly piece of furniture that contained instructions only in the Spanish or French that you had studied in high school. If you discovered that a competing product was sold with English instructions, you might quite naturally return the first purchase unassembled and buy the competing product instead.

Perhaps the clearest example of customers' preferring their own language comes from Norway, which has two official languages, both of which are Norwegian. The first, called *bokmål*, is used largely in Norwegian cities, newspapers, and books (thus its literal meaning, "book language"). It bears a strong resemblance to Danish. The second, *nynorsk* ("New Norwegian"), is used mainly in the mountainous areas of the Norwegian countryside, a few novels, and some television and radio programs. It is closer to Old Norse and bears some similarities to Icelandic. Bokmål and nynorsk are close enough that one can usually get the gist of the other, even without much training. All school districts in Norway choose one language as the primary language of instruction but must teach the other as a secondary language as well.

Even though all Norwegians can understand both languages, Norwegians show a clear preference for packages in their primary language. As a former resident of Oslo, Norway's capital, where bokmål is primary, I recall watching what

TABLE 20.3	**Checklist for Localization**

- Acronyms to be expanded, translated, or left unchanged (ISO, UN)
- Acronyms not to apply abroad (EPA, INS, IRS)
- Bilingual glossary to be created and added for translated text
- Body language in illustrations (gestures)
- Calendar differences (Arabic, Hebrew)
- Capitalist terms to be avoided in some countries (profit, bottom line, competition)
- Capitalization rules
- Cartoons (acceptable, inappropriate)
- Characters to be avoided ($, @, #)
- Color (meaning in other cultures)
- Company name spelling (transliteration)
- Consistency of technical terms (if previous localized text exists)
- Conversions (currency, measurements)
- Country codes per ISO
- Country names (e.g., People's Republic of China versus Mainland China)
- Cultural differences
- Currency format (if converted from dollars)
- Date format and presentation
- Dress conventions (bikinis, jeans, miniskirts, shorts may not be acceptable in some cultures)
- English terms not to be translated
- Examples (generic, offending)
- Figures ("billion" is expressed as 1,000,000,000 in the United States but as 1,000,000,000,000 in the United Kingdom)

- Gender conventions
- Graphics
- Hyphenation
- Icons
- Illustrations
- Measurement systems and units (length, area, mass [weight], volume)
- Non-Western alphabets (Chinese, Japanese)
- Number display (decimal signs, thousand separators)
- Paper format and sizes (A4)
- Personal names (don't use first names only)
- Political (not geographic) names (Beijing, not Peking)
- Provisions for text expansion
- Sacred symbols
- Sex (e.g., images of physical display of affection)
- Signs or symbols (public buildings, road signs)
- Sort order
- Standards (DIN, ISO)
- Telephone (no toll-free 800 line)
- Telephone: proper codes to and from foreign countries
- Temperatures (Celsius or Fahrenheit)
- Time formats and notations
- U.S. concepts (racial discrimination, Social Security, technical communicator)

Source: Fred Klein. May 1997. Beyond technical translation: Localization. *Intercom.* 32–33. Reprinted by permission of the Society for Technical Communication.

happened when grocery stores began running short of milk cartons labeled *melk*, the bokmål word, and had to restock the coolers with cartons labeled *mjølk*, the nynorsk word, originally intended for stores in the small mountain villages. Oslo customers would buy up all remaining *melk* cartons before touching the *mjølk* cartons, as if the Norwegian dairy cows produced inferior milk when it was labeled *mjølk*. The lesson for editors seems clear: customer sales and relations improve when companies localize. In most circumstances, translating into the readers' language is worth the time and money it takes.

Writing to Facilitate Translation: Minimize Ambiguity

Planning for translation before writing and editing requires knowledge of how translation occurs and of specific strategies to facilitate translation. Writers and editors can make choices in their native language that will make translation more accurate and minimize ambiguity.

To increase the efficiency and accuracy of the translation, writers can try to eliminate the ambiguity that comes from unfamiliar words, complex structures, and culture-specific metaphors. Errors in grammar and punctuation compound the problems of translation. The following guidelines can aid in writing and editing a document destined for translation.

- **Use short sentences and substantial white space.** Many languages use more characters to say the same thing that an English sentence says. If the translation will preserve format, including page and screen breaks, you may have to edit to shorten sentences so that when they are translated they will still fit on the screen or page.
- **Avoid jargon and modifier strings.** The principles that produce clarity are even more important when the text will be translated. There may not be equivalents in the translation language for unfamiliar words and jargon. Strings of modifiers raise questions about what modifies what.
- **Create a glossary of product terms.** Definitions help translators find equivalents for unfamiliar or product-specific terms.
- **Eliminate culture-specific metaphors.** Sports and political metaphors are common within a culture, but not all cultures play the same sports, and political systems vary. Varner and Beamer observe that military and sports metaphors are common in Western countries—*strategies, price wars, planning attacks,* and *digging in* have become the language of competition, but their military roots make Americans seem aggressive (pp. 74–75). Even an apparently neutral metaphor such as *keyboard* may not be meaningful in Asian countries, where hundreds of characters take the place of much shorter Western alphabets.
- **Avoid acronyms and abbreviations.** An organization is a social construction; unless an organization is international, its acronym may not be recognized in countries other than the host country. The translation program (like spelling checkers) will not recognize many acronyms. Abbreviations pose similar problems. High-tech terms, like CPU and RAM, known by their acronyms better than by the words they stand for, are exceptions to this guideline.
- **Avoid humor and puns.** Much humor is at the expense of something valued by someone else, and puns depend on specific meanings of languages. Both are difficult to translate well and are risky to use.

To see how such general advice may be applied to specific instances, we can look at a set of instructions by technical writing student Mike Freeman, at the time a senior majoring in architecture technology at the University of Memphis (Figure 20.1). In each example, excerpt 1 is from his original set of instructions, while excerpt 2 shows the change he made to prepare the instructions for translation. The changes are in italics to help you locate them. Although the changes are not entirely ideal for translation in every case, they are noticeably easier for translators to work with than the original text is.

Translation Quality

Sooner or later, if you are involved with translation, you will likely find yourself asking the following question: What should I do if my company is pressuring me to compromise the quality of a translation? The question brings out a classic dilemma. Caught between the horn of budgets and deadlines on the one hand and the horn of translation quality on the other, an editor managing a translation project can often feel perplexed, stuck, or—in worst cases—gored. In the short term, appeasing your bosses' demands for spending less money in a shorter period of time by sacrificing the quality of a translation can seem the best solution. In the long term, however, such a solution can reveal that much more money and time will have to be spent than would have been the case if a sufficient amount of both had been allotted from the first. After all, if the customer can't make sense of the documentation, finds it insulting, or misinterprets it to the point of misusing the product it accompanies in a way that leads to injury or death, your company will lose sales, perhaps permanently. Explaining how this is so may not be easy if your superiors are not familiar with the complexity of language and translation. However, an editor's duties include advocating the readers' concerns. Readers who are disgruntled with the quality of a document's translation can easily transfer their negative opinion to your company and its products. In the long run, advocating a high quality translation benefits the readers, your company, and you.

At first glance, the importance of translation quality may seem a given. After all, if a translated document appears shoddy to a customer, then the customer is likely to look askance at the company that produced it. If the document is an instruction manual, then the poor quality of translation can render the product it accompanies unusable. However, the problem for most technical editors in the United States, where foreign language instruction is minimal to nonexistent, is knowing whether the translated document has even been translated into the proper language, much less whether it is poorly or superbly executed. Usability testing, through the employment of in-country reviewers, is usually an integral component of the translation process—yet it is also often easiest for a company to skip in the face of short deadlines or budgets. Even when in-country review is thorough, you as the technical editor will still probably have to rely on the translation company to tell you what the results of the review or usability testing reveal.

To achieve high quality, translation companies often use some version of a quality procedure. When you contact a translation company, you'll be assigned to a project manager. This person will assess your project's type and difficulty, then assign a translation team to the project. If the project is highly specialized, say, a surgical device, the project manager will find a translator with special training, in this case a physician who is also trained as a translator. When the translator is finished, a translation editor will examine it and make any changes deemed necessary. The project manager then recruits one or more in-country reviewers. These

Example: Clarifying the Language
1. The following set of instructions is designed to show *and give* examples of how to use a Summagraphics 1212 four-button digitizer.
2. The following set of instructions is designed to show *you* examples of how to use a Summagraphics 1212 four-button digitizer. *The digitizer replaces the mouse in the AutoCAD program to save time.*

Example: Substituting Infinitives for Gerunds
1. Objective: *Achieving* greater speed with less effort is the main purpose of the digitizer.
2. Objective: *To achieve* greater speed with less effort is the main purpose of the digitizer.

Example: Inserting the Subordinating Conjunction *That*
1. Notice there are no rollers on the bottom of the puck.
2. Notice *that* there are no rollers on the bottom of the puck.

Example: Clarifying Unclear Antecedents
1. Electronic pulses start within the tablet and locate the cross-hairs on the puck to determine its location. . . . Notice the resemblance between the mouse and the puck, when *it's* within the screen-pointing area.
2. Electronic pulses start within the tablet and locate the cross-hairs on the puck to determine the puck's location. . . . Notice *again* the resemblance between the mouse and the puck, when *the cross-hairs are* within the screen-pointing area.

Example: Simplifying the Language, Steps, and Explanation
1. To draw a line using the command blocks, place the puck cross-hairs in the command block labeled "line." Now move back to the screen-pointing area. With the top left button on the puck, pick where you want to begin and end your line. Notice how much faster you can perform commands, *by picking on the command block, as opposed to using the pull-down menus and scrolling for line, then scrolling for segment, then drawing your line.*
2. To draw a line using the command blocks, place the puck cross-hairs in the command block that is labeled "line" *and press the top left button. You can* now move back to the screen-pointing area. With the top left button on the puck, pick where you want to begin and end a line. Notice how much faster you can perform commands.

Figure 20.1   Examples of Preparing a Text for Translation

The second version of each statement shows changes to facilitate translation.

Source: These instructions, written and prepared for translation by Mike Freeman, then a student at the University of Memphis, appeared in the *Journal of Business and Technical Communication* 11.3: 349–350.

Example: Adding a Glossary
1. No glossary.
2. Glossary:

AutoCAD   AutoCAD is a Computer-Aided Design program that allows a person to draw and design with the aid of a computer.

Command blocks   Command blocks are the square blocks that are on the tablet.

Puck   The puck is the device that is used to communicate with the tablet. The puck is a rectangular device with four buttons.

Pull-down menus   Pull-down menus are the options that are listed horizontally across the top of the screen. When you point and click on the pull-down menus, another list will appear vertically.

Screen pointer   The screen pointer is the small arrow on the computer screen. You control the screen pointer with the mouse or the puck.

Tablet   The tablet is the large flat device that contains the command blocks. You slide the puck on the tablet to perform functions.

Figure 20.1 *(continued)*

will be persons who resemble the intended readership—physicians, for example—but who are not translators themselves. They will point out where the text is confusing or is rendered in such a way that not many native readers would absorb its concepts. The project manager will submit their review to the original translator to make appropriate adjustments. When done, the project manager will often suggest that the client company test the text with its own reviewers. Often they are the company's sales force working in the target language area. This procedure is lengthy, labor intensive, and sometimes time consuming, but it yields a highly usable text that is unlikely to be misinterpreted. Editors who rely on a sole translator are taking chances, possibly in ways that can be hazardous to the readers of their documents.

Machine Translation

The first step in translation is sometimes a computer translation. Machine translation can give human translators a speedy headstart; however, by itself, it is inadequate for readers. Machine translation matches the words of the source language with their equivalents in other languages, and it parses sentence structure, albeit somewhat crudely. Words in different languages rarely have a one-to-one relationship. Indeed, even between different dialects of the same language this can be the case, as we have seen above with "truck" and "pavement." So far, computer programs are not yet sophisticated enough to replicate human judgment in assessing a word's shades of meanings and connotations and to use them unerringly in their appropriate contexts. The best machine translation programs depend on tightly controlled terms that adhere to the definitions stored in a firm's computer glossaries. Currently, the most widely used computer-aided translation (CAT) tools for professionals are Systran, officially adopted by the European Commission (Cronin), and Trados, partly owned by Microsoft and compatible with its popular Word program. When used well by proficient human translators, they allow "content management": reusing or adapting an approved, previously composed and translated text in a new document.

Human translators take a text through several stages to capture the exact meaning of the source text as they render the target text. Robert Bly, a famous poet, is also well known for his literary translations. His book *The Eight Stages of Translation* reveals how a text gradually takes on its full meaning as a translator passes it through each stage. The first stage is a literal translation, so crude that it would be nearly incomprehensible to most readers. (Users of web translators, like Systran's Babel Fish or AltaVista <world.altavista.com>, are familiar with such results first-hand.) Each succeeding stage makes the text more idiomatic and brings the text closer to the connotations intended by the author. Today, most machine translations are capable of achieving only stages one or two. A few of the most powerful and expensive can reach stage three.

Editors can take advantage of machine translation in two ways:

1. by allowing human translators to focus on the upper stages of translation;
2. by storing information that will probably never be used but which may need to be checked for its content.

In the first case, machine translation can save time and money, especially if the project is large and requires translation into several dialects. A boat manufacturer that wishes to sell its vessels in Spanish-speaking nations may find machine translation useful, especially if the documents for its products are updated every few years. As part of the content management process, the computer can keep track of passages that may remain unchanged and flag those where a change will be necessary. The computer can then render an early stage translation of the passages requiring alterations. Translators in such countries as Argentina, Mexico, Spain, and Venezuela can then render the early stage translation into their local dialects. This process results in less cost and earlier delivery of the translated documents.

In the second case, editors can use machine translation to their company's advantage by storing large volumes of documents, most of which will probably never be used again, in a literal translation. This literal translation will allow them to get the gist of the text's meaning if they discover later that they may indeed need to use the document. A product liability case can serve as an example. An automobile manufacturer may have assembly plants in many countries. Personnel records are kept in the language of the local workforce. In most cases these records need not be translated. However, if a product defect is traced to a particular plant, lawyers at the company's headquarters may want to comb records to see if labor relations may have played a role. Having all personnel records translated for use by humans in court would result in monumental costs and take years to complete. Machine translations, while unusable in court, can suggest what the gist of the records' meaning is. An editor, attorney, or others assigned to read these very rough translations can then decide whether or not to have records translated that appear to be relevant. This approach is far more economical than requiring high quality translation of all documents.

Other Localization Tips

Double-byte languages will raise the cost and increase the time of translation. Because computer keyboards were first crafted by American engineers familiar only with a 26-letter alphabet, languages with many diacritical marks, like Hungarian, or languages that use ideograms instead of an alphabet, like Chinese and Japanese, require more than one keystroke to enter a letter or ideogram into the computer. Today they are often handled through Unicode. If you are an editor in charge of a translation project, ask the translation company if any of the languages you are contemplating are double-byte. Be prepared to hear that translating documents into these languages will take more time and money than languages that use the Roman alphabet with few diacritical marks.

Converting measurements to metric is generally considered the job of you, the editor, not the translation company. Practically speaking, the United States stands alone in not using metric measurements—meters, liters, and grams. (A few items in Great Britain and Canada are still sold in pounds or yards, and distances in Britain are still posted in miles.) Consequently, virtually any document destined for localization or translation must have measurements converted to metric. Because anyone with a calculator can make such conversions, translation companies often view such a task as the client's responsibility. Translation companies will do the conversions if requested but will charge a hefty premium as a deterrent to future requests.

A4 paper size is standard in most of the world. Although equivalent to the 8½ by 11-inch paper used in the Unites States, A4 is slightly longer and narrower. A surprising number of localization horror stories tell of editors' making sure that every localization feature was handled impressively until the document was faxed or typeset abroad. Then the text would not fit on the page because someone had forgotten that 8½ by 11-inch paper margins do not match A4 margins. A similar scenario is common when editors do not realize that few languages can be expressed in as short a space as English can. Many languages take 30% more space on a page than does the same passage in English. Dutch and sometimes German can take up to 40% more. Thus, editors using simultaneous translations, in which more than one language appears on a page, must calculate much more space for the non-English languages. Those who use consecutive translation, in which each language appears in its own section, must plan on including additional pages for the non-English languages. The same is true for character counts in electronic documents. A field space on a web page that permits no more than 250 characters will create havoc for translators if the English in that field already takes up 249 characters. Without the necessary space, something will have to be lost in the translation, usually part of the meaning.

Visuals, including cartoons, work better in some languages and countries than others. While Americans find illustrations helpful, if not absolutely necessary, Japanese often consider illustrations critical. Their absence may make a text seem incomplete. Indeed, many serious publications in Japanese, including technical documents, are illustrated with cartoons. By contrast, if cartoons were included in a German translation, German-language readers would not consider the document serious or legitimate.

Finally, localization entails leaving decisions about what makes sense for local audiences to those, like translators and reviewers, who know the local culture and language. Advertising slogans that work in American English, like Nike's "Just Do It" or Pepsi's "Uh huh!" elicit blank stares when translated directly into other languages. In such situations it is much better for editors to allow the local sales force or translators to come up with a slogan that works locally than to insist on absolute consistency between languages. Similarly, it is better for editors to recognize that their perceptions and ways of comprehending the world are shaped by their own culture than to demand a rigid adherence to a culturally bound document. I am aware of one U.S. firm with offices worldwide that insisted that its employee survey include U.S. categories of ethnic background: white, black, Hispanic, Native American. The Hispanic category made no sense to employees at the company's office in Spain. Nevertheless, the company insisted that the categories needed to be the same around the globe so that the computer at its U.S. headquarters could tally the results and calculate the survey's statistics reliably. What the company got back were many incomplete or mismarked survey forms; quite a few of the employees outside the United States refused to mark any category. The translation team had warned the company that this might indeed happen, but to no avail.

Researching Social and Cultural Information

Even a person who is sensitive to and respectful of differences will not intuit all the ways in which language may evoke unintended responses. Acquiring information about any unfamiliar audience requires research.

Some sources are readily available: printed materials in the genres in which you will work, international employees of your company, professors and international students, and customers. The sources listed at the end of this chapter provide numerous helpful explanations and details. For products for international distribution, a consultant in international communication or the country in which your products will be distributed may be hired. Asking questions at the beginning of document development will help prevent problems, but because you won't be able to anticipate all the questions to ask, the review of the draft should include a review for globalization and localization.

Editors interested in keeping up on issues in international editing may subscribe to *TC-Forum*, a quarterly magazine published online by Intecom at <www.tc-forum.org>. To subscribe to *TC-Forum*'s listserv, on which the most current international editing issues are discussed and explored, send an email request to subs_tc-forum@tc-forum.org.

Using Your Knowledge

Two values that enhance editorial effectiveness are respect for differences and sensitivity to the experiences of readers. Research into the values, customs, and languages of intended readers in various cultures is necessary to supplement what an editor can sense intuitively. In addition, by anticipating translation and localization, editors can aim for words and structures that will minimize ambiguity.

Further Reading

Andrews, Deborah, ed. 1996. *International Dimensions of Technical Communication.* Society for Technical Communication.

Artemeva, Natasha. 1998. The writing consultant as cultural interpreter: Bridging cultural perspectives on the genre of periodic engineering report. *Technical Communication Quarterly* 7.3: 285–299.

Bosley, Deborah, ed. 2001. *Global Contexts: Case Studies in International Technical Communication.* Longman.

Bly, Robert. 1983. *The Eight Stages of Translation.* Rowan Tree Press.

Cronin, Michael. 2003. *Translation and Globalization.* Routledge.

Crystal, David. 2003. *English as a Global Language,* 2nd ed. Cambridge University Press.

Hoft, Nancy L. 1995. *International Technical Communication: How to Export Information About High Technology.* Wiley.

Klein, Fred. May 1997. Beyond technical translation: Localization. *Intercom* 32–33.

Maylath, Bruce. 1997. Writing globally: Teaching the technical writing student to prepare documents for translation. *Journal of Business and Technical Communication* 11.3: 339–352.

Munday, Jeremy. 2001. *Introducing Translation Studies.* Routledge.

van Parijs, Philippe. 2000. The ground floor of the world: On the socio-economic consequences of linguistic globalization. *International Political Science Review* 21: 217–233.

Phillipson, Robert. 2003. *English-Only Europe?* Routledge.

Varner, Iris, and Linda Beamer. 2005. *Intercultural Communication in the Global Workplace.* 3rd ed. McGraw-Hill/Irwin.

Discussion and Application

1. Prepare a document that you have written for translation (or exchange papers with a classmate for this task). In the original, identify features and phrases that might cause confusion. Use the guidelines in this chapter to determine the types of expression that may be difficult to translate.

(See Tables 20.2 and 20.3, Figure 20.1, and the section on editing to avoid ambiguity.) Then compose alternatives. Make a list of the types of changes you made.

2. The following paragraph incorporates language that is common and familiar in discussing business in North America. Underline the metaphors of war and sports. If this paragraph were sent to international managers in a global corporation, what difficulties in translation might such metaphors create?

> A department committee will meet for strategic planning to identify goals for surviving in light of the new policies of retrenchment. We will have to dig in to fight the misunderstanding of management about the department needs and goals. If our current funding is reduced, we will not be able to play on a level playing field with the competition. It appears that the vice president is trying to do an end run around our department manager.

3. Use a machine translation (MT) tool, such as the one found on the World Wide Web at altavista.com, to translate a passage into a language of your choice. Then use the MT tool to translate the passage back into English. Identify the phrases and sentence structures that differ from everyday, idiomatic English. Make a list of portions of the text that might cause confusion, misunderstanding, or incomprehension.

4. Find a British newspaper, such as *The Times* or *The Guardian*, or a British magazine, such as *The Economist.* (These publications and more maintain their own websites.) Locate an article written by the publication's own staff or the British news agency Reuters. (Avoid U.S.-based wire service stories, such as those from the Associated Press or *New York Times* News Service.) Using this chapter's information on British spelling, grammar, and vocabulary, "translate" the article into the English used in your region. This exercise can be repeated with other English-language publications, for example from Australia or Hong Kong.

21 Editing Websites

The Web has become the source of choice for many people seeking quick access to information. Your college or university provides information through its website; products that you may purchase have websites to provide information on their use; government websites offer consumers information about health, nutrition, laws, technology, and more; private organizations use websites to provide information or promote their point of view about a variety of issues, such as the environment or transportation policy. These organizations create websites full of information because they know someone will come looking for it.

In many ways, the Web has changed the way people look for information and how they read. The variety of websites is as vast as the variety of print materials. These changes create new challenges and opportunities for editors. What can you contribute as an editor? The brief answer is that you contribute a habit of anticipating reader needs. You also contribute knowledge about organization, style, visual design, and consistency, just as you do when editing print documents. The process of analysis and evaluation that you have learned for comprehensive editing can work for editing websites too. But websites are distinct genres (types of documents) in themselves, not just extensions of print. Furthermore, online storage of information allows new possibilities for its delivery, including updating multiple references from a single source.

This chapter uses websites to illustrate the kinds of things that editors consider in editing online documents. An informational website is distinguished from one used to transact business or purchase products and from entertainment sites. It has the serious purpose of making information available to users who have to complete tasks or get information.

This chapter applies the editing principles and processes you already know to websites, but it shows where new principles and processes apply. The chapter begins by considering how websites differ from print and the ways in which readers use websites. Understanding genres and planning for their development and use are starting points for editors. It approaches the editing process by the method you already know how to use in comprehensive editing: planning to meet user needs for information, organization, navigation, style, screen design, and consistency.

Key Concepts

Websites differ from print. Print organizes information in a specific sequence (though readers can access the information selectively). Websites include separate files of information whose connections are revealed through menus of topics and hyperlinks from one piece of information to a related one. Readers have more power to impose their own structure, but they also need help from the interface to perceive the scope of the information. Because of the worldwide availability of many websites, these documents need to accommodate a variety of users: novices and experts, native-language and second-language users, readers who seek concepts, readers who seek instructions, and so forth.

Measures of effective websites parallel the measures of effective print documents: content is complete and right for the users; content is organized so that users can find what they need; visual design provides cues to document structure; style is crisp; terms are used consistently; identifying information facilitates searching. Usability, the ease with which users can find and understand the information they need, is a measure of whether these features are used effectively.

Websites as Content Repositories and Databases

Websites of any complexity are composed from separate files. They rarely are single "pages." Websites may consist of hundreds or even thousands of separate files. This structure enables a user to search for specific information and ignore the rest. But before users can search effectively, they need a sense of what is available, where to find it, and what terms identify it. An online document reveals its structure through its interface: the textual or graphic information that enables a user to connect to the content in the document.

Editors ask not just what readers will find in the document but also how they will find it. A printed book reveals its scope and structure in a tangible way. Readers search books by the index, table of contents, headings and running headers, and just by perceiving the structure of the information as it is organized and bound. Editors of websites pay attention to search functions, such as the document interface and menu items, search terms, and hyperlinks.

In this online content repository, the files may be "static"; that is, they do not change unless a web manager or help author manually replaces, deletes, or adds files. Increasingly, however, online documents are updated dynamically. They may enable users to select preferences about what information will be displayed. If you go to a site for travel arrangements, for example, you can choose what airlines and airports to feature when you load the page. Your preferences will be stored in an online database. The content that is foregrounded in the site that you open differs from the content on the same site that someone else opens. You may have to log on so that the software can identify you, or it may identify you from information that it stores in your computer.

A website may also draw from files in an online database. If the owner of the database changes the records in the database, the website automatically changes. This possibility of linking multiple documents to a database is the basis for the

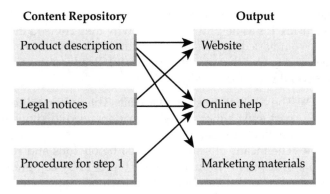

Content Repository   **Output**

Figure 21.1   Single Sourcing for Online Documents

If the online documents are linked to files in an electronic content repository, an update of
the source file automatically results in an update of all the documents to which it is linked.

concept of "single sourcing": the creation of one source (the database) to which
multiple documents (such as a website, online help, and marketing materials) are
linked. By updating the database, the database owner automatically updates all
the documents that draw on this single source (see Figure 21.1). This technology is
an important way to keep information up-to-date and consistent throughout the
information pieces of an organization.

As editor, you probably won't have to design and maintain this technology,
but you should understand its concepts and possibilities and know where the information you edit will be used. You can also make editorial choices to facilitate
the searching and updating that mark usable online documents. These choices include appropriate tagging of document parts (see Chapter 5) and identifying information, especially titles, for the separate files. Part of the document plan
should include a plan for regular updates of the information.

Reading and Searching Online

Understanding readers and the ways they will use online documents guides your
editorial decisions about content, organization, navigation, visual design, and
style. A number of reasons bring users to their computers, but users come to informational websites to get answers to questions like these: What are the symptoms of this disease? How can I crop this picture? What is the evidence for global
warming? Readers do not come expecting to read at the screen for a long time,
and because they are so goal driven, they want their information quickly. As
Ginny Redish (June 2004) observes, most readers scan rather than read websites.
They "read to do," not "read to learn" (p. 5). To facilitate their searching, they
may wish to control the appearance of the screen and make choices about which
information is presented to them.

Because websites are often available worldwide, it is hard to classify readers
according to levels of knowledge, vocabularies, and cultural backgrounds. You

have to assume that they bring a mixture of backgrounds and interests and competencies. It's as helpful to think of why they come to the site as of who they are.

What varied readers have in common is a wish to get to the information they seek and to find that information useful. Reader strategies and backgrounds point to several goals for websites:

■ The content must be available. The writer, editor, and subject matter expert must anticipate what readers need to know and what questions they are likely to ask.

■ The means of searching must be obvious and reliable. Users may search index terms from a menu of topics or by comprehending the organization of related items. Writers and editors anticipate the methods of using the site, not just the content needed.

■ The text should usually be short enough to fit on a screen. Readers in a hurry do not want to read long paragraphs and scroll through screens. Writers and editors cut unnecessary words. If the text requires substantial development, a version suitable for printing should be available.

■ Empowering users to set preferences or to locate the information that suits their interest and expertise can make the document useful to a variety of readers.

To find out more about the specific readers who are likely to use the documents you are editing, you might consult the document plan and customer service representatives. Even better, look for opportunities to observe potential readers at work using websites to complete tasks. You will learn about how they do their work, but observing will also give you images of individuals using the site. "Reader" will have a face and name and not be an abstraction. All of this knowledge about potential readers will help you make editing decisions.

Planning and Developing Websites

Informational websites are complex enough to require a comprehensive document plan at the beginning of development. That plan gives direction to the various people involved, including editors, who need to schedule their own time and tasks as well as know who is responsible for developing content. Editors can help with this planning along with content experts, designers, technical experts, clients, and a project manager. In Chapter 1, you saw that Charlene was part of the team that planned the online safety training. Her role in that project provides a good model for an editor's participation throughout planning and implementation.

The document plan shows at the outset of development the purpose, uses, and scope of the document. It includes a statement of specifications for the project content and updates, templates for the components, and a style guide. A separate project plan tells how the project will be developed and on what timeline: assignments of tasks, project milestones, reviews and testing, schedules, and budgets. You will benefit from participating in these management decisions because if there's slippage by others, you will be the one caught by failure to meet final deadlines.

Defining the Concept and Specifications

In planning, team members must first develop a shared concept of what the website is meant to accomplish and for whom. If it is supporting the use of a product, will it provide conceptual information or just procedural information? Will it include tutorials? FAQs? Will the information be accessible through an index of terms? Will users be able to type in search terms, or must they select from an index that is provided? How often will the document be updated, and who will be responsible?

With content experts, designers, and technology experts, you can help to develop specifications for the project, templates for the components, and a style guide. Specifications include the types of information and how the information will be organized. Specifications might also include any plans for translation or localization, provisions for ensuring accessibility, and provisions for updates.

Charlene's team (Chapter 1) developed a storyboard visually outlining the topics and their connections. They identified content components and their organization. With this level of detail, responsibilities for developing content can be assigned. The storyboard provides the editor with the big picture of the project, including content, scope, and people responsible. The information also enables a project manager to schedule due dates. Those dates influence the editor's schedule.

Templates

Templates for the pages or sections include two types of information: visual design and the topics and their structure for related types of pages. The template will incorporate styles (see Chapter 5) for the display for the various textual parts, such as headings, menu items, paragraphs, list items, and links. The template will also include a plan for screen layout: where various elements, such as the menu, will appear. Visual consistency helps users search because they know where to look.

A map of the structure for related types of pages encourages writers to provide complete information and in the same order. For example, if you were developing a department website for a university, the information for each degree program or specialization might begin with a description of the goals for that program. Each section might include criteria for eligibility and application, degree requirements, course descriptions, a faculty list, and examples of student achievements or projects. (These divisions would represent anticipated student questions: What is this field about? How do I sign up for the major? What courses will I have to take to complete the program? Who will teach the courses? What are the other students like, and what do they do?) The template showing structure would enable different program directors to develop content for each module consistent with that of the other modules.

Style Guide

The more you know about the document contents, the better you can complete some editing tasks up front and save time later. A style guide establishes terminology, including capitalization and spelling, to encourage all members of the project team to use terms consistently. Terms include product and part names, technical terms, and references to the company. Definitions and examples can clarify how to use the terms. Anticipating synonyms for the terms that users might think of will lead to good search functions. If some of the terms can be abbreviated, include the acceptable abbreviation.

If there is no company style guide to indicate general style preferences, such as for cross-references, numbers, and punctuation of lists, include those as well.

Conventions for illustrations and logos can be described. Because some users may be visually impaired and will depend on text readers to hear the content, the style guide might require alternate text for each illustration. For example, the alternate text for Figure 21.3 (p. 374) might read "Scan-select-and-move-on page for the Firefox browser, illustrating a menu of website sections and help topics."

A style guide may also include file-naming conventions for each type of document. A systematic way of naming files will enable management of large collections files. For example, in the files for the website for this book, the Discussion and Application files all have teDA as the beginning of their file name (te is for technical editing). That convention makes it easy to group and to find the files.

Making Content Work for Readers

Because readers search websites for information, the most important evaluation criterion for editors is to ensure that accurate and complete content is available and that it meets the needs of users. Your document plan and storyboard should represent the best judgment of team members about what content needs to be provided and at what level of detail. When you edit for content, you check what is there against the plan.

As with any content, the heuristic of asking *who-what-when-where-why-how-so what* is a starting point to determine content. Imagining that a reader is asking these questions will guide you to the reader's experience and needs for information and away from just describing your product or organization.

Failures in the content occur for two contradictory reasons: too much information, not enough information. Too much information may be excessive background or explanations that bury the details the reader seeks.

Too little information often occurs when instructions tell what to do but not how. When the content is instructions, it's better to err on the side of too much information. An informed user can skip sections, but a beginner usually cannot fill in gaps except by guesses or intuition, and their guesses could be wrong. To edit content for instructions, you read asking "how." If the answer is not apparent, you may need to request more information from the writer or content expert.

Another gap in content may be identifying information. Some development teams neglect to articulate what's obvious to them (but not to readers): the purpose and scope of the website, who they are, and where they can be found. Each website page should display the name of the site and/or sponsoring organization. Templates for different types of pages can include such information automatically.

Identifying information in the code also increases the chances that search engines, such as Google and Yahoo!, will lead to the site. The title, a necessary part of HTML and XML code, should be specific, as is this one:

<title>Web Content Accessibility Guidelines 2.0</title>

Search engines scan titles. If someone types "accessibility" into a search engine, this site will almost certainly appear.

Organization

As you saw in Chapter 17, principles of organizing for print documents include organizing by task and organizing by the inherent structure of the information. Those principles work for the content sections of websites. However, an editor of websites is concerned with another layer of structure: how the different files may be classified in terms of their purpose.

Ginny Redish (June 2004) identifies four types of pages in most websites (p. 6); see Figure 21.2.

- **Home page:** identifies the site, establishes "brand recognition," sets the tone, tells users what the site is about, gives a big picture of the possibilities on the site, starts users down the right path quickly.
- **"Scan, select, and move-on" pages:** provide more menu options than the home page can accommodate; convey information without requiring users to read paragraphs of text. Figure 21.3 is an example of such a page.
- **"Scan and get information" pages:** provide the content for the site; users may encounter paragraphs, but they will appreciate short ones with headings that identify contents so that they can scan for the information that interests them.
- **Forms pages:** allow interaction or necessary sharing of information.

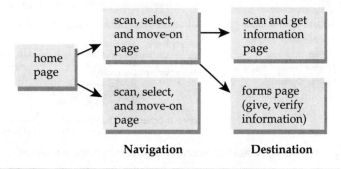

Figure 21.2   The Four Main Types of Pages in Most Websites

Source: Janice C. (Ginny) Redish. June 2004. Writing for the Web: Letting go of the words. *Intercom,* p. 6. Used by permission of the Society for Technical Communication.

This structure may guide the development of individual pages by requiring you to know their different purposes. Users will find it easier to navigate your site if you don't mix purposes.

As part of the updates for a website, you may cast your editorial eye on the organization of FAQs or items in a knowledgebase. These pieces of information are often organized in the order in which someone asks a question. But that order makes it hard for new users to find what they are looking for. They might scroll through lots of irrelevant information to find what answers their question. (Or they may give up.) At the very least, you can group the items by topic.

Good organization facilitates navigation because it links related information. Users might find their specific objective if they get into a section of a site where such information is likely to be located. The need to organize information also forces web and help designers to identify the main topics and perhaps thereby to clarify how the document will achieve its goals.

Navigation and Searching

"Navigation" is a metaphor for locating information on a website. One associates navigation in its literal meaning with wide open spaces, such as oceans or skies. The directional signals are not entirely clear from the environment, so the navigator uses tools to identify location and to choose the right direction. The minimal navigation tool on a website or in online help is a menu of the topics of the information available there. In Figure 21.3, the main menu items are at the top of the page.

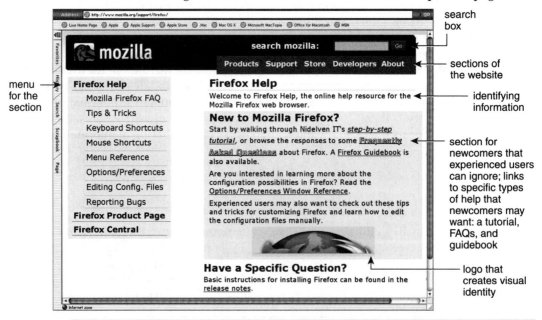

Figure 21.3   A Scan, Select, and Move-On Web Page
(www.mozilla.org/support/firefox)
Copyright © 1998–2005 The Mozilla Organization. Used by permission.

Scan, select, and move-on pages prioritize navigation. Note that Figure 21.3 provides users with multiple ways of finding information. If users know a term, they can type it in the search box and go directly to information on that term (though they may find more choices than they want). They can navigate to different parts of the site using the top menu. Because they are now in the "support" section, they can pick from the topics in the left-hand menu. The links in the shaded portion in the middle will take new users to a tutorial, FAQs, and guidebook (for reference). On part of the page that you cannot see in Figure 21.3, the page leads to an online chat and web forum, other sources of information provided by a community of users.

The page is not flashy. Most of the information is text. The designers have aimed to help users move on, not to demonstrate their skill with design that calls attention to itself. The page is visually well organized, and users can find information in predictable places.

A variation of a scan, select, and move-on page is a site map. A site map (an outline of the divisions and subdivisions of a website) reveals the structure of a website. Its function is similar to that of a book's table of contents, enabling a user both to get to specific information and to see the scope of the site. It helps users navigate the site by letting them see where things are. For example, is shipping information included in the customer service section or in its own section? Are the department's courses listed under "undergraduate programs" or in a section called "courses"? If users can see the structure of the site, they can make good guesses about where to find information.

As editor, you will also value the site map for checking that the information is complete (topics are not missing) and for evaluating the structure (topics are grouped in logical ways).

Screen Design and Color

Screen design refers to the way the information will be positioned on the screen and what combinations of type and graphics will represent the information. Effective screen design creates a page that supports the purpose of the document, is easy to read, minimizes scrolling, and allows readers to find information easily. Meeting these goals requires awareness of the users' hardware and software, creating legible type, and following the goals of consistency and simplicity.

Effects of Hardware and Software on Screen Design Choices

Readers and their equipment have some control over how a screen will display. The software that enables users of the Web to find and display documents that are stored on other computers is called a *browser*. The reader's browser, not just the writer's file, determines what the reader sees from what is coded. It interprets the HTML or XML code. Most browsers are graphic, but some users set their preferences to see only text because it loads more quickly or because they are using screen readers to accommodate low vision. Different browsers for the Internet display the same information with different fonts and sizes. The length of the displayed text line adjusts to fit the available space of the monitor or window unless it is coded in fixed sizes, which may result in waste of space.

The operating speed of the computer and modem or other network connection determines how quickly information will load. Color, graphics, patterned backgrounds, sound, and animation all slow the loading time.

Legibility of Type

Another screen design challenge is that type on screens is generally harder to read than type in quality print reproduction. The wide screen of a typical computer monitor invites a line length that violates easy reading, and HTML codes do not encourage the range of display options, such as multiple columns and side headings, that word processors and page layout programs permit to achieve an optimum line length in print. Cascading style sheets and XML can give designers more power over output (see Chapter 5). Italics are hard to read on many monitors, and screen resolution is probably worse than resolution in print. These variations of print and screen explain, in part, why paper documents coded for the Web without redesign for the medium may succeed in making information available but rarely succeed in making information readable.

To solve the problem of line length, some designers use tables to section the screen. These sections increase their control over placing text and graphics for maximum readability. The sections can be used to create columns, side headings, side-by-side text and graphics, and vertical navigation bars. Although they make the screen more readable for people with normal vision, people who depend on screen readers may hear confusing text because their screen readers move across the page, not in sections.

The difficulty of reading from screens argues for simple backgrounds rather than patterns that minimize the contrast between characters and background. To maintain contrast, designers also choose colors of different intensities for text and background. Dark characters on a light background will be easier to read than background and text created with two colors of the same intensity, such as red and black.

Font choices affect legibility. Simple shapes of characters formed with broad strokes offer advantages, and several have been designed for the Web, including Verdana (sans serif) and Georgia (serif). In your code you may specify a font, but it's smart to include a category option, such as serif or sans serif, in case the user's browser does not support the fonts you have named.

Visual Consistency and Simplicity

Given the challenge of creating readable text and graphics for computer screens, design principles such as consistency and simplicity can be even more important for online design than for print. Placement of different types of information follows increasingly stable conventions. Functional information (menus, toolbars, search boxes) typically spans the top of the screen. Structural and content information (topics) may go across or down one side. These functional and structural divisions help users keep their place in a long document and enable searching. Variations interrupt users from their main task of locating and understanding the content.

Consistency in the use of type, icons, and color to signal types of information and structural divisions also increases the usability of a document. A document style sheet and templates for design encourage consistent choices just as they do for print. Variations in different parts of text, such as headings and paragraphs, should be apparent from typeface, type size, and spacing, but each of these parts should be displayed consistently.

Color can enhance comprehension and motivation, or it can contribute to a confusing and busy look. Arbitrary shifts of background color or other inconsistent uses of color will confuse because readers expect shifts to mean something. On the other hand, using different colors consistently to distinguish types of information could enhance comprehension. Perhaps, for example, all of an organization's research activities appear on pages with blue banners at the top and its advocacy activities appear on pages with green banners.

Simplicity follows from determining what is essential in contrast to what may be only decorative or novel and focusing on usefulness to readers rather than on cleverness of the designer. A simple design that is easy for users to comprehend functions better than an elaborate design that may be breathtaking on first viewing but clumsy to use.

Style

"Let go of the words." That's the consistent style advice of website and online help experts (see Redish, *Writing for the Web*, in Further Reading). That advice follows from the comparative difficulty of reading from the screen and the likelihood that users are seeking specific information. They may be impatient with plodding through a long explanation or description of product features.

But don't be brief at the expense of necessary information. Readers can skim or skip too much information, but they can't fill in the gaps of missing information. If the text seems to get too long, cutting words and content is one solution. Other solutions are to divide the text into smaller units or provide more signals about the structure of the long text so that readers can skip to the part that interests them.

Follow the other guidelines for style explained in Chapters 15 and 16. Considering international users, also refer to guidelines in Chapter 20.

Usability Testing and Accessibility

No matter how good you are at editing, you will not discover all the possible ways in which users might try to use the online document and possibly fail. A usability test gives representative potential users some tasks to do using the website or help system. By observing the way users negotiate the website or online help, the tester can identify some ways to improve the information and navigation. For example, if you were testing the site represented by Figure 21.3, you might ask users starting on the home page to find a tutorial for using the software. The test would include other similar tasks that typical users are expected to do at the site.

Accessibility means that users, even those with some impairments, can use the website. Impairments might be visual, cognitive, or motor. Some assistive technologies are available to help users. These include screen readers, enabling people with visual impairments to hear what is written. For these technologies to work, the HTML or XML coding has to be valid. The code can be validated up to a point with an automatic tool such as WAVE 3.5 at the *WebAIM* site. The *WebAIM* site in Online Resources provides a number of tips for accommodating possible disabilities in some users. One general tip is simplicity. Clutter and confusion overwhelm and interfere with assistive technologies as well. Other tips are to provide alternate text for illustrations and to avoid complex data tables.

Maintenance and Updates

After a website is tested, revised, and published, it can begin to be out of date almost immediately. The product that the website supports may change, requiring parallel changes in the site. New research may require updates of the information on a medical information website. In spite of your testing, you may find through user feedback that some information is not clear or is incomplete and needs to be revised.

Some problems like these are inevitable, but it's not inevitable that an online document will get out of date. A schedule for updates, process of review, and assignment of someone to the task of updates will provide administrative support for keeping the information valuable.

Using Your Knowledge

One thing is certain about technology: it will change. As it changes, your knowledge about it needs to grow. The more you learn, the more valuable you will be to the organization for which you work. To stay ahead of the curve, you can take advantage of professional development opportunities offered by professional associations or your company. And you can be a user (as well as an editor) of informational websites, online help, and books on the topic of information technology.

Types of documents (genres) will also change as technology changes and as we learn more about user responses. Genre knowledge as well as technical knowledge will help you be a good editor of these materials. To develop this genre knowledge, you could begin by reading some materials in Further Reading.

One thing will not change: your broad preparation for editing will serve you well in editing online documents because it offers the systematic procedure of evaluating content, organization, visual design, style, and consistency in terms of user needs. Others on the project team may focus on content or on the technology, but you offer the valuable perspective of users negotiating the interface and locating the information they need. You make editorial choices to enable these users to get their jobs done.

Further Reading

Anderson, Steven L., and Charles P. Campbell. 1998. Editing a web site: Extending the levels of edit. *IEEE Transactions on Professional Communication* 41.1: 47–56.

Farkas, David K, and Jean B. Farkas. 2001. *Principles of Web Design*. Longman.

Hackos, Joann T., and Dawn M. Stevens. 1997. *Standards for Online Communication.* Wiley.

Hammerich, Irene, and Claire Harrison. 2002. *Developing Online Content: The Principles of Writing and Editing for the Web*. Wiley.

Microsoft Corporation. 2004. *The Microsoft Manual of Style for Technical Publications*. 3rd ed. Microsoft Press. The second edition is available online from www.microsoft.com/downloads/.

Nielsen, Jakob 2000. *Designing Web Usability: The Practice of Simplicity*. New Riders.

Price, Jonathan, and Lisa Price. 2002. *Hot Text: Web Writing That Works*. New Riders.

Redish, Janice C. (Ginny). June 2004. Writing for the Web: Letting go of the words. *Intercom*. 4–10.

Weber, Jean Hollis. 2004. *Is the Help Helpful? How to Create Online Help that Meets Your Users' Needs*. Hentzenwerke.

Online Resources

Accessibility portal: www.webaim.org. This *WebAIM* site provides numerous articles and links on different disabilities and ways to design to accommodate them. It also offers an accessibility tool for checking how well a site complies to the accessibility guidelines.

Style guide: Lynch, Patrick, and Sarah Horton. *Web Style Guide*. 2nd ed. Also available for purchase in print. info.med.yale.edu/caim/manual/contents.html.

Usable Information Technology: www.useit.com

Web specifications: World Wide Web Consortium: www.w3.org

Web Content Accessibility Guidelines 2.0: www.w3.org/TR/WCAG20/

Discussion and Application

1. Examine examples of sites suggested below on the Web to determine
   - probable target readers
   - likely purposes of the readers and goals of the site sponsor
   - organization (main divisions, parts—connected or segmented, sequential or by categories)
   - navigation devices: what, where located, types of information
   - screen layout
   - visual design (text or tables, paragraph length, headings)
   - use of color or icons to identify the section
   - style (sentence style and illustrations)
   - accessibility

   Then evaluate each site to determine whether the choices have been effective. Summarize your findings overall by noting differences in choices and their match to intended uses.

   > copyright page: www.copyright.gov
   > state government site for your state
   > site for your college or university

2. Look for FAQs in a commercial website. They are a resource for users who have similar questions, but they work only if users can find the question that matches their own. Grouping related items facilitates searching (see Chapter 17), but it's easier for the site manager to add on without grouping. What is the pattern of organization? If the organization seems as though it would help users find the right information, articulate the qualities of this pattern that work. If you find an unorganized group of FAQs, suggest a way to organize them to suit user needs. If a keyword search option is available, test it to see whether it leads to useful information.

3. Test a page of a website for accessibility using the WAVE tool at www.wave.webaim.org/index.jsp. You might find it most interesting to test a small, noncommercial site, such as the site of a student organization or department.

22 Legal and Ethical Issues in Editing

Laws and codes of ethics aim to protect the good of a society as well as to protect individual rights. In technical publication, intellectual property laws govern copyright, trade secrets, and trademarks. Laws also require organizations to accept responsibility for product safety. Editors work with other members of product development teams and legal experts to verify adherence to these laws and the representation of them in the text through warnings and notices of copyright, trademarks, or patents.

Codes of ethics are less formally encoded than laws, as are the sanctions for violating them, but, like laws, they aim to protect individuals and groups from harm and to provide opportunities by creating a work environment in which individuals can achieve. They may be encoded as statements of professional responsibility, such as the "Ethical Guidelines" by the Society for Technical Communication, or derive from cultural values of right and wrong, such as the values that individuals should not harm other persons and that they have responsibility for maintaining the quality of the environment.

Legal and ethical issues pertain both to individuals and to organizations. Editors can be most effective as individuals if corporate policies establish commitment to legal and ethical behavior and if corporate procedures allow for review of products and documents by a variety of knowledgeable people, not just the editor or even just the legal department.

This chapter reviews legal and ethical issues that pertain to technical publication and suggests corporate policies to encourage ethical behavior.

Key Concepts

Law and ethics balance the rights of individuals in a society with the needs of the society as a whole for information, ideas, products, and the expectation of health and safety. Intellectual property laws protect creations in various media, including print and digital, written texts, music, photographs, and software. Countries interpret these rights differently. Laws also govern contracts and product safety. Ethical issues include misrepresentation and environmental responsibility.

Legal Issues in Editing

Editors share responsibility with writers, researchers, and managers for protecting documents legally and for ensuring that they do not violate intellectual property, product safety, and libel laws. Editors verify that permissions to reprint portions of other people's publications are obtained, and they review instructions and warnings for potentially dangerous products. Using the symbols ©, ®, and ™ indicates copyrights, registered patents, or trademarks.

The editor can fulfill legal responsibilities more readily and thoroughly if the organization's policy manual and style manual include policies and guidelines about intellectual property, product safety, and misrepresentation.

Intellectual Property: Copyright, Trademarks, Patents, Trade Secrets

The law of many countries, the United States among them, recognizes that one can own intellectual property just as one can own land or a business. These protections are intended as incentives to develop ideas or products that may improve the quality of life for the community. According to the 1787 U.S. Constitution, "the Congress shall have power . . . to promote the progress of science and useful arts, by securing for limited times to authors and inventors the exclusive right to their respective writings and discoveries." Intellectual property includes original work of fiction or nonfiction, artwork and photographs, recordings, computer programs, and any other expression that is fixed in some form—printed, recorded, or posted on the Web. These expressions are protected by copyright law. Other intellectual property includes work protected as trademarks, patents, and trade secrets. The owner of intellectual property has the legal right to determine if and where the work may be reproduced or used.

Copyright

The United States Copyright Act of 1976 protects authors of "original works of authorship." Copyright means literally the right to copy. The Copyright Act gives the owner of a copyright the right to reproduce and distribute the work and to prepare derivative works based on the copyrighted work. Reproducing work copyrighted by someone else violates the copyright law. As editor, you verify permission to use copyrighted material and take steps to protect documents published by your company.

Ownership

Copyrights belong to the author who created the work unless the author wrote the work to meet responsibilities of employment. In the case of "work for hire," the employer owns the copyright. Most technical writers and editors work for hire, and the manuals and reports they write are the intellectual property of their employers. Some publishers require authors to surrender the copyright to them. Works by the U.S. government are not eligible for copyright protection; they are in the "public domain" and can be used by other people without permission.

Collections with contributions by multiple authors are generally protected by a single copyright, but the sections may also be copyrighted individually.

In the United States, copyright extends 70 years beyond the copyright owner's death. Works written for hire are protected 95 years beyond publication or 120 years from creation, whichever is shorter. When the copyright expires, works are in the public domain and may be reproduced and distributed by others. (See www.copyright.gov for more details.)

Copyright Notice, Registration, and Deposit

Copyright is automatic in the United States as soon as a work exists in fixed form, even if the work is not published for wide distribution. (Your course papers and projects are copyrighted.) Protection does not require a notice or registration. However, registration with the Copyright Office gives maximum legal protection. Registration requires sending an application form (Form TX for most technical documents), a fee ($30 in 2005), and two copies of the document to the Copyright Office in the Library of Congress.

For the best protection, a published work should include a notice of copyright. In a book, the notice usually appears on the verso page of the title page and is often called the copyright page. The notice includes the symbol ©, abbreviation "Copr.," or the word "Copyright"; the year of publication; and the owner's name. Here is an example: © 2006 by Pearson Education, Inc.

You can get more detailed information on U.S. copyrights from the Copyright Office, either in print or on the website (www.copyright.gov). Particularly useful publications are Circular 1, "Copyright Basics," and Circular 3, "Copyright Notice." You can download application forms from the website.

International Copyright Protection

Copyright in one country does not automatically extend to another. Use of works in a country depends on the laws of that country. Countries that have signed one or both of two multilateral treaties, the Universal Copyright Convention (UCC) and the Berne Convention, offer some protection to work published in member countries. Circular 38a lists countries that maintain copyright relations with the United States. Some countries offer little or no copyright protection for work published in other countries.

Permissions and "Fair Use"

Because work is protected by copyright, permission must be obtained to reproduce all or part of someone else's work. Usually, the writer requests permission from the copyright holder, but editors verify that permissions have been acquired before the document goes to print. The permission ought to exist in writing, whether in a letter or on a form. The correspondence will establish exactly what will be reprinted and where. The copyright owner may charge a fee for use of the material. If permission is denied, the material cannot be used. Permission must be acquired for each use. Permission to use copyrighted material in one publication does not give one the right to use it elsewhere.

A request for permission to reprint copyrighted material should include the following information:

- title, author, and edition of the materials to be reprinted
- exact material to be used: include page numbers and/or a photocopy
- how it will be used: nature of the document in which it will be reprinted; author; intended audience; publisher; where the material will appear (for example, quoted in a chapter versus cited in a footnote)

Fair use allows some copying for educational or other noncommercial purposes. For example, you may photocopy an article in a journal to study for your research paper. But your professor cannot copy whole sections of a book for the class so that the class won't have to buy the book because doing so would deprive the book's publisher of sales of its intellectual property. In a work setting, it is best to be cautious about copying, especially if you will distribute the work widely and profit, even indirectly, from this distribution.

Copyright prohibits duplication of software for multiple users unless an organization has purchased a site license. The terms of use are often listed on the software package.

Copyright and Online Publication

The Internet and Web have provided the public with wonderful access to information. An ethic of openness, sharing, and access has encouraged individuals and organizations to offer documents, statistics, and other information to be freely used. This ethic and the ease of copying material distributed through the Internet or Web makes some people think that it is all right to do so, but material on the Web and even email messages are protected by the same copyright laws that protect print, especially when the information has commercial value. Because publishing on the Web makes material available without purchase, you are not depriving an author of sales if you copy material from the Web for your personal use and knowledge. Such use is probably fair use. But if you distribute something you found on the Web to make money or to people who would otherwise read it on the Web and see the advertisements that accompany it, or if you copy the words and use them as your own without citation, then you are violating copyright law.

Cyberspace law is a new area that is still developing. For the present it is prudent to assume that the laws for print publication apply to online text. You need to get permission to use material from the Web in other contexts. You may even need permission to link from your website to another site, especially if you wish to link to pages deep in the site, bypassing the home page where the advertisements may appear.

Trademarks, Patents, and Trade Secrets

Trademarks are brand names, phrases, graphics, or logos that identify products. The rainbow-colored apple is a trademark of Apple Corporation, as is the name iPod. If the marks are registered with the United States Patent and Trademark Office (PTO), no one else can use those particular marks to represent their own products. Editors look for proper representation of trademarks that may be referred to in the text. The symbol ® next to the mark means that the mark is formally registered and certified with the PTO. The symbol ™ indicates a mark registered on a state basis only or one that has not been officially placed on the Principal Register in the PTO. Trademarks are capitalized in print. Dictionaries identify words that are registered trademarks. Product literature is a good way to find out about whether to use the ® or ™. A typical procedure is to use the symbol on first use of each trademark or to list all trademarks used in a publication in the front matter. Constant repetition of the symbol in the text may become distracting, and the law does not require it.

Patents protect inventions in the way that copyright protects expressions. Their significance to editors is that the text records patent registration with the symbol ® just as it does for trademarks.

Law also protects trade secrets, such as the specifications for a new product or a customer list. Employees owe a "duty of trust" to current and former employers. It is illegal for a company to hire you to find out what a competitor is planning, and it is illegal for you to give trade secrets of a former employer to a new employer or of your current employer to anyone else. Documents produced under government contract projects may be classified and require secrecy. For either private or government work, you may have to use special protections to keep information in your computer files secure. The duty of trust represents one exception to the First Amendment protection of free speech.

Product Safety and Liability

According to U.S. law, companies and individuals must assume responsibility for safety of their products as they are used or even misused by consumers. The products themselves must be designed to be as safe as possible, but because design itself cannot ensure safety, instructions for use and warnings must be complete and clear. The instructions and warnings must cover use, anticipated misuse, storage, and disposal. Manufacturers cannot avoid responsibility with disclaimers (statements that the manufacturer is not responsible for misuse or accidents).

Instructions, Safety Labels, and the Duty to Warn

The first strategy of documenting safe use of a product is to write clear and complete instructions. Clarifying procedures where safety is an issue should be part of document planning. Editors may wish to request a summary of hazards so that they don't have to rely on the instructions alone to identify them. In reviewing a draft, editors rely on the principles of organization, style, visual design, and illustrations, as discussed in Chapters 14–21, as well as an attitude of vigilance and care. Ambiguity must be clarified. For example, if a procedure calls for "adequate ventilation," an editor may query whether "adequate" means "fresh" air or if circulating air will suffice. Taking the perspective of the reader, the editor may note some gaps that a writer misses.

If there are hazards of using products, manufacturers and suppliers must warn of the risks unless the product is common and its hazards well known. Instructions do not constitute warnings, nor can a warning substitute for instructions. A warning calls attention to a particular procedure verbally, visually, or both. Some standards for identifying different categories of risks and for symbols and colors to identify them have been developed by the American National Standards Institute (ANSI), the Occupational Safety and Health Administration (OSHA), and the International Organization for Standardization (ISO). For example, the word *danger* and the color red are used only when serious injury or death may result. *Warning* (orange) and *caution* (yellow) warn of less serious risks.

Safety labels should be attached to products where users will see them before and as they use the product. This principle may seem obvious but is not always followed. For example, a printed warning to thaw a turkey slowly in the refrigerator to avoid food poisoning was placed inside the frozen cavity of the turkey where it could not be found until the turkey was already thawed!

The Editor's Legal Responsibility

Like everyone else who is involved in product distribution, writers and editors have some responsibility in the eyes of the law for safe use of a product. They can be named in product liability lawsuits, though the usual practice is to name the company. A sense of professional responsibility as much as respect for the law encourages editors to be careful in the legal edit.

Libel, Fraud, and Misrepresentation

Libel is a defamatory statement without basis in fact that shames or lowers the public reputation of an identifiable person. People who can prove libel may win damages from a publisher. The possibility of being sued for libel worries editors of fiction and periodicals more than it does technical editors because people are more likely to be discussed, referred to, or otherwise cited in works of fiction and in periodicals than in technical documents. Nevertheless, all editors should read alertly for facts to verify the accuracy of any negative statements that may be made about individuals.

Fraud and misrepresentation deceive the public. Misrepresentation may occur in labeling products or making claims about them or in claims about credentials of individuals or about data.

Ethical Issues in Editing

Editors and others in technical communication have responsibilities that extend beyond meeting the requirements of the law. These responsibilities derive from some values that seem universal, no matter what their philosophical or religious bases, such as the responsibility not to harm others. This principle enables people to live together in groups and with some trust in others. It may reflect the wish for survival ("I won't hurt you if you won't hurt me") as well as more generous and lofty principles of mutual responsibility. Cultural and professional values help people anticipate the consequences of actions and make judgments in particular cases.

Ethical controversies arise when the goals of expedience, profit, or convenience of one stakeholder conflict with the best interests of other stakeholders. In technical communication, a variety of stakeholders and some of their interests include the following:

> *Users:* safety, access, quality, ease of use, accuracy, privacy
> *Clients:* economy, time to produce, marketability, quality, safety, confidentiality
> *Employer:* ability to do business
> *Profession:* reputation of the profession, fair wages for others in the profession
> *Environment:* safety, health of the people who share the environment

Users, Clients, and Employers

Users count on safety, privacy, and accuracy of information. They also value ease of access, ease of use, and quality products. Values in society prohibit discrimination on the basis of age, physical impairments, class, race, and gender. For example, the accessibility guidelines for the Web require a text alternative for an illustration. Software can recognize and read the text to a user with visual impairments. Providing that text costs a company very little in production time. But a company that needs a profit to stay in business may cut other corners in document production. For example, they may skip the usability test or even the editing in order to get the product to market sooner and to save development costs. The result may be clumsy and frustrating instructions. Is such behavior unethical? It's easy to oversimplify. It's not ethical to take shortcuts if doing so endangers a user or promises to enable a user to complete a task and then prevents that very completion. But if the risk is only the user's inconvenience, the ethical behavior is less clear. People make judgments and choose tradeoffs every day without violating ethical guidelines. But misrepresentation, safety compromises, and breaches of privacy are not negotiable.

Acting ethically may be even more demanding in some situations than acting legally. For example, a contractor who read the specification for "stainless steel nuts and bolts" and used a cheaper, less durable metal for the bolts acted within the law (arguing that "stainless steel" modified only "nuts") but did not act ethically, considering that the bolts caused the structure to fail.[1] The contractor's interest in profit conflicted with the users' interest in safety, the owner's interest in a fair return on investment, and the interest of an environment not burdened with unnecessary waste. The contractor may have technically acted within the law but did not act ethically.

Misrepresentation of Content and Risks

Because of the serious, informative purpose of technical communication, readers expect it to be truthful. They may approach advertising with enough skepticism to question claims that are made, but because readers expect technical communication to be accurate, writers and editors have a particular responsibility for accuracy.

---

[1]This example was described by John Oriel, 1992, p. 172, in Editing engineering specifications for clarity, *Conference Record, IPCC 92—Santa Fe*, RT 9.5/168–173.

Misrepresentations deceive readers and may cause them to make mistakes in procedures, interpretations, or decisions.

Intentional misrepresentations, such as faulty data in research or false claims about ingredients or features, are illegal, but misrepresentation is also an ethical issue because there is no certainty that instances of misrepresentation will be identified and penalized by the law.

Misrepresentation may be unintentional, as in a graph in which the segments are not mathematically defined or perspective that inadvertently emphasizes one piece of information. Editors have an ethical responsibility to understand the structure and style of visuals as well as words so that they can correct unintentional misrepresentation.

Misrepresentation includes omission as well as inaccuracy. Omissions are especially significant when there is danger of harm. Risks of purchasing or using a product should be documented. The editor may have to ask the writer directly what the risks are and not just wait to be told.

Professional Codes of Conduct

Technical communicators, like physicians, lawyers, professors, and clergy, consider themselves professionals rather than staff or laborers. Such classification brings privileges of status and material benefits—and responsibilities as well. Professionals accept responsibility for their work rather than deferring to a superior. It is unprofessional in any setting to "pass the buck," or to blame someone else for a problem for which the professional person bears some responsibility. The editor's ethical responsibilities increase along with other responsibilities—a comprehensive editor is more responsible than a copyeditor. Yet all editors have a part in ensuring the quality and integrity of documents.

Figure 22.1 shows the ethical guidelines approved by the Society for Technical Communication. Like statements by other professional associations, these guidelines affirm professional competence as an ethical requirement for communicators. This competence refers to "truthful and accurate communications"—writing and editing well and pursuing accurate and understandable expression.

The guidelines also affirm adherence to laws, including laws regarding confidentiality, as well as to honesty and fairness. Thus, the guidelines draw on general values (honesty and fairness), laws, and the expectations for professional performance in a particular field.

One point of conflict and confusion in the ethics of technical communication relates to the professional value of "objectivity." Objectivity seems to represent truth uncontaminated by the bias of opinions. The goal of objectivity is that the person (the "subject") should get out of the way and let "facts speak for themselves" (the facts are presumed to be objects separate from the subject). Should the communicator remain neutral, taking no position but rather translating technical information accurately for the public? Furthermore, urged to adapt to an audience, should the communicator defer to an audience's wishes? These slippery words—*objectivity, neutral, translation, adaptation*—appear in various discussions that attempt to define the values and ethics of technical communication.

STC Ethical Guidelines for Technical Communicators

As technical communicators, we observe the following ethical guidelines in our professional activities. Their purpose is to help us maintain ethical practices.

Legality
We observe the laws and regulations governing our professional activities in the workplace. We meet the terms and obligations of contracts that we undertake. We ensure that all terms of our contractual agreements are consistent with the STC Ethical Guidelines.

Honesty
We seek to promote the public good in our activities. To the best of our ability, we provide truthful and accurate communications. We dedicate ourselves to conciseness, clarity, coherence, and creativity, striving to address the needs of those who use our products. We alert our clients and employers when we believe material is ambiguous. Before using another person's work, we obtain permission. In cases where individuals are credited, we attribute authorship only to those who have made an original, substantive contribution. We do not perform work outside our job scope during hours compensated by clients or employers, except with their permission; nor do we use their facilities, equipment, or supplies without their approval. When we advertise our services, we do so truthfully.

Confidentiality
Respecting the confidentiality of our clients, employers, and professional organizations, we disclose business-sensitive information only with their consent or when legally required. We acquire releases from clients and employers before including their business-sensitive information in our portfolios or before using such material for a different client or employer or for demo purposes.

Quality
With the goal of producing high quality work, we negotiate realistic, candid agreement on the schedule, budget, and deliverables with clients and employers in the initial project planning stage. When working on the project, we fulfill our negotiated roles in a timely, responsible manner and meet the stated expectations.

Fairness
We respect cultural variety and other aspects of diversity in our clients, employers, development teams, and audiences. We serve the business interests of our clients and employers, as long as such loyalty does not require us to violate the public good. We avoid conflicts of interest in the fulfillment of our professional responsibilities and activities. If we are aware of a conflict of interest, we disclose it to those concerned and obtain their approval before proceeding.

Professionalism
We seek candid evaluations of our professional performance from clients and employers. We also provide candid evaluations of communication products and services. We advance the technical communication profession through our integrity, standards, and performance.

Figure 22.1 STC Ethical Guidelines

(www.stc.org/ethical.asp)

Reprinted by permission of the Society for Technical Communication.

One approach to answering these questions is to consider whether facts can ever speak for themselves and whether anyone can ever be neutral. As you learned in Chapter 2, readers interpret based on their prior experiences and knowledge, and different readers will find different meanings in the same texts. The object can never be entirely separated from the subject, or the person who observes and interprets the object. Furthermore, the acts of arranging information in sentences or on a page influence interpretation. As you saw in the chapters on style, for example, the independent clause is more emphatic than a dependent clause, and a writer chooses to emphasize a point by placing it in an independent clause. Presumably the writer will aim to reflect the content in this choice of structure, but still the writer is making a choice. Writing is never entirely neutral, even when the subject is materials and procedures. The claim of neutrality may be avoidance of commitment, an excuse not to get involved.

The nature of the expertise that enables people to call themselves "professionals" also influences answers to these questions. Expertise generally requires specialized education and perhaps experience. As a consumer of services by professionals, such as your physician, you expect information about options even if you claim the right to make the ultimate choice. You respect your own limitations enough to listen to the options and welcome them.

Likewise, your client or manager may depend on your expertise. It would be unethical to withhold it because in so doing you might jeopardize their ability to make the best decision. For example, your client or manager may request the use of a particular type of software that you know is less efficient than another or may not know the environmental consequences of petroleum-based inks for printing compared to soy inks. If you remain silent, they may never know the options. Or you may resist making recommendations in a feasibility study because the reader has higher status in the organization than you do. In all those cases, your professional expertise carries with it a responsibility to inform a client or reader, even when the reader may not want to hear your information. Your judgment represents the collected wisdom and knowledge of the profession as represented by your expertise, not something arbitrary and personal.

Ethics in technical communication involves more than reporting accurately what a subject matter expert says. It requires more than passively telling the truth. It also requires a proactive position in offering information and helping to guide the decisions of people who lack the specialized knowledge of the communicator and therefore do not see all the consequences of choices.

Environmental Ethics

Environmental ethics, like all other ethics, are based on the principle of responsibility for the social good as balanced against short-term prosperity. As consumers of paper, electricity, and inks, technical publications groups use products that harm the environment and, by extension, human health. The use of these products is ethical according to cultural values, but misuse is not because the danger to health and safety increases disproportionately to the benefits of use. Corporate policies can encourage product choices that are environmentally friendly. Policies as well as individual practice can minimize waste such as from excessive photo-copying and printing.

The paper and publishing industries spill industrial toxins into the air and water. Many of those toxins are dioxins used to bleach paper. Dioxins cause cancers and birth defects. Recycling paper and using recycled paper cuts down on waste because the manufacture of recycled paper is a less toxic process than the manufacture of new paper, and it also consumes fewer trees. Uncoated papers are more readily recycled than coated papers, and white paper is more readily recycled than colored. White, uncoated paper is thus preferable except in special circumstances. Processes that minimize multiple draft copies should be encouraged.

Other choices of materials and processes can also minimize environmental and health consequences. Inks made from soybean oil are made from a replenishable agricultural product rather than from petroleum, and UV-curable inks eliminate solvents in manufacturing. Conservative use of electricity and plastic is environmentally as well as economically responsible. Electricity used to run computers, printers, and photocopiers may be generated by fossil fuels, which contribute to global warming. Plastic wrappers for manuals and other packaging will probably become landfill and may be superfluous. Thus, choices about papers, inks, and other consumables affect not just whether readers get the information they need but also general health and welfare of the community. As with all ethical issues, responsible corporations and individuals look beyond immediate needs and convenience to longterm and broad consequences of choices.

Bases for Ethical Decisions

Ethical decisions can be based on principles (such as honesty), law, professional standards, and assessment of consequences.

In a survey of technical communicators, Sam Dragga identified eight types of explanations for decisions about the ethics of a communication decision (see Further Reading). The most common reason was *consequences*: what happens to readers as a result of that choice. The extent to which a choice is unethical is proportional to the degree of deception or injury to the reader.

Other explanations are document *specifications* and *common practice*. Specifications refer to document descriptions, such as number of pages. Common practice establishes some expectations. If it is common practice in an employee evaluation to emphasize achievements and deemphasize limitations, then the potential deception of doing so seems less grave than misrepresentation of data in a research report. Readers may not be deceived by emphasis if practice leads them to expect it.

Technical communicators generally rejected the explanation that the burden for accurate interpretation rests on readers rather than writers.

If an important basis for ethical choice is consequences, then editors and writers must consider consequences broadly—longterm as well as immediate consequences; to secondary as well as primary readers; to the environment as well as the corporation; and so forth.

Establishing Policies for Legal and Ethical Conduct

An editor or writer working for a company is only one person among many who are responsible for products and their documentation. Although any professional person should accept personal responsibility for the safety and welfare of others, ethical behavior in a corporate setting is also a social and cultural issue. To increase chances of the organization and its employees following high legal and ethical standards, and to minimize the potential for conflict of an individual employee against the organization as a whole, the company should have policies defining what constitutes ethical communication, procedures to include training of staff in ethical issues, and document review for adherence to the policies.

A first step for an employee is to review existing policy statements and the house style manual. If potentially useful sections are missing, or if some require revision, a proposal to research and revise the manuals could be developed. Such revisions should be the responsibility of a group of people representing different components of the organization, including technical communicators, engineers, managers, and lawyers. Developing or revising these policies as a group helps to create the sense of group responsibility for adherence to high standards of legal and ethical behavior.

A style and publications procedures manual might cover these policies and procedures:

- identification of trademarks
- registering copyrighted materials
- requesting permission to use material copyrighted by others and forms or letter templates for doing so
- visual and verbal identification of warnings and hazards
- accurate representation of verbal and visual information
- ethical use of resources and procedures to minimize waste
- review of ethical and legal issues in each document. The review should be the responsibility of various employees, including but not limited to the editor.

The employee policy manual should include commitments to

- legal and ethical conduct by the organization and its employees with examples specifically related to publications, including commitment to accurate representation of a product's value and limitations and instructions and warnings that accommodate the needs of users
- organizational as well as individual responsibility for legal and ethical conduct

Such statements affirm corporate responsibility and encourage individuals to act ethically by creating an environment in which ethical actions are expected and respected. They provide guidance for making ethical decisions on a case-by-case basis. They may prevent problems from occurring and support editors if problems do occur.

Using Your Knowledge

The stories that make the news about ethical failures in businesses and other organizations are usually stories of egregious failures that compromise the well-being of numerous people. Most ethical and legal decisions, by contrast, are tiny and daily. Few reflect life-death issues nor even issues that will result in lawsuits. But daily ethical and legal behavior creates work groups, businesses, and a society in which individuals can thrive.

As a responsible member of the profession of technical communication with a commitment to other members of the profession as well as to readers and users, as an employee, and as a good citizen who accepts responsibility for the quality of life around you, you can take actions to increase the chances of making legal and ethical decisions.

1. Be informed. Read "Copyright Basics" at www.copyright.gov (or the comparable information for another country in which you are editing) to start.
2. Respect the stakeholders. Thinking about the effects of your work on multiple stakeholders will guide you to ethical and legal decision making.
3. Encourage your company to develop policies that will ensure a culture in which it is expected that people will follow the best practices of working within the boundaries of legal and ethical behavior.

Further Reading

Copyright Basics (Circular 1) and Copyright Notice (Circular 3). Copyright Office, Publications Section, LM-455, Library of Congress, Washington, DC 20559. Also available at www.copyright.gov.

Dombrowski, Paul. 1999. *Ethics in Technical Communication*. Longman.

Dragga, Sam. 1996. Is it ethical? A survey of opinion on principles and practices of document design. *Technical Communication* 43.3: 255–265.

Fishman, Stephen. 2003. *The Copyright Handbook: How to Protect & Use Written Words*. 7th ed. Nolo.

Phillips, Jerry J. 2004. *Products Liability in a Nutshell*. 6th ed. West Publishing Company. See especially Inadequate warnings and instructions, and misrepresentations, pp. 207–234.

Velotta, Christopher. 1987. Safety labels: What to put in them, how to write them, and where to place them. *IEEE Transactions on Professional Communication*. PC 30.3: 121–126.

Discussion and Application

1. For this activity, work in a discussion group with two or three classmates. Or, if your instructor directs, write out a response to one or more of the situations.

   For each of the situations that follow, discuss:
   - whether and on what grounds the situations might have legal or ethical dimensions
   - policies that might minimize conflict between individuals or groups within the organization regarding the document options or choices
   - options for action

   a. You are writing online help for the operating system of a new computer. As with any system, failure to follow certain instructions can result in loss of information. You try to call the user's attention to these places by writing WARNING and highlighting the word with color. Your manager objects to the emphasis with the argument that too many warnings will make the product seem deficient. The legal department confirms that there is no breach of law if the warnings are less frequent or less emphatic.

   b. U.S. law requires the manufacturers of cigarettes and beer to include a warning on each package or container about the hazards to health of using the product. Some companies comply with the law by printing the warning in all capital letters using condensed type. These typographic choices make the warning hard to read.

   c. User manuals for electronic equipment frequently include a page or two of warnings at the beginning of the manual. There may be as many as thirty warnings about such hazards as shock from using the product in the bathtub, heat damage to the equipment from improper ventilation, and dangers of frayed cords. Using principles of organization from Chapter 17 and theories of reading from Chapter 2, discuss the effectiveness and ethical significance of grouping multiple warnings at the front of the manual. What would be the legal and managerial reasons for doing so?

   d. A booklet for prospective investors represents the increasing value of an investment over time with a line graph that shows a sharp increase in value. Narrow segments on the horizontal axis and shading of the area beneath the line exaggerate the steepness of the line. Although the booklet is a marketing tool, its understated (not slick) style and factual information suggest the goal of informing rather than persuading. Does its marketing purpose make the exaggeration ethical?

2. Check user manuals, appliances, and power tools that you own to see how manufacturers have met their duty to instruct consumers in safe use and to warn them of dangers. Evaluate whether the instructions and warnings might be improved. Prepare a brief report of your findings.

3. Technical communication emphasizes "audience adaptation"—adjusting content, style, and design of documents to the needs, backgrounds, and uses of readers. An extension of this principle might be to give clients what they ask for. Discuss whether there are situations in which audience adaptation might go so far as to breach ethics in technical communication.

23 Type and Production

As the link between the development and the publication of documents, editors know about type and production. Type refers to the shape and size of the letters on the page. Production refers to the preparation of documents in the form in which they will be published: print or digital. A production editor may supervise publication after development editors and copyeditors have prepared the text. However, all editors should know what happens to the document once the text is established, because choices about production influence early editorial decisions and schedules.

This chapter provides an overview of type and production and the services that commercial printers can provide. Its purpose is to enable you, as editor, to communicate with graphic designers and printers. The orientation of the chapter is to print, though the principles also apply to online documents.

Key Concepts

The size and look of a font as well as letterspacing, linespacing, and wordspacing affect how easy it is to read type. Fonts that work for print may not be best for reading online. Novice designers should be conservative in their use of type but should also look at the space of the page differently than they do for course papers or typescripts. Fullscale commercial printing requires that illustrations be digitized or converted to halftones: dots of varying sizes. Color printing requires separation into the four colors of cyan, magenta, yellow, and black. In fullscale printing, books are usually printed on large sheets of paper and then folded and trimmed. Editors must manage their project schedules to allow for the steps in production.

Working with Type

Letters within the English alphabet can be recognized in a variety of shapes and sizes. A *typeface,* or a *font,* is a collection of characters with distinguishing shapes and a name, such as Times Roman or Helvetica.[1] Typefaces and sizes and the spacing between words and between lines of type affect the ease of reading. Typefaces also express different qualities of character, such as professional, avant garde, or elegant, that shape the way readers respond to a document. Using type well is an art, but the principles underlying the uses of type are also based on research.

Fonts and Their Uses

Fonts (typefaces) are chosen for readability. They are generally divided into four categories:

- **Display**. Display fonts are used for posters, signage, and sometimes book covers. Display fonts are used when information will be read at a distance—for example, a poster on the wall for an upcoming workshop will require a display font. Fonts such as Aachen Bold and Arial Black are typically used for display.
- **Decorative.** Decorative fonts are the elaborate fonts and script fonts such as Nuptial Script, which is commonly used for wedding invitations. These fonts should be used only for limited amounts of text as they are difficult to read.
- **Serif.** Serifs are cross strokes at the ends of the main strokes of letters. (See the illustration that follows.) The serif fonts are the most commonly used for the body text in documents. The serifs may increase the ease of reading on paper because the horizontal lines of the serifs propel the eye across lines of type. The serif styles are also familiar to North American readers, and with familiarity comes comfort in reading. Times New Roman and Garamond are popular serif fonts and are good choices for the body of most print documents. Georgia has been developed for online documents.
- **Sans Serif.** The sans serif fonts (without serifs) are generally used for headings in documents although many of the sans serif fonts are quite readable for body copy. Sans serif fonts are an excellent choice for the body text when you need to conserve space in a document. Helvetica is a commonly used sans serif font. Sans serif fonts are used widely on the Web, including Verdana, developed for on-screen display.

You can see examples of serifs in both the lowercase and capital letters in the words "Times Roman" below. Both vertical and horizontal cross strokes appear in the capital letters. A sans serif style omits the cross strokes. The strokes for the capital letter H in "Helvetica" simply end.

| Times Roman | Helvetica | TIMES ROMAN | HELVETICA |

The examples below show a few of the many typefaces available. Serif styles include Bookman, Garamond, and Schoolbook.

| **Bookman** | Schoolbook | Avant Garde |
| Garamond | **Arial Black** | Verdana |

---

[1]In this chapter, "typeface" and "font" are used as synonyms, reflecting current practice, which has been influenced by the language of word processing and page layout software. Traditionally, a font was a complete set of characters in one size of a typeface.

You can see how Arial Black could be difficult to read in paragraphs, but it can be attractive and effective for display type. Verdana was developed for the Web to accommodate the low resolution of monitors and the organization of space by pixels.

Font Selection

Thousands of fonts are available, both as freeware and for purchase. When selecting a font for a publication, choose a complete font, one that contains all the characters, such as nonbreaking ellipses, accents, and typographer's ("smart") quotes, that your publication will require. If the text will be translated, the font should include the characters for the languages into which the text will be translated. Also, make sure the font will print reliably. Many of the inexpensive or free fonts look fine on the monitor but do not print properly on a laser printer or on the high-end printing equipment used by commercial printers. A good guideline is to choose a PostScript font from a well-known font supplier such as Adobe. If you are going to use special fonts, check with your printer regarding compatibility. If you will work with a professional typesetter, that person can show you a type specimen book with examples to help you choose.

The resolution is lower for online displays than for print. This means that fine details are harder to represent online, and fonts that look good in print may be hard to read online. Fonts that have been developed for low-resolution online display, such as Georgia and Verdana, use wide characters with a tall x-height (the part of the character that is as tall as the letter x).

Type Size

Commercial printers use the pica as a standard measure. The replication of a pica ruler below compares picas and inches. It shows that six picas equal approximately an inch. Type is measured in points, with a point equal to one-twelfth of a pica. Thus, 12-point type will fill slightly less than a pica of vertical space, and six lines of 12-point type will fit in approximately one inch of vertical space.

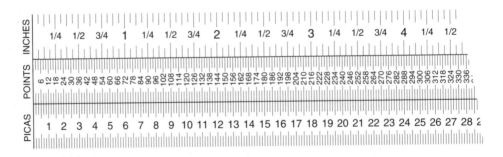

The measures are not exact, however, and 12-point type in one typeface will be larger than 12-point type in another typeface. In the examples that follow, different typefaces in the same size fill different amounts of vertical and horizontal space.

This line is set in 12-point Times Roman.

This line is set in 12-point Helvetica.

This line is set in 12-point Bookman.

The parts of the letters have different proportions in the typefaces. For example, Helvetica and Bookman letters have a proportionately large x-height, the part of the letter equivalent in height to the lowercase x. Thus, the ascenders (the projections above the body of the letter, as on the b and h) and the descenders (the projections below the body of the letter, as on the p and y) are proportionately shorter than in Times Roman.

The capital letter M is shown below in various sizes in the typeface Times Roman.

MMMMMMMMMMMMMM
6 8 9 10 12 14 18   24   30   36   48   60   72

The choice of typeface and type size can make a difference of a number of pages in a long document such as a manual or book. It can determine whether a story fits into one column in a newsletter or spills into a second column. According to a type specimen book, 10-point Times Roman will fit 2.6 characters per pica, whereas 10-point Helvetica will fit 2.4 characters per pica. A page that is 42 picas deep, with each line being 25 picas long, has 1,050 picas of space available (42 × 25). In Times, 2,730 characters would fit on the page (1,050 × 2.6), but in Helvetica, 2,620 characters would fit (about 20 fewer words). The numbers may not sound so different for one page, but a book of about 180,000 words (such as this textbook) would require about 20 more pages set in Helvetica than in Times.

Most blocks of text—the body copy as opposed to the headings or display copy—are set in 10-point or 12-point type. Type smaller than 9 points is difficult to read and is reserved for footnotes, indexes, and other material that readers will read selectively and in small quantities. Type that is 14 points or larger is reserved for children's books, large-print books for visually impaired readers, and display copy.

Headings may be set in the same size as the body copy if they are set in boldface. Boldface alone will let readers distinguish visually the headings from the body. For headings of different levels, you may distinguish the levels with space alone. For example, a level-one heading could be centered but in the same size as a level-two heading set on the left margin.

If size is the only variable in the headings, you will have to skip at least two sizes between levels if readers are to recognize the levels. Thus, if a level-one heading is left justified in 14-point type, a level-two heading can be set in 10-point type.

Type size on websites is often relative rather than absolute, and it can be controlled by the browser and user (see Chapter 5). HTML defines font sizes on a scale from 1 to 7, with 7 being the largest and 3 being the default size. Each increment is about 20% larger or smaller than the size next to it; thus size 4 is about 20% larger than size 3. Designers should prepare pages to handle a range of font sizes and check them on different monitors.

Leading, Letterspacing, Wordspacing, and Line Length

Various options are available for the treatment of individual letters, words, and lines of type—including leading (space between lines), wordspacing and letterspacing, and line length.

Leading and Linespacing

Type would not be readable if the descenders in one line crashed into the ascenders of the line below. Thus, good typesetting includes leading (pronounced *ledding*) between the lines. The term derives from the days when type was set from metal molds. Strips of metal were placed between the lines of characters. Even if the type is set solid (without leading), there will be a tiny bit of space between the lines. Most type for body text in print is set with one or two points of extra leading. Thus, 10-point type may be set on a 12-point line. The expression "set 10 on 12" and the marginal notation "10/12" both direct the typesetter to use 10-point type on a 12-point line.

If you check the format/paragraph menu in Microsoft Word or other word processing program, you will see an option for specifying linespacing. Although an option for "double" or "single" spacing is available, remnants of typewriter days, professional typesetters describe linespacing in terms of points. The default spacing for "single" inserts an appropriate amount of leading, but if you were going to adjust the leading, you would specify an appropriate linespacing using "exactly" and the point size. For example, you might choose 10-point type from the font menu but specify linespacing of exactly 12 points from the paragraph menu.

The following examples show type set with the right amount of leading, with too little leading (solid), and with too much leading. The leading affects the ease of reading. The first paragraph is set in 10-point Helvetica on a 12-point line. The extra two points of leading are especially important for sans serif type in order to help readers distinguish lines and read comfortably.

Right Amount of Leading

Leading, the space between lines of type, helps the eye move horizontally. Too little leading increases reading difficulty because the type seems crowded and the letters cannot be distinguished. Too much leading makes the text look childish. The right amount of leading is comfortable to read and helps the eye find the correct line when it moves from the right back to the left margin.

The next paragraph is set in 10-point Helvetica on a 10-point line (no leading). The crowding increases reading difficulty and the chance that readers will skip lines when they look for the next line at the left margin.

Too Little Leading

Leading, the space between lines of type, helps the eye move horizontally. Too little leading increases reading difficulty because the type seems crowded and the letters cannot be distinguished. Too much leading makes the text look childish. The right amount of leading is comfortable to read and helps the eye find the correct line when it moves from the right back to the left margin.

The final paragraph is set in 10-point Helvetica on an 18-point line. The extra space, besides being wasteful and looking unprofessional, could interfere with comprehension in a long document by spatially separating related ideas.

Too Much Leading

Leading, the space between lines of type, helps the eye move horizontally.

Too little leading increases reading difficulty because the type seems crowded

and the letters cannot be distinguished. Too much leading makes the text look

childish. The right amount of leading is comfortable to read and helps the eye

find the correct line when it moves from the right back to the left margin.

The longer the line, the more leading is needed (to help the eye move horizontally). Sans serif type generally requires more leading than serif type, and larger type sizes require more leading than smaller type sizes.

Line depth is always measured baseline to baseline. The baseline is formed by the bottom of the letters excluding descenders. The T-square feature of some pica rulers makes accurate measurement of line depth easy.

Letterspacing and Wordspacing

The horizontal spacing as well as the vertical spacing affects readability. Typesetting uses proportional spacing, meaning that each letter receives space proportionate to its width. By contrast, most typewriters and older computer printers give each letter the same width so that an *i* takes as much space as an *m* (monospacing). Proportional spacing makes type easier to read.

Letterspacing refers to how close together the letters are set; wordspacing refers to the amount of space between words. As with leading, too much or too little spacing interferes with reading. The following examples show letterspacing that is just right, too tight, and too open. All three paragraphs are set in the same size (10/12-point) and face (Times Roman) of type.

Correct Letterspacing

Correct letterspacing helps readers recognize words because they can easily identify the shapes of the words. It also affects the number of times the eye stops in its move across a line of type. Too much letterspacing requires the eyes to make more stops than desirable. The eye has to make a wider sweep for fewer words. The crowding of condensed type also increases reading difficulty.

Too Tight Letterspacing

Correct letterspacing helps readers recognize words because they can easily identify the shapes of the words. It also affects the number of times the eye stops in its move across a line of type. Too much letterspacing requires the eyes to make more stops than desirable. The eye has to make a wider sweep for fewer words. The crowding of condensed type also increases reading difficulty.

Too Open Letterspacing

Correct letterspacing helps readers recognize words because they can easily identify the shapes of the words. It also affects the number of times the eye stops in its move across a line of type. Too much letterspacing requires the eyes to make more stops than desirable. The eye has to make a wider sweep for fewer words. The crowding of condensed type also increases reading difficulty.

You can trust a reputable typesetter to letterspace correctly. If you are composing type using page layout or word processing software, use default spacing rather than playing with the computer's capacity for expanding or condensing type.

Line Length and Margins

The amount of space between words and, to some extent, between letters is partly determined by line length and justification of margins. The left margin is almost always straight in body copy; that is, the beginning letters of each line align vertically, so the text is left justified. If the lines of type align on the right margin, the text is right justified (or right aligned). If the lines align on the left but not on the right, the text is set ragged right.

Left-Justified Text (ragged right)

In 2001, U.S. industries released 6 billion pounds of toxic chemicals into the air and water. The paper, printing, and publishing industries released 3% of the total.

Right-Justified Text

In 2001, U.S. industries released 6 billion pounds of toxic chemicals into the air and water. The paper, printing, and publishing industries released 3% of the total.

Centered Text

In 2001, U.S. industries released 6 billion pounds of toxic chemicals into the air and water. The paper, printing, and publishing industries released 3% of the total.

Left- and Right-Justified Text

In 2001, U.S. industries released 6 billion pounds of toxic chemicals into the air and water. The paper, printing, and publishing industries released 3% of the total.

To achieve both left and right justification, extra space must be inserted between words and letters. High-quality typesetting or page layout software can achieve the spacing in such a way that the eye often won't recognize the extra space, but even with good tools, the wordspacing may be obviously exaggerated. This space calls a reader's attention to the type rather than to the content and thus interferes with reading. The likelihood of exaggerated spacing is greater with short lines, such as newspaper columns, and with right-justified text. Many designers, therefore, prefer ragged right text. Perhaps the page as a whole looks a little less tidy (or, from another point of view, less rigid). The choice of ragged right, however, acknowledges the goal of readable type; in this case, the reader's ease in moving across lines of type supersedes the desire for a straight right margin.

Optimal line length is relative to type size. As a general guide for print, lines include 1 ½ to 2 ½ alphabets, or 39 to 65 characters. The optimal line for 12-point type will thus be longer than the line for 10-point type. Lines produced with this guideline will include 9 to 10 words. Another guideline is to make the line length 2 ½ times the type size—25 picas for 10-point type. If the line length is short and words are long, end-of-line hyphenation may be necessary to prevent excessively short lines.

On the computer screen, readers like short lines. They don't like to scroll vertically, but they may not even think to scroll horizontally, across the screen. Some website designers divide the screen into columns so that lines don't spread across the screen. A designer can specify absolute widths, but because you don't know what the reader's screen will display, it's safer to specify proportional widths and let the reader make adjustments in the size of the window.

Design Tips for Beginning Designers

If you will design pages for print without the guidance of a professional graphic designer, you can create good pages by following a conservative approach to page design and by aiming for consistency and simplicity. One of the first steps is to use the language of typography, not of typewriters. Instead of "double spacing," you may set a heading with 12 points of space before and 3 points after. Then your headings won't float between paragraphs but will attach visually to the paragraphs that they identify. You will educate your eye to look differently at the 8 ½ × 11-inch piece of paper than you do as a student typing double-spaced term papers. Because the page is too wide for a readable line of type, you may create columns or side headings or use graphics on one side to create lines of the right width (remember that this width is about 2 ½ times the point size, about 4 ¼ inches or 27 picas for 10-point type).

All designers, whether beginners or professionals, save time and create consistency by using a grid and by establishing templates for different document types using the styles feature of page layout and word processing programs (see Chapter 5). A grid organizes the space on the page. The simplest grid defines margins for one column of text, but even beginning designers can work with a page divided vertically in thirds. The left column becomes a place for side headings, illustrations, and comments. The two right columns may be joined to create a space twice the width of the left column. In this space, most of the text will appear. In the document templates, you can embed commands for spacing, type style, font, and size as well as indention and other features into styles and then apply all those commands simply by selecting the style.

Using characters that emulate typesetting will give your work a professional appearance—solid lines instead of two hyphens for em dashes, curly quotation marks, and nonbreaking ellipses. You should probably type just one space after a period (though some companies will prefer the style of two spaces). You can discard the typewriter style of emphasis (underlining), in favor of boldface or an increase in type size. You might use rules under or above headings or titles, but you will place them one to three points away from the text so that they don't break the descenders or ascenders of the type.

For a clean look, avoid mixing typefaces. It's safe to use a sans serif style for the headings and a serif style for the body text, but leave more elaborate design choices to professional designers. Stick with well-known typefaces, such as Times Roman, Garamond, and Helvetica.

Indentions and gutters between columns can be narrower than you might think—¼ inch will suffice (though you will need more for the fold of a brochure). Prefer left-justification to centering. On a brochure, align tops of panels, just as you left-justify type. Leave more space at the bottom than at the top so that the text doesn't seem to fall off the page.

And, include identifying information on your documents—including a date and a way to contact the company or you if necessary.

Working with Illustrations

Production editors often handle illustrations separately from the text. Simple line drawings created with graphics software may be part of the text file, but photographs may require professional attention. If they are available only as prints, they may need to be scanned professionally to create halftones of sufficient resolution that the print will resemble the original. (See definitions in the following section.) Many consumer-level scanners produce excellent results with black-and-white photography, illustrations, and small photographs. However, if you wish to use a photograph so it fills an entire page or use a photograph for a poster, you require a professional scan of the image. This professional scan ensures that enough digital detail is captured in order to reproduce a color image crisply on a printed page. The commercial printer can also help with halftones and advice on resolution.

Halftones

Photographs, or continuous tone art, include shades of black and white or full color. But printing presses use solid color inks. Before photographs can be printed, the shades in the artwork must be converted to dots of various sizes. Large dots densely grouped will appear as a dark shade when printed, while small dots widely spaced will look like light shades. The conversion from continuous tones to dots that look continuous is a *halftone*. The file of a digital photograph contains its

Figure 23.1   Halftone and Magnified Section

The illustration on the left is printed on a 133-line screen.

information in dots, but a printed photograph must be scanned or converted through a photographic process. Figure 23.1 illustrates a magnified halftone to show the dot composition.

For color photographs, a process called *color separation* is necessary to identify the four colors that are used in printing: cyan, magenta, yellow, and black (CMYK). These colors in combination can produce the full spectrum of colors. A halftone for each of the colors is prepared either by photography or by scanning. The document runs through the press four times, one time for each of the ink colors, with each new color being superimposed on the others. (See the discussion of printing that follows for more information on four-color printing.)

The creation of a halftone adds some expense. The cost is reasonable, but when you are getting bids for your document, you will need to specify the number of photographs to be included and how they will be supplied in order to get an accurate bid. Always check with your commercial printer and communicate the necessary specifications to the graphic designer.

Resolution

In order for digital illustrations and photographs to print properly, the images require sufficient digital information for the method of reproduction. The number of possible dots in a given area is called resolution. With high resolution, there are more dots per inch (DPI) than with low resolution. A color image of 72 DPI is called a low-resolution file and is fine for display on a website; however, it is not suitable for commercial printing. There is not enough digital information for a clear, crisp reproduction on paper. Black-and-white photographs or line art can be low-resolution files, though the more digital information, the better.

The DPI always needs to be considered in conjunction with the number of lines per inch, or LPI, used by the printer. Newspapers, for example, are printed at 70 or 80 LPI while glossy magazines have a line screen, or LPI, of 133. The higher the dots and lines per inch, the finer the quality of reproduction. The illustration in Figure 23.1 is printed with a 133-line screen.

It is important to consider the LPI and DPI when working with digital cameras. The higher you set the resolution on the digital camera, the fewer images you will be able to store on it. If you are producing a web and a print version, you will need both a low-resolution (72 DPI) and a high-resolution (300 DPI) version of your photographs. You will be able to save the high-resolution file to a lower resolution in your photo manipulation software.

Your printer can advise you on the required DPI for your job. Generally, a good formula is LPI × 2 = approximate DPI required. Color photographs for a printed document with a 133- or 150-line screen should be about 250 to 300 DPI.

Correction of Photographs

Most photographs you see in glossy magazines and brochures have been digitally corrected. Black-and-white photos are adjusted so there are no overwhelming areas of dense black nor stark white. Instead, the image is adjusted to provide a continuum of tonality. Color photos are also corrected in order to fix problems with oversaturation and color balance. This type of work is best left to a graphic designer. This service does add to the cost of a print job, but the finished product will be more satisfactory.

Photographic Releases from Subjects

Editors should obtain a signed release form from any individual in a photograph. On this form the individuals give their consent for you not just to publish their photo but also to digitally alter it or reuse it in other publications. If your company has a legal department, it will be able to provide such a consent form. Otherwise, prepare a letter stating where you will use the photograph and for what purposes and how you may alter or reuse it. The permission that the individual signs covers only the stated conditions.

Choosing Paper

Production editing requires decisions about paper as well as preparation of the text and illustrations. Paper is a factor in the desired resolution of images, for example. Paper, like typeface and layout, affects readability. Paper may be too transparent and allow shadows of the print from the reverse side to show. Paper may not accept ink well and the type may smudge, or it may be too absorbent and the type may bleed. It may have a shiny coating that glares. Criteria for paper selection include size, weight, opacity, and finish as well as color. Post-consumer recycled content and potential to be recycled are also criteria for choice.

Much printing for large quantities is done on large sheets of paper that are later trimmed to the proper size or on large rolls of paper. The standard size of "book" paper in the United States is 25 × 38 inches. (Sizes vary slightly in coun-

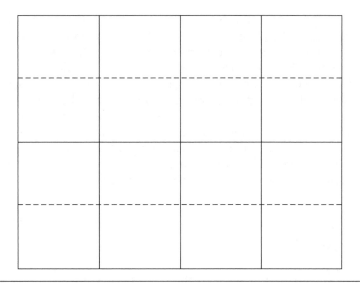

Figure 23.2   Book Paper, 25 × 38 Inches, Showing Cuts for Pages

tries using the metric system.) As the diagram in Figure 23.2 shows, a sheet of paper this size will yield eight 8 ½ × 11-inch folios (a page that can be printed front and back) or sixteen 6 × 9 folios. The broken lines show additional cuts for the sixteen folios. Extra space is available for trimming.

The standard size for "bond" paper is 17 × 22 inches. This paper yields four 8 ½ × 11 folios. "Cover" paper measures 20 × 26 inches.

Standard paper sizes determine standard page sizes. It is possible to print a book in a size other than 6 × 9, 7 × 9 ¼, or 8 ½ × 11, but the per page cost will increase because of the waste of paper. Standard paper sizes also determine desirable book length, generally a multiple of 16 pages. Each group of pages printed and bound together is called a signature. The most common signature size is 8 pages or 16 pages. Thus, a book of 97 pages rather than 96 will require a whole extra sheet of book paper for one page of text. Editing to condense slightly will save money.

The basis weight of the paper is derived from the weight of a ream of the paper in its uncut size. Thus, a ream of book paper may weigh 60 pounds while a ream of the same weight bond paper weighs 24 pounds. The following table shows equivalent weights of different kinds of paper.

Book Paper	Cover Paper	Bond Paper
basis: 25″ × 38″	basis: 20″ × 26″	basis: 17″ × 22″
45 pounds	25 pounds	18 pounds
50	27	20
60	35	24
70	40	28
80	45	32

For books and textbooks, 50- or 60-pound book paper is generally satisfactory. Classier publications may use 70- or 80-pound book paper.

Opacity and finish, as well as weight, affect the character of the document, success of the printing, and the extent to which the paper supports or interferes with the document's use. Opacity refers to how easy it is to see the ink through the page. Shadows from the reverse of the page distract readers. Paper weight is just one factor in opacity. Finish, or treatment to the surface of the paper, affects how well the paper accepts ink and whether the paper seems elegant or businesslike. Finishes, arranged in order of increasing smoothness, are known by these names: antique, eggshell, vellum, and machine finish. Coating makes the paper smoother and minimizes the spreading of inks and the blurring that results. Photographs will reproduce more crisply on coated paper. The raised patterns on some fine stationeries are created by embossing.

Understanding the Production Process for Print Documents

Production requires the creation of production ready pages, either through desktop publishing or through commercial typesetting and page makeup; printing either through digital photocopiers or by offset lithography; and binding. The production process is represented by the flowchart in Figure 23.3.

Desktop Publishing and Digital Printing

The world of printing has evolved rapidly in the last two years and continues to do so. Many companies that produce technical documents bypass commercial services by using desktop publishing to develop originals of the pages and by printing and binding in house. A production department may provide help with page layout so that the writers can spend their time developing content.

High-end digital photocopiers are excellent machines for producing a small quantity of a publication at a reasonable cost. These digital machines can do both black-and-white and color printing and are able to collate and bind publications easily. Commercial printers also provide digital printing services. Offset printing still remains a cost-effective method for large quantities.

Fullscale Commercial Services: Typesetting, Page Makeup, and Offset Printing

For documents that use many photographs or that will be printed in large quantities, or if there is no production department, it may be cost effective to use commercial services for composition and offset printing. The editor may provide camera ready originals and ask only for printing and binding. Alternatively, the editor may also contract for composition (typesetting and page makeup) and preparation of illustrations. If the document will be printed by offset lithography, the pages will be grouped into signatures (8 or 16 pages) for printing on book paper (large sheets). Metal plates are made from the signatures and then affixed to the printing press. Once the sheets are printed, they are folded, trimmed, and

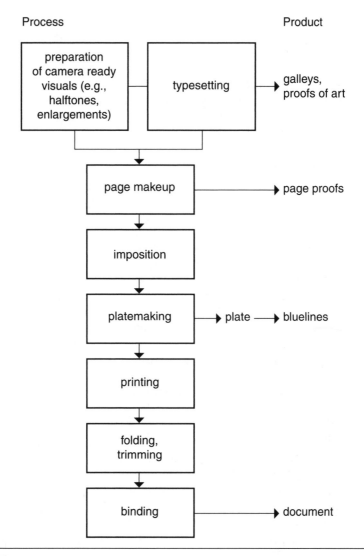

Figure 23.3   The Offset Printing Production Process

bound. During this process, the editor will be in contact with the printer to ensure that the job is proceeding on schedule and that the document is being printed correctly. There may be several stages of proofreading. More detailed information on this process follows.

Typesetting and Page Makeup

If the editor is satisfied that the text is reasonably correct and will not require substantial changes after typesetting, the text may be typeset directly into pages. The first proofs will be page proofs. But if corrections might change the page breaks, it is more cautious to begin with galleys, long sheets of text in the chosen typeface, type size, and line length. The galleys are proofread and corrected before they are made up into pages. Corrections could change the amount of type on a given page, and the pages would have to be redefined.

Once the galleys have been corrected, pages are made up, including placement of illustrations. The page proofs are then checked against the proofread galleys to ensure that no copy was omitted and to confirm the accurate correction of galleys, the correct sequence of pages, the proper placement of illustrations, and meaningful page breaks.

Imposition and Platemaking

When the page proofs have been proofread and corrected and illustrations approved, the proofs and illustrations are arranged into signatures. The pages must be arranged so that they will be in sequence when the signature is folded and trimmed. They must be arranged so that top and bottom margins of pages in the book or other document will be the same. Figure 23.4 shows one sequence of pages for a book. Note that the pages across the top are upside down. The page will be folded in half lengthwise after printing so that all the pages in the bound book will face up. The process of arrangement and alignment is called *imposition.*

A photograph is made of the signature. A thin metal plate is exposed to the negatives in a processor, and the image is transferred to the plate. A proof copy may be made of the document at this point showing not just the separate pages (as in page proofs) but their sequence and alignment as well once the signature is folded. These proofs are called blueprints or bluelines because the print is pale blue. Bluelines provide the final chance to check the document before printing.

Offset Lithography

Different types of printing presses make their impressions on paper in different ways, but most large-scale commercial printing is done by a process called offset lithography. "Offset" refers to the fact that the image is transferred from a metal plate onto a rubber blanket and then onto the paper. "Lithography" means writing on stone. In art, it is a way of reproducing illustrations from plates formed by greasy crayon on stone.

A basic principle of chemistry explains printing by lithography: oil and water do not mix. The ink used in printing is oil-based. The metal printing plates are treated chemically so that the plate will accept water while the type image repels water. The plates are dampened with water. Then, when the ink is applied, it adheres to the type but not to the space behind and around the characters. The metal plate is fixed to a cylinder on the printing press. It rotates while a second cylinder, covered with a rubber blanket, rotates against it in the opposite direction. A third cylinder, the impression cylinder, presses the paper against the blanket cylinder. The inked image is transferred from the metal plate onto the blanket and from the blanket onto the paper. Because the blanket is flexible, it conforms to rough surfaces on the paper. Figure 23.5 illustrates how this process works.

When the printing is complete, the sheets are folded to define the separate pages in the signatures. Then, because the edges of the paper will be folded or uneven, the folded signatures are trimmed around the three outside edges.

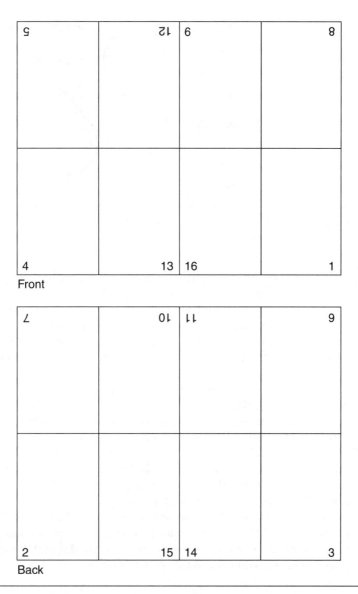

Figure 23.4 Arrangement of Pages after Imposition

The pages are arranged so that page 2 is directly behind page 1.

Color Printing

The least expensive printed documents use only one ink color. When a second color is added, the expense nearly doubles, because the paper may have to run through the press separately to apply each ink color. The press will have to be cleaned and set up for each new ink color, adding to the labor cost. Some presses can print more than one color at one time, but because they are more elaborate than the single-color presses, they cost more to operate.

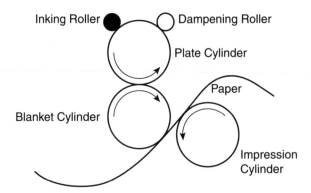

Figure 23.5   Offset Lithography: Schematic

When the document includes color photographs, commercial printers use a four-color process of printing to produce the optical illusion of all the colors that the eye can see with just four colors of ink. The four colors of ink are cyan (greenish blue), magenta (bluish red), yellow, and black. This color model is referred to as CMYK, for cyan, magenta, yellow, and black. A halftone image is created for each of the four colors, with the colors separated by scanning or filters on a camera.

The document is printed first with yellow ink; then it is run through the press again with magenta, cyan, and black. The black is used to correct for deficiencies in ink pigments that muddy the basic colors and to increase the overall image contrast. The press operator must be careful to align the printing images for all the colors to achieve *registration*, or the precise superimposition of one over the other. Otherwise, the final print will have streaks of yellow, magenta, and cyan rather than the desired rainbow of hues. The extra procedures, time, and skill required for color printing explain its cost.

Computers usually represent colors by a different scheme, RGB (for red, green, blue), from the CMYK model. If you will supply color photographs in digital form, make sure you are using the color model that the printer will use. Your page layout software should let you choose.

Binding

Decisions about binding need to be made before pages are laid out. Binding requires extra space on the page, and different types of binding require different-sized margins. Bindings also vary in durability, and they determine how easily a book will stay open or stand on a shelf.

Manuals are frequently bound in a three-ring binder or with wire spirals or plastic combs. These bindings require 5/16 inches of space along the binding edge. When you plan the page, you should allow the 5/16 inches plus the margin of space around the words. In desktop publishing, you will probably create

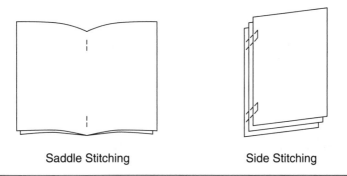

Saddle Stitching   Side Stitching

Figure 23.6   Common Methods of Binding

different righthand and lefthand pages, with more space on the inside margin for binding.

Other types of binding are saddle stitching, side stitching, perfect binding, and library binding. Saddle stitching is appropriate for booklets. The pages are laid open, and staples are forced through the middle. A saddle-stitched booklet stays open easily. Side stitching is similar to saddle stitching, but the pages are closed before staples are placed near the edge of the fold (see Figure 23.6). Side stitching is used when the bulk is too great for saddle stitching. It requires about 3/8 inches of space for binding.

Perfect binding is used for paperback books. The signatures are collected, and the spine is roughened. The cover is then affixed with an adhesive. If the adhesive dries out too much, it cracks when stressed.

The most durable binding is a library binding, a type of hardcover binding. The durability results partly from the sewing together of the signatures and from reinforcements on the spine. The rounded back allows the cover to open and close properly.

Because the signatures are folded together, the inside pages and outside pages of the signature will not be even. The pages on the inside of the fold will "creep," or extend further than the pages on the outside. Thus, the pages will have to be trimmed. The trimming uses a small part of the outside margin of the page.

Working with Commercial Printers

Commercial printers are always very glad to provide helpful suggestions on how to make your print job run more smoothly or on how to design a document in the most cost-effective manner. They can make paper recommendations or advise you as to how slight variations in the size of your finished product can maximize the use of a sheet of paper. They can advise you about the resolution of images you provide and about the type of paper that will work best with photographs.

Printers can also advise on ways to save money by indicating the best places to insert color photographs.

A production editor will benefit from developing a good working relationship with a printer. To do that, you will need to provide necessary information and complete your responsibilities on time during the process. Before you take or send materials to a printer, you will get a bid on the job and arrange for delivery of the materials by a certain date.

Obtaining a Quotation from a Printer

If you are unfamiliar with the work of a specific printer, ask to see samples and/or references. A reputable printer will be glad to provide these. Always get three quotations. In order to get a price quotation on your print job, a printer will need the following specifications. The specifications allow a fair comparison of bids. They also become the basis of a contract for services.

- **Quantity.** How many books, brochures, or other items do you want?
- **Finished size.** How big is the final product, for example 8″ × 10″?
- **Pages.** How many pages are in your document?
- **Illustrations.** How many continuous tone and line drawings are there in the document? How will you supply them? As digital files? As prints to be scanned?
- **Inks.** Is this just a one-color job (for example, black), or does it contain full-color (CMYK) photographs?
- **Paper (interior pages).** What is the grade/name/weight of the paper, for example, 70 lb Luna Gloss?
- **Paper (cover).** Are you using a heavier paper for the cover? For example, 100 lb Luna Gloss?
- **Bleeds.** Do the pages bleed? That is, does the color or art extend right to the edge of the page? If it does, the printer needs to use a bit more paper to bleed, or extend the live area, so the document trims properly.
- **Binding method.** Is this job saddle stitched (stapled) or perfect bound (glued)?
- **Supply of materials.** Indicate if your files are electronic and what software they were created in. If some materials are provided as paper copy, indicate it here.
- **Return date.** Ask how long it will take to print, bind, and deliver the job to you based on when the printer receives the final files.

In addition, the bid sheet may include policy statements about overruns in printing (whether you will be responsible for buying them), author's alterations (whether they will be charged per line or otherwise), and subletting (whether the bidder may sublet work to another printer).

Delivering Materials to the Printer

Electronic files are the most common means of sending a document to a commercial printer. Depending on the printer, you may courier disks to the printing plant or send files electronically via email or FTP. Usually, a printer will want the collateral materials including the printed pages.

A Printout of the Entire Document, Including Covers

This print copy is used for reference and to ensure that your files are printing out on their equipment exactly as they printed out on yours. If there are any discrepancies, they will be resolved early in the process and before costly proofs are generated. Keep a copy of the printout for yourself in case you have to refer to it.

Disk(s) with All Files

Include not only your page layout files but all your electronic graphics, logos, and illustrations. Include a cover sheet listing all files on the disk and the names and version numbers of the software used.

Other Media

If the printer is going to scan your photographs/illustrations and place them in the document, include these items in your package with identifying marks, such as the figure number. Indicate clearly where the scanned files are to be placed in your document.

A Memo from You

Include a memo identifying all materials sent and cite the bid quotation number and any relevant job specifications. This memo provides a good opportunity to confirm delivery dates and any other benchmark dates in the printing process. If you are the main contact person for this print job, always include sufficient contact information—there may be a problem that requires immediate action on your part, and your job may be delayed if the printer cannot reach you.

Using Your Knowledge

Knowing the specialized vocabulary of graphic designers and printers enables you to communicate with them and make informed decisions about type and printing. When you are in doubt about type or have to design your own print documents, choose 10- to 12-point type, in the Times Roman or Helvetica font, and set it on a line length of 2 ½ alphabets, or about 27 picas wide. For a book, use 50- or 60-pound paper. For materials that will be read online, choose Verdana or Georgia and relative type sizes. These are safe choices, not always the most sophisticated. Schedule enough time for printing or online development to allow for all the steps of preparation, printing or launch and testing, and proofreading.

Acknowledgment

Laura Palmer, a graphic designer from Vancouver, British Columbia, provided information for updating the sections on printing and illustrations.

Further Reading

Beach, Mark, and Eric Kenly. 1999. *Getting it Printed.* 3rd ed. North Light Books.

Bruno, Michael H., ed. 2004. *Pocket Pal: A Graphic Arts Production Handbook.* 19th ed. Graphic Arts Technical Foundation.

Craig, James, and William Bevington. 1999. *Designing with Type: A Basic Course in Typography.* 4th ed. Watson-Guptill.

Williams, Robin. 2004. *The Non-Designer's Design Book.* 2nd ed. Peachpit Press.

Discussion and Application

1. Find examples of typography that encourage reading or discourage reading. Evaluate the typography in the context of the document. Do the bad examples seem to be accidental or intentional? What could be done to improve the typography? What are the positive features of the good examples?

2. Type a paragraph in Times Roman, then copy it and reformat it in two additional typefaces, including one in sans serif style, such as Helvetica. Compare the amount of space each one takes. Characterize the type: Is it inviting? Sophisticated? Masculine? Elegant?

3. To illustrate how the paper is folded on the signature, create a miniature sheet of book paper by using notebook paper. Divide it into sections, and number the sections according to the plan in Figure 23.4. (Tip: page 2 should be right behind page 1.) Then fold the paper in half so that the two shorter sides touch and the front is outside, with page 1 in the lower left corner. Fold in half again, with this fold at right angles to the first. Fold a third time at right angles to the second fold. Trim the top, bottom, and right folds. Your pages should be numbered in sequence from 1 to 16.

4. Examine your textbook and other bound documents you possess. Identify the type of binding used.

24 Project Management

With Heather Eisenbraun

In order to edit well, editors need to anticipate needs for and uses of the text and to make the right choices about organization, style, design, and visuals to enable or to persuade users. These points dominate most of the chapters in this textbook.

Although expert editorial decisions are essential to the quality of documents, good management and a good process of document development also influence quality and your opportunities to do your best work as an editor. Last-minute work invariably compromises the work of editors and can jeopardize other parts of the project.

Project management is the process of planning, scheduling, communicating expectations, and tracking progress in order to achieve a finished product on time and within the budget. Management of a publications project requires the balancing of three variables: time, resources (that is, the number of writers and editors), and scope or quality. Given a project, a manager must determine which variables are fixed and which are flexible. For example, if it's most important to finish a project by a certain deadline, a manager can either add more resources to the project or consciously lower the quality. Often, writers and editors will be faced with sacrificing at least one of the variables. They may feel torn between doing the very best they can and doing the very best they can with the time and resources they are given. The priorities should become clear in planning.

Editors participate in organizational management as members of project teams, and the production editor has specific responsibilities for moving a project toward publication in print or online. As individuals, editors must also manage their own time, often juggling several simultaneous projects, coordinating with the writers and technical experts, and meeting deadlines.

This chapter begins with a review of the impact on document quality and on editorial roles of a well-managed document development process. It then considers the key features of both organizational and individual management: planning, record keeping, estimating, scheduling, tracking, and evaluation. Chapter 22 reviews legal and ethical issues of management, Chapter 23 covers contracting for services from commercial printers, and Chapter 12 introduces ISO 9000 certification.

Key Concepts

> The quality of a document, print or online, depends not just on good writing and editing but also on a well-managed process for developing the product. One management model used widely in technical communication is the life-cycle model. Thinking of a project from inception to conclusion as it develops into different phases over time gives a whole-project focus. Good management requires assessment of the project scope and uses, task analysis, time and budget estimates, scheduling, supervision and collaboration, record keeping, and tracking of progress. Managers must communicate expectations throughout the cycle. They should evaluate not just the effectiveness of the document or other product but also the effectiveness of the process used to manage development.

The Case for Managing the Document Development Process

Workplace publications projects are often complex and developed by a team of people whose activities must be coordinated. An example is the computer safety training described in Chapter 1 that Charlene's group developed. In that case, the publication is the project, the "deliverable" that the client expects to receive. Another example is user documentation described in Chapter 1 that Paula's group developed to support BMC's software products. In that case, the reference manual is just one of the deliverables of the project. The publication goals have to be synchronized with the product development goals.

The complexity of developing these publications requires planning and coordination. When people work in teams and depend on others to deliver material, review drafts, sign off on documents, or collaborate in any way, they must understand expectations for themselves and others at the start of the project. There could be multiple layers of management and planning, with the publications manager reporting to the project manager.

Strategies for managing development of documents have developed because unmanaged projects compromise on quality and risk failure to meet deadlines. The unmanaged project is usually a linear one, with writing and editing occurring at the end of project development. When editors enter a project at the end, they are often limited to basic copyediting and have little opportunity to influence document structure and style and product design. Thus, their expertise is incompletely used. As the last people to work on the document, editors might be squeezed against impossible deadlines. The quality of documents suffers, and supervisors, clients, and customers lose confidence in people who cannot deliver their products by established deadlines.

External incentives beyond customer satisfaction and product quality require effective management. If a company wishes to sell products in the European Economic Community, it must achieve ISO 9000 certification, which requires documented management procedures (see Chapter 12).

The Life-Cycle Model of Publications Development

JoAnn Hackos has developed a model for document development based on the computer software development process. She identifies five phases in this process:[1]

1. Starting the project—information planning
2. Establishing the specifics—content specification
3. Keeping the project running—implementation
4. Managing the production phase—production
5. Ending the project—evaluation

Hackos also classifies publications organizations according to how well they manage projects, from "ad hoc" organizations that have no standards beyond those for individual projects, to organizations with procedures that are repeated for all projects, to optimizing organizations marked by self-managed teams and close contact with users. The procedures all require planning, estimating, and scheduling. Although the examples in the Hackos model relate to software documentation, the principles and process can be applied in other settings.

Planning and Content Specifications

Planning means setting goals and priorities, deadlines, and schedules for meeting the goals by the deadlines. It also means assembling a team and assigning people to tasks. A project plan includes a statement of the project goals, identification of the full range of tasks to be accomplished, and a schedule for completing the project on time.

The people who will be involved in the project develop the plan collaboratively. Planning must consider the needs of stakeholders, people with an interest in the outcome of the work, not just those who will do the work. In the optimizing environment that Hackos describes, teams are self-managed, but to accomplish that goal, everyone needs to know what the specifications are, who is responsible for what components, and when the milestones will occur. The milestones are points between start and finish when certain achievements should be attained. At milestones, the project team meets to review accomplishments and to revise the project, if necessary. A progress report to a supervisor, client, or external funding source may be required. It is also important to identify handoff or synchronization dates.

An editor might be the best person to make recommendations about the document specifications to the project manager. If the editor is trained in usability and audience analysis and is a regular member of the product team (that is, not called in at the last minute to edit the documentation), the editor probably knows

---

[1]JoAnn Hackos. 1994. A model of the publications-development life cycle. *Managing Your Documentation Projects*. 25–43. Wiley.

what is best for the project. The project manager ultimately makes the decision, but it is common for the editor to write a documentation proposal to present along with the project manager's project proposal. The documentation tasks may include conceptualization of the document with consideration of readers and purpose, research and generation of a draft, technical review, comprehensive editing, copyediting, usability testing, and production.

Any publications plan requires some description of the content to be developed. For a reference task like Paula's, the team may develop multiple deliverables—possibly a reference manual, a tutorial, and online help. Each one of those products requires specifications of topics to be covered.

A company that develops similar materials for multiple products will have some outlines of comparable projects that can be used to determine what is required for a new project. Clients, such as the clients for Charlene's project, may specify the topics that they need to have covered. An external funding source, such as an organization that issues a request for proposals, may also specify the topics to be covered.

Estimating Time and Developing Budgets

A project manager will need to know how much time the editing will take based on estimates of project scope. If people, including editors, estimate intuitively, they usually underestimate the time required for writing and editing. Accurate time estimates keep the project on schedule and help to define editing expectations.

Freelancers and contractors, and even some salaried employees, need to estimate costs of the job as well as time. You may communicate your estimate to the client in terms of cost per hour, per page, or per job. However, clients deserve to know the total estimate. Time estimates are essential for cost estimates.

Estimates for document development require agreement on the type and level of editing required. Estimates also require the document specification and records from similar projects in the past. You can review sample pages from the present document, but if you have management procedures in place, you probably have records from similar projects in the past and can use those standards. Before you can estimate how long a job will take, you need to know what responsibilities you will be expected to complete.

Classification of Editorial Tasks and Responsibilities

Good estimates begin with a clear sense of expectations. Because editing is a complex process, it does not have a single meaning. A manager or client who thinks editing means proofreading may assume that an editor can "edit" a 200-page document or a 90-file website in a day. Such an assumption would not allow for even minimal copyediting.

Time estimates follow from agreement about tasks. Several decades ago, Robert van Buren and Mary Fran Buehler developed the "Levels of Edit," a taxonomy of editing tasks that they used to reach agreement with clients about expectations in order to manage their editing tasks. Table 24.1 expands their taxonomy to reflect comprehensive editing and production. It defines editorial tasks and responsibilities and the desired outcome of each for the document and for the readers. It is the basis of the editing agreement in Figure 24.1.

When an editing job is assigned, you can use the classification in Table 24.1 to clarify the editing tasks that you are expected to complete, and you can estimate the required time and cost for your services on the basis of those expectations. Clients or co-workers should understand just what they will receive and understand time requirements in terms of the editing services that they request.

The copyediting and comprehensive editing tasks are typical for editors responsible for the text and visuals. The coordination and production tasks may be performed by a production editor, but in some small companies, editors responsible for establishing the text assume the full range of responsibilities. Preparation of an index, product testing, and training of users are responsibilities beyond the usual expectations for editors, but editors often perform these tasks as well.

An editing agreement, such as the one in Figure 24.1, can prompt you to consider all the tasks that will be involved in the editing job and to clarify expectations with the person who requests editing. Many projects may be estimated informally, but it may be policy within your organization to have such an estimate signed by both the editor and the person requesting editing so that each individual will understand the expectations from the start and agree to meet them.

Although individual projects can vary in what they require, some standards have been offered in the literature on managing technical publications. Thomas Duffy found that experienced editors—with a median of 11 years of experience— could edit 28.8 pages a day for comprehensive editing and 38.4 pages per day for copyediting.[2] Joyce Lasecke offers these formulas for developing different types of materials.[3] Editing should require less time than developing these topics, but it will help if your company develops some standard metrics for these different types of editorial tasks.

Type of Topic	**Guideline Hours**
Step-by-step procedures	4–5 hours per procedure
Glossary terms and definitions	0.75 hours per term
Reference topics such as explanations of concepts, theories, overviews, and product information	3–4 hours per topic
Error messages (problem statement and recommended solution)	1.5–2.5 hours per message
Time to create or modify graphics other than screen captures	0.5 hours per graphic object

[2]Thomas M. Duffy. May 1995. Designing tools to aid technical editors: A needs analysis. *Technical Communication* 42: 267.

[3]Joyce Lasecke. November 1996. Stop guesstimating, start estimating! *Intercom.* 6.

TABLE 24.1 **Editorial Tasks and Responsibilities**	
Task	**Responsibility (Purpose)**
Basic Copyediting	
spelling, grammar, punctuation	correctness
consistency: verbal, visual, mechanical, content	consistency
match of cross-references, callouts, TOC, etc.	consistency
completeness of parts: sections, visuals, front matter, back matter, headings, etc.	completeness accuracy
accuracy of terms, numbers, quotations, etc.	correctness, consistency,
illustrations	completeness, accuracy
copymarking for graphic design	correct typesetting and page layout
Comprehensive Editing	
document planning	suitability for readers, purpose, budget
style: tone, diction, sentence structure	suitability for readers, purpose; comprehension
organization	comprehension
visual design	comprehension, usability
illustrations	comprehension, usability
cultural edit	suitability for readers according to their national and ethnic backgrounds
legal and ethical edit	comply with legal, ethical, and professional standards
Coordination, Management	
communicate expectations, establish scope	ensure that the team works toward common goals
document tracking, transmission	keep production on schedule
checking on the project, correspondence	collaboration; make sure everyone still agrees on goals
scheduling editors, proofreaders, usability tests	keep production on schedule; allow time for quality control
getting permissions, applying for copyrights	meet legal requirements; protect the writer's or company's rights
Production	
negotiating with compositor, graphic designer, printer, technical staff for online coding: bids, schedule	quality, economy; meet deadlines
graphic design	reader use, comprehension; document attractiveness
Desktop Publishing	
preparation and maintenance of templates	consistency of design; save time
keyboarding text, corrections	production-ready copy
preparation of illustrations	production-ready copy

TABLE 24.1	*(continued)*
Task	**Responsibility (Purpose)**
HTML/XML	
coding	consistency, readability
checking links	accuracy, ease of navigation
testing display on different browsers	access, usability
testing download time	access, usability
Indexing	
preparation of an index	locate information
Product Testing	
running program, using product	usability, readability, completeness, safety
organizing and conducting usability tests	usability, readability, completeness
Training of Clients, Co-Workers	
training in writing	writer competency
training in program, product use	user competency

The following estimates are used as benchmarks for performance as well as for estimating large projects, but the numbers could range widely for each activity.[4] Length, technicality of the material, number of graphics, and care with which a document has been prepared by a writer will all affect the actual time it takes to complete work.

Writing new text	3–5 hours per page
Revising existing text	1–3 hours per page
Editing	6–8 pages per hour
Indexing (all tasks)	5 pages per hour (user guide pages, not index pages)
Production preparation	5% of all other activities
Project management	10–15% of all other activities

The time required to complete a project may include team meetings, email, negotiations with vendors, and reviews with clients as well as time for reports and record keeping. In addition to these categories, include a cushion of time for "project creep," the tendency of projects to expand beyond their original specifications. This amount might be 5–10% of the total project estimate. As you keep records of projects in your company, include a place for comparing original project specifications with final specifications to get a sense of how much projects expand in your company.

Record Keeping

Estimates for documents already developed should be based on records from past projects. If you know, for example, that you can copyedit six online help topics in

---

[4]Lola Frederickson and Joyce Lasecke. 1994. Planning for factors that affect project cost. *Proceedings, 41st International Technical Communication Conference.* Arlington, VA: Society for Technical Communication, 357.

Client _____   Phone number _____

Address _____   Email _____Fax _____

Document: title _____   Type _____

Date submitted _____   Date due _____

Length in pages/screens _____   Form submitted: paper copy_____

digital copy _____

Visuals: total number _____ photos _____ tables _____ graphs _____ other _____

Other nonprose material_____

Editing required

_____ spelling, grammar, punctuation   _____ document design

_____ consistency   _____ style: tone, diction, sentence
                                    structure, globalization

_____ match of cross-references,
          figure numbers, etc.   _____ copymarking for graphic design

_____ completeness of parts   _____ preparation of production-ready pages

_____ accuracy of terms, numbers, etc.   _____ preparation of production-ready visuals

_____ visuals   _____ HTML/XML coding or checking

_____ organization   _____ HTML/XML links check

Other related tasks (reference check, product testing, index, permissions requests, etc.)

_____

_____

Online editing acceptable? yes _____ no _____

Estimates of time: hours _____ working days _____

Milestone (review) dates:

Handoff date:

Intellectual property: The editor of this document may use it as a portfolio piece, available online as well as in print. Other uses will require separate permission.

Conditions (if any):

This estimate is based on document specifications and editing tasks as shown here. It is binding only so long as the specifications and editing tasks remain constant and the document is available for editing on the date cited. Any changes will require a new estimate.

_____   _____
editor   date   client   date

Figure 24.1   Editing Agreement

an hour by a writer whose work you know, you use that figure for estimating the copyediting of an entire job. Likewise, comprehensive editing may require an hour for each two pages of typescript. Good estimating begins with good record keeping. For each job you do, you should record the job type and completion time. Soon you will have averages on which to base future estimates. Your records could be organized according to categories such as these:

Date	Job	Scope	Time
Jan. 3	proof newsletter	4 finished pages	30 minutes
Jan. 3–4	edit proposal: comprehensive	32 typescript pages	8 hours
Jan. 10–11	edit help topics	30 topics	5 hours

You can begin record keeping even as a student. When you are assigned an editing project, estimate the time it will take. Then keep records and compare the estimate with the reality. Don't base the estimates just on the length of the document. Consider the nature of the document (illustrations, formulas, and complex structure will all increase time). Also consider its condition (errors increase time). These records will give you rough bases for future estimates.

You should also see some increase in productivity over time. For example, if comprehensive editing requires an hour per page on your first project, your third or fourth project may require, by the end of the term, only half an hour for a comparable page.

Your records can also demonstrate productivity. Your department may be required to assess your achievements periodically in quantitative form. To many outsiders, the work of editing is invisible, and to quantify it in terms of pages or documents distributed shows productivity in a form that makes sense to other managers. Collectively, the records of individuals within a document production group form the basis of estimates for large projects. For example, you may learn that the documentation for a particular project will take 800 hours, including 150 hours of editing. This information helps project managers estimate the entire project, including documentation, rather than expecting the documents simply to appear. The collective records may also justify the hiring of new employees.

Sampling

Averages imperfectly predict the time required for a future job because of variations in writers' levels of skill, in subject matter, and in the condition of the typescript. If the document is complete when you receive it for editing, averages should be accompanied by reviews of sample pages. Sampling is especially important if you are working with a new writer or project. To estimate based on sampling, first skim the entire document to determine the number of pages of text and the number and type of illustrations, the amount of technical material, the extent of reference material, and any other features that will help you assess the scope of the editing task. Then edit sample pages, perhaps the first few pages of two chapters and pages with technical information from two parts of the document. On the basis of the time it takes you to do this work, estimate the entire editing job.

If the document has just been scheduled for development, the project manager should provide specifications for the document. These projections should be in writing in case the expectations develop to include, say, a reference manual as well as a tutorial without a proportionate increase in time.

A schedule must be presented as part of the job estimate. See the section on scheduling.

Setting Priorities

In spite of all these efforts to predict and control time, editors invariably have too little time. What do you do when you estimate 100 hours for a task but only 40 hours are available? There's no easy answer to this question. Part of the solution is to establish priorities. The manual that accompanies the $600 software package on which your company stakes its reputation deserves more of your time than the in-house employee newsletter. If you must cut corners, cut them on the lower-priority documents.

It has been said that writers and editors never finish their work; they simply abandon it. Sometimes you will have to let go of documents knowing you have not yet done your best work on them just so that you will not sacrifice more important tasks. No document is ever perfect, and in that sense writing and editing are never complete.

Document Scheduling and Tracking

Document development is usually a longterm project controlled by a specific distribution date. Projects are typically measured in terms of weeks, months, or even years rather than in hours or days. Furthermore, document development is typically collaborative, with the achievements of one collaborator determining how quickly and well another collaborator can finish his or her tasks. Editors are particularly vulnerable to inadequate scheduling or delays because the bulk of their work often comes at the end of the development cycle. They may be expected to make up for the delays of others so that the document can be distributed according to the original schedule. A thorough and realistic schedule established at the beginning of the development cycle, with periodic due dates rather than a single due date at the end, distributes the responsibility for keeping a project on schedule on all collaborators rather than disproportionately on editors. Efficient development requires a detailed management plan at the beginning of the development cycle and periodic tracking.

A good project manager will check the status of her project twice as often as she can allow the project to slip in time. For example, if a project has a deadline that must be met by + or – four days, the project manager should check on the progress of all tasks no less than every other day to identify issues or assess new risk areas.

Scheduling Due Dates

To establish due dates, managers frequently work backward from the necessary or desired distribution date and stagger due dates. For example, 100 copies of an annual report may be due on September 2. The document consists of 25 individual research reports to be prepared by 8 staff members. Its estimated length is 300 pages of 8 ½ × 11 pages prepared by the production staff. Because the reports follow a standard format, an electronic template setting margins, typeface, and headings is available for the researchers. This template results in relatively uniform reports that need minimal formatting by the production staff. The report will be printed and bound by a quick-copy shop.

Some dates must be established by phone calls to contractors, especially, in this case, the printer. Other dates may be negotiated by consulting with the contributors. The following production schedule is based on the assumption that the production staff will spend no more than six hours per day on this task and the copyeditor no more than four hours per day because of other concurrent assignments.

Task	Completion Time	Beginning Date
Mail date		August 23
Printing and binding	2 days	August 20
Page layout, proofreading	@ 2 hours × 25 = 50 hours	July 18
Copyediting	@ 4 hours × 25 = 100 hours	July 16

Copyediting must begin at least twenty-five workdays before the report is delivered for printing and binding. If concurrent projects place other demands on the copyeditor, the beginning date may be pushed backward. To maintain a regular production pace, a copyeditor might request one report draft each day beginning five weeks before the report is scheduled to be delivered for printing and binding. However, document tracking will be easier with fewer deadlines; thus, the reports may be collected in groups of five each week.

The schedule of assignments needs to be distributed far enough in advance of the first deadline to make it possible for researchers to complete their first reports on their due dates. They will work on subsequent reports while the first reports are being edited and proofread. Researchers should be consulted for their preferences about due dates. If research projects are already completed, the reports may be due early in the schedule. The later dates should be reserved for projects that are still being actively pursued. Some researchers may wish to have all their reports due at one time to facilitate their own time management.

The schedule could be as simple as the following one, which lets each researcher see at a glance when reports are due. For simplicity, researchers are identified here by letter, and report topics by number.

Researcher	Report Due Date				
	July 16	**July 23**	**July 30**	**Aug. 6**	**Aug. 13**
A	1		2	3	
B	4	5		6	7
C	8		9	10	
D	11		12	13	14
E		15	16		17
F				18	19
G	20		21		22
H		23, 24, 25			

Without such planning, researchers are likely to wait until the last minute to write their reports, and the editor and production staff will face a 150-hour task to be completed in just a few days. The job is not likely to be finished on time, its quality is certain to suffer, tempers will be strained—and the organization may lose funding for future research.

More complex projects, especially those requiring comprehensive editing and technical review, those with illustrations or complex page layouts, and those being commercially typeset and printed, will require more complex scheduling.

Scheduling Reviews

The editorial process is rarely completed by an individual contributor. Often documents must be reviewed and approved by various content experts, project managers, marketing departments, and legal advisors. The editor is usually the person to route the reviews to the list of reviewers and resolve the edits that arise from those reviews.

When scheduling reviews, the editor must make sure that reviewers are able to meet those expectations. A reviewer might be on vacation during the time the editor plans to send the document to review. Some reviewers intend to return reviews on time but get too busy to do so. An editor who works with chronically late reviewers will allow for the possibility of a missed deadline. An editor who works with remote developers must consider holidays in other countries. For example, a developer in China shouldn't be expected to return reviews during the Chinese New Year.

Tracking the Document through Development and Production

When there are multiple documents or sections of a document, or when the production is complex, involving typesetting and illustrations as well as editing, some systematic way to track the document through the production sequence will aid in management. You need to know both how the whole project is developing and how each individual project is progressing. You will probably maintain a file on each project, including correspondence, if any, and notes of meetings establishing due dates, plus the versions of the document as it moves through editing and production.

If you share documents on paper copy, you may also wish to attach a tracking sheet to the document itself so that, when the document lands on your desk, you can tell at a glance exactly what needs to be done next. In the following example, the sheet will identify the report and the steps completed in production. As each step is completed, it is dated and initialed by the person who does the task.

Report # _____   Researcher _____

	Date	Initials
Received		
Copyedited		
Writer proof		
Editor's check		
Final revisions		

In addition, you will maintain a file on the whole project, indicating its progress by showing how the individual projects are progressing. The simple form below assumes that the procedure is for the copyeditor to edit electronically. After he enters his corrections and notes queries, the report goes back to the writer for proofreading and emendations that the queries initiated. It is then returned to the copyeditor for another check. Finally, it goes to the secretary, who prepares the production-ready form.

Report	Received Final	Copyedit	Writer prf	Check	Final
1	7/16	7/16	7/18	7/20	7/23
2					
3					
etc.					

The form records each action on the project and each time it changes hands. The form helps the editor to keep production moving and to locate any missing documents. A more elaborate project, with acquisition, graphics, and typesetting stages, would require a more elaborate tracking form.

You can find a variety of project scheduling and management software, from simple tracking software to keep up with project milestones and tasks to sophisticated systems that manage documents and workflow. For example, let's say a product team needs to produce a document and that document will be authored by a writer, passed to an editor for editing, returned to the author for technical review, and then sent to the project managers for approval before printing. A writer can author the document in one of many different electronic formats and submit the document to the document management system. Because of the predefined workflow, the system can email the editor to alert him to begin editing the document. When the editor finishes editing, he will resubmit the document, and the system will alert the author that she can now review the document before it is sent for approval. When the author approves the document, with changes or without, the project and marketing managers are alerted via email to go into the system and approve the document. After they approve the document, the system can alert all of them that the document workflow process is complete. At that point, the person responsible for printing the final product sends it to the printer. If the company has a printing department, it's even possible for the printers to be alerted that the document is ready to be printed.

Project management software is often expensive, and it can't replace good estimates and frequent communication with the rest of the team. Corporate-wide project management systems are highly customizable.

Version Control

As a document is developed, reviewed, and edited, it will soon exist in multiple versions. These versions need to be identified so that reviewers will always work on the most recent version. Team members can waste time changing material that has already been modified by someone else or working on sections that have been eliminated from a revision. Likewise, good revisions could be lost because a subsequent revision reverts to an earlier version. Version control is especially important when multiple people have access to the same electronic files and privileges to make changes within the files, but even an editor working alone may maintain multiple files of the same document, including an archive copy, a copy that is being reviewed by a subject matter expert, and a third copy on which the editor continues to work.

Three strategies are valuable in version control: using configuration management software, using the "save version" feature of word processing software, and establishing a naming convention for versions. The naming convention for the electronic files is the most important strategy for version control when the reviews and editing are on paper copy.

However you manage version control, it's a good idea to preserve a copy of the original document with no changes. If the edited copy becomes corrupted in any way, the original remains intact so that the work on it doesn't have to begin from scratch. And the original provides a source for checking the accuracy of editorial changes, particularly style changes. It can be used for comparison with a later version.

Configuration Management Software

Configuration management (CM) software is a file storage and management system. A file server stores all of the files and documents, and system administrators can make back-ups of the files to make sure no one loses work. The CM software provides an interface that allows you to browse the documents in the system. You can check documents in or out, add or delete documents, or revert to a previous version of the document. Some CM tools will allow you to graphically compare (or difference, "diff") two versions of the same file or lock files that you are currently modifying so no one else may check them out and make changes. The benefit to this sort of system is that it allows multiple users to modify the same file, saving versions of each, without overwriting anyone's changes because of a source control mistake.

CM software is available in all ranges of prices, from freeware and shareware to very sophisticated, expensive systems that companies use throughout the enterprise and with all ranges of features.

Using Software to Identify Versions

If you are editing electronically, you can use the version control feature of your word processing software to keep records of the various interventions into the text. For example, Microsoft Word gives you the option of choosing "versions" from the file menu when you save. In the save window, you can type a comment to define how the document has been changed. The comment might read "archive copy," "technical review," or "graphics inserted." The file then contains a record of the date and time of the version, who saved it, and the comment.

You can retrieve earlier versions of the file through the versions option. This feature provides a way to maintain ready access to the original, or archive, copy. When you choose "versions" from the file menu to open a document, you can open the original in the same window as the edited version and compare the differences.

Each version should represent a milestone in development, such as the original, the version for technical review, the version for usability testing, and the final version. If you define "versions" for each minor revision, you will soon be overwhelmed with the number of versions, mostly indistinguishable from the others. For multiple versions of essentially the same document, some system of file naming will accomplish the goal of keeping track.

File Naming Conventions

Extensions to the file names for documents can identify which version is the most recent and comprehensive. These extensions could be letters, numbers, or dates or a combination of these that makes sense within the company. In the example that follows, the identifying name is "styles," and the date represents the last revision.

Styles09.10.05

Evaluation

Editors need to evaluate their document development and document production process periodically to determine how they might be improved. Evaluation should include some assessment of how well the product meets the expectations of users, whether through usability testing, user feedback, or records of calls for help to product support, or, in another setting, whether the grant is funded and the article is published. But evaluating the product will tell only part of what you need to know to keep improving the processes of publications management. For this goal, you will need to know whether the planning, estimating, scheduling, and reviewing processes have been effective, whether you have used software tools efficiently, and whether the people involved in the process have been happy about it.

Record keeping during the process will provide good evaluation data. For example, you can compare actual times to complete tasks with estimated times. If there is a significant difference, were the estimates based on faulty expectations, did the project grow in size, or were there personnel problems? How did the promises of dates for project milestones compare with actual delivery dates? Did people turn in work late? If yes, do people need some help with time management, or were they given too much to do?

Take the opportunity of project completion to discuss with co-workers what worked well in the project and what needs to be changed. A formal meeting with a planned set of questions engages all group members in problem solving and planning and helps everyone feel invested in the process and outcome.

Setting Policy

Editorial policies may govern documents, collectively and individually, and editorial procedures. In editing the text, editors will establish mechanical style. Specific documents will require other, more comprehensive policy decisions. For example, the editor of the proceedings of a conference will need to decide whether to publish printed papers or transcripts of the oral presentations.

Editors may also need to establish policies about the scope and quality of publication, including the use of illustrations in documents (such as whether to include illustrations, whether to allow photographs as well as line drawings, whether to make writers responsible for production-ready copy), the means of creating production-ready copy (electronic publishing, typesetting), the use of color in printing, and paper and binding. The policies may differ for various types of documents, such as those for in-house distribution and those for mass public distribution. Policies can always be revised as new situations arise so they can be helpful without being rigid.

Policies may also establish editing procedures. For example, each publishing organization should have a policy about editing online, clarifying what changes may be made directly and what ones need prior approval. Other policies establish who reviews the drafts and when. Quality control may require a policy of two proofreadings by someone other than writer and editor at the page-proof stage.

Just as style sheets increase editing efficiency, so can established publication policies increase efficiency by minimizing the amount of time required for decisions.

Using Your Knowledge

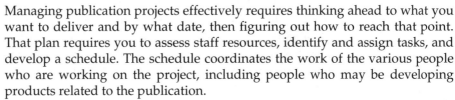

Managing publication projects effectively requires thinking ahead to what you want to deliver and by what date, then figuring out how to reach that point. That plan requires you to assess staff resources, identify and assign tasks, and develop a schedule. The schedule coordinates the work of the various people who are working on the project, including people who may be developing products related to the publication.

Nothing is more important than communicating expectations and checking in periodically to make sure that everyone still agrees.

Acknowledgment

Heather Eisenbraun, a documentation group manager at National Instruments in Austin, Texas, provided substantial assistance in revising this chapter. I am grateful for her experience and help.

Further Reading

Allen, O. Jane, and Lynn H. Deming, eds. 1994. *Publications Management: Essays for Professional Communicators.* Baywood.

Burdman, Jessica. 1999. *Collaborative Web Development: Strategies and Best Practices for Web Teams.* Addison-Wesley.

Dicks, R. Stanley. 2004. *Management Principles and Practices for Technical Communicators.* Longman.

Hackos, JoAnn T. 1994. *Managing Your Documentation Projects.* Wiley.

Van Buren, Robert, and Mary Fran Buehler. 1991. *The Levels of Edit.* 2nd ed. Society for Technical Communication. Rpt. from Pasadena, CA: Jet Propulsion Laboratory, publication 80–81.

Wysocki, Robert K., and Rudd McGary. 2003. *Effective Project Management: Traditional, Adaptive, Extreme.* 3rd ed. Wiley.

Discussion and Application

1. You are the editor of an anthology of articles by different writers in different locations on the subject of computer documentation. The anthology will be published by a small commercial press. The press will print

and bind the anthology from copy you provide. Make policy decisions on the following issues with the goal of creating a useful, high-quality anthology that also can be produced economically and efficiently. Identify the bases for decision making.

a. How will the articles be acquired? How can you get quality articles and a broad coverage of topics? Will there be a general solicitation in journals that potential contributors read? Will some people be invited to contribute, and if so, will acceptance of their articles be offered in advance or will their articles be subject to peer review? Will only original articles be accepted, or will reprints also be accepted?

b. How will you prepare copy ready for reproduction? Will you provide templates to writers so that they prepare their versions in the final form? Will they submit text electronically that the production staff prepares?

c. Printing economy dictates 8 ½ × 11-inch pages. Will pages be laid out in two columns, or in one column with wide margins? Two columns would probably be more readable and would allow more words per page, but they would be harder to produce because of controlling for column breaks as well as page breaks and because multiple writers will prepare the chapters.

d. How consistent must the individual chapters be? If one includes a list of references for further reading, must all of them?

e. How and where will contributor biographies be printed: at the beginning of the anthology or with each chapter?

f. What kinds of illustrations and how many per chapter will be allowed? What will be the responsibilities of authors for providing digital copies?

2. For the same situation, make management decisions on the following issues with the goal of getting the anthology ready for production by the date the press has specified.

a. When will chapters be due? Will they be due on the same date, or will you establish variable dates? If the dates vary, which ones should come first?

b. In what order will you edit the chapters? Will you edit them in sequence, from one to twelve, or might there be reasons to proceed in a different order? What factors will determine the order?

c. On the basis of policies you have established for the book, what information will you have to communicate to the contributing writers about preparation of typescripts? Consider, for example, documentation style and length. Make a list of points to cover in document specifications.

d. Suppose one contributing writer has a due date of the first of the month, and two weeks later you still have not received the chapter. What can you do?

e. Suppose one contributing writer has submitted an electronic file that cannot be converted to your word processing system without extensive intervention in the text. Should your own production staff type the conversions (which means removing that person from another important project and also increasing the costs of producing the anthology), or should you ask the writer to revise and resubmit the file? Your answer may not be the same for all writers involved. What will determine your choice?

f. Outline a form for tracking the progress of the anthology overall, including acquisition, acknowledgment of the manuscript, acceptance or rejection, and other production stages.

3. Open a document in your word processing or page layout software. Locate the method for saving versions, and experiment with the method. Or, in a group project, save files in a location that everyone can access. As you work on the document, write comments, and save a version identifying the nature of your work on the file.

25 Client Projects

Your instructor may assign a project for which you must locate a client beyond the classroom and edit according to the objectives that you and the client agree on. This project provides an opportunity for you to demonstrate and apply what you have learned in the editing course. The principles you have applied to sample documents can sound and feel different when you have to look a person in the eye and recommend some modifications of the document that person has written. Your project, in which you are more independent than in projects your instructor provides, can become a useful portfolio piece for job applications.

This chapter covers some guidelines for locating a project, negotiating an agreement between you and your client, completing the editing, interacting with the client, and presenting the final version orally. These guidelines may be useful as well if you attempt freelance editing jobs.

Key Concepts

A client project can provide editing experience and result in a good example for a job search portfolio. A project plan or proposal requires you to set editing objectives and methods for achieving them. An editing agreement establishes the expectations for editor and client. Editor and client may communicate at a conference to review editing progress. Planning and organizing the conference and using goal-oriented language will lead to productive conference outcomes. An oral presentation of results focuses on editorial problems and solutions.

Selecting a Good Project

Your instructor may locate the clients and projects or send you to find your own. If you locate your own project, your instructor will provide some guidelines for what constitutes a good project. Chances are you will be encouraged to look for genres of professional practice: instructions, policy manuals, proposals, business plans, websites, brochures, or newsletters. This textbook has not prepared you to edit fiction or poetry. If your instructor approves your editing of an academic paper for another student, it may be with the condition that you get the approval of that student's instructor. You may be able to find a document that requires editing from the Web or another source, but unless you have access to the writer and can negotiate with that person about editorial changes, you will not have the full experience of editing.

Employee policy manuals can provide excellent projects, and you may find such a manual at a place where you have a part-time job or where you volunteer. Sometimes professors are happy to have help with class information or handouts. You may find instructions on fliers in your campus computer support facility or library or the campus health center, or a collection of related fliers that lack a common design and pattern of development. A service organization may be writing a proposal and need advice about how to strengthen its argument. Such projects may share the history of having been developed over time, with additions by various writers until some coherence in style and purpose has been lost. An editor can help by rediscovering how the parts fit together.

Be cautious about committing to a project that might be more demanding than time allows. If you take on such a project, you may have to manage it to complete it in phases, with one phase counting for your project.

Project Plans and Proposals

As you saw in Chapter 24, work on projects is more likely to reach good outcomes that satisfy clients in the time allowed if the project develops according to a plan and if the editor and client communicate at the beginning, at project milestones, and at the conclusion.

At least two readers may be interested in your project plan: your instructor and your client. Your instructor wants to make sure that you have a feasible project that meets the expectations of the assignment and that is likely to result in a useful deliverable (defined on p. 420) for the client. Your client wants a deliverable that meets whatever need he or she has for the document. Your client also wants to influence its outcome and should participate in defining the goals of editing.

The main beneficiary of advance planning is you. You will define goals and a method for achieving them. By communicating your plans with your client and instructor, you will avoid possible misunderstandings about what the project entails. By establishing a schedule, you will avoid last-minute, frantic work and sleepless nights.

Your plan should include at least the following sections: analysis of the document's purpose and readers and evaluation of its current strengths and needs; the objectives in editing; the deliverables; the responsibilities of team members in a multi-person project; the schedule and milestones; the budget. Depending on the expectations your instructor and your client establish, you might write two different documents for the two audiences, using much of the same information in both. Your instructor may want a proposal that reveals your understanding of the editing process and makes a convincing case for the need to edit. Your client may be satisfied with a letter that outlines the goals and process and accompanies a contract (discussed later in the chapter).

Analysis and Evaluation

As you learned in Chapter 14, comprehensive editing begins with an analysis of document goals and an evaluation of how well the document achieves these goals in its current form. This assessment is the basis on which you can develop specific editing objectives. You can use the heuristic of considering content, organization, visual design, style, illustrations, and copyediting, all evaluated on the basis of your comprehensive analysis of purpose and readers and whether the document is in the right form to meet the purposes for the readers.

Your instructor may be more interested than the client in the details of this analysis. If your instructor requires you to write a proposal as a basis for approving the project, the analysis will be the part of the proposal that establishes a need for editorial intervention and the match of the editing tasks with assignment requirements. It also establishes your qualifications to do the work (because you understand it). For the client, an abbreviated analysis, perhaps in letter form, will establish that you understand the editorial problems. You also establish terminology that you and the client share for the work that you will do.

Objectives and Deliverables

The objectives of editing grow out of your analysis. For example, some employee manuals develop a hostile tone that conveys an assumption of employee irresponsibility. One of your objectives might be to edit the style and content to emphasize expectations rather than punishments. You might have observed in analyzing the original document that long lists of tasks are hard for employees to learn. One of your objectives will be to group related tasks.

The deliverables are the products you will offer the client. Will you print pages that can be reproduced? Will you also provide digital copy that can be updated? Will you edit content for a website assuming someone else will do the coding for display, or will you also code the content?

Schedule and Responsibilities

The convincing part of a project plan or proposal is the part that tells what you will do to accomplish the objectives and when you will do it. Categories of tasks include editing of the text, production (preparing the pages, print or online, as they will be reproduced), and meetings with the client. Some projects will involve research beyond careful reading of the text. Some projects will include usability testing as part of the review.

If you will collaborate with one or more classmates, each task you list on your plan should include the name of the person responsible. The client may also have some responsibilities, such as providing some information or files.

In the following example, two students are collaborating on the revision of instructions for assembling a desk built in the workshop of a local prison.

Date	Task	Responsibility
10/17	Project plan	Zack and Eva
10/19	Meet client, editing agreement	Zack and Eva; client
	Get AutoCad drawings	Client
10/19	Site visit: observe a user assembling the desk	Zack and Eva
10/20	Update product description; revise section on drawers	Zack
	Reorganize tasks	Eva
	Page design	Eva
10/24	Meet client; mid-point review	All
10/27	Revise text and graphics	Zack and Eva
11/2	Handoff for usability testing	Zack and Eva

The schedule should include meetings with the client. A preliminary meeting is essential to establish a shared understanding of what needs to be done. A midpoint meeting is desirable. It is a milestone (see Chapter 24) that holds you accountable for completing work. It also encourages the communication that keeps editors and clients working together cooperatively. You can get a preliminary response to work accomplished and verify that the project should continue as planned or be modified according to the results so far. In planning, you cannot always accurately forecast project needs and constraints. For example, you may have planned to combine two sections of a manual in a way that seemed theoretically feasible at the time of the plan, but now that you have worked on the material, you see that the idea isn't so good after all. The meeting is a time for you and the client to agree to this change in plans.

Budget

Chances are you will complete your classroom project as a service to your client, recognizing that you are still an apprentice. The value to you is a project for your class and class credit, the experience, a possible portfolio piece, and a possible letter of recommendation. The client may be doing you a favor by giving you his or her time for reviewing the project and its goals with you. But you may have a client who offers to pay for your time, especially for a complex project with commercial value that demands more than the classroom assignment requires. Even for a service project, you could also have some expenses for travel to a location for observations or for supplies or for production. Will you absorb these out-of-pocket expenses, or will you ask the client to reimburse you? Any expenses and responsibility for them should be explicit in the project plan or proposal. An agreement about who pays for these expenses can be stated in the contract.

Establishing a Contract

A contract is a legal document that summarizes the agreement between the client and the service provider (you). Signatures on the contract confirm that the parties have agreed to the terms about what services will be provided and the deliverables that will result.

You can use the editing agreement in Figure 24.1 (p. 426) as a starting point for developing an appropriate contract for this project. The intellectual property clause is important to establish your right to use the project as an example in your portfolio. This statement gives you the privilege to use the completed work for a specific purpose, not for any or all purposes. You may have to modify the statement to accommodate constraints on the privilege. For example, a client may give you permission to use the document only if all identifying information is removed or may allow you to print it but not post it on the Web. Some companies may object to giving you this privilege at all if you are working on information that is for internal use only. Knowing this restriction at the beginning of the project will give you the option of finding another project, doing this one anyway, or negotiating for some limited use of the materials.

Conferencing with the Writer or Client

A good working relationship with a client or writer requires face-to-face meetings or, in the case of long-distance editing, contact by phone, letter, or email. These conferences may be between editor and writer or between editor and client or with multiple stakeholders. In this section of the chapter, the client is assumed to be the writer, not a third party who is an intermediary between writer and editor. Even if you negotiate a project with a client who is not the writer, at some point you should meet the person who might be responsible for implementing editorial suggestions.

The way you manage yourself in these meetings will set the tone for your relationship. Conferences may occur for planning, review at milestone points, and at handoff. The review conference, during which the edited document is examined or plans for further editing are made, is the most sensitive because of the potential for implied or real criticism. Documents may be passed back and forth between writer or client and editor without a conference, and it saves time to do so. However, if the editing will be extensive, it may be better to plan a conference to talk about the changes than to surprise a client with a heavily marked document. When you are editing a long document with multiple sections, it will be smart to schedule a review conference following your editing of the first section. The conference will provide an opportunity for communication between client and editor to clarify expectations. You will both work more efficiently for the remainder of the project.

Some companies with product teams use a variation of the conference called an "inspection meeting." All the people working on the product gather to review the edited version of the document and to negotiate the next steps in development. The inspection meeting combines technical review with editorial review. Such a meeting reinforces the team concept of development instead of the one-to-one relationship between writer and editor. And besides, it is easier to resolve disagreements when all the people who might have suggestions are in one place at one time instead of depending on sequential and sometimes contradictory comments.

Even if a meeting has the potential to generate tension, it is preferable to exchanging heavily edited documents without discussion. It allows for give and take—the opportunity to question and clarify and to make nonverbal gestures of accommodation, such as nodding your head and smiling. You can make these conferences go well by organizing your comments, by using tactful and goal-oriented language, and even by arranging the furniture in a way as to suggest cooperation (see "Furniture Arrangement").

As you plan for the conference and its organization, think of the overall goals. The purpose of the conference is to clarify the next steps in document development. All conferences should result in some agreement and understanding among collaborators regarding goals, tasks, responsibilities, and schedule.

Using the review conference for instruction in principles of writing distracts from the main conference purpose and from the focus on the document at hand. Instruction also demotes a writer to the role of student rather than colleague and collaborator. Save instruction for another setting, or limit it to responding to questions a writer asks. If you repeatedly edit one writer for persistent, correctable problems, it's quite reasonable to suggest a time for instruction. But don't mix instruction with the development and production of a specific document.

Conference Organization

A conference is a form of communication, just as a printed, graphic, or electronic document is. Thus, the conference should adhere to principles for effective communication, such as organization. Before you meet with the client or writer (or for an inspection meeting), develop a plan for what you want to discuss and the order in which you will approach topics. Notes identifying topics and pages to which they apply will help the conference proceed efficiently and with a minimum of fumbling. An organized conference should increase the writer's confidence in you as editor and manager. It will reflect your comprehension of the whole project as well as your respect for the writer's time.

An overview statement, like the introduction to a document, identifies the topic and goals of the conference and the order in which you will proceed. You could organize by topic, perhaps asking questions about content and then offering comments about style. Working top down, from content and organization to the details, will keep the conference focused on substantial issues that require conversation. Or you could proceed through the document from beginning to end, chronologically (though not line by line). The conference could reasonably end with suggestions and a schedule for revision.

If you meet with several reviewers in a document inspection meeting, the importance of organizing the conference increases. Use their time efficiently by defining in advance the issues about which the reviewers may disagree or about which you need help. Work toward consensus on major issues and leave the minor issues alone. You can address issues such as capitalization and punctuation later without using everyone's time.

Review of the Edited Document

In a review conference, do not feel that you have to call attention to every emendation in the document. You do not have to initiate a conversation about all the punctuation changes, for example, but you should be prepared to explain them if the client or writer should ask. Your goals are to verify that your editing is correct and consistent with the overall document goals and, working with the writer, to establish the next steps in project development, whether revision or approval for continuing production. It is inefficient to plan a line-by-line review of the document in conference. A writer who is determined to consider each emendation may appreciate a copy of the edited document before the conference. The writer can then review it line by line on his or her own time.

A heavily marked typescript may intimidate a writer and create defensiveness that can spoil the conference. If company policy gives you the privilege of editing electronically, you can easily prepare clean copy incorporating the editing. But keep records of the editing for style and organization so that you will be able to query about the accuracy and seek approval for substantive changes. Also keep the original version so that you will be able to restore it if necessary.

If policy and the writer's trust permit you to share clean rather than marked copy, three positive results can be achieved. First, a clean copy is less threatening than a marked one, so the egos of the writer and editor are less likely to interfere with the work of producing an effective document. Second, it will be easier to spot additional editing needs in a clean copy than in one in which the marks distract from reading. Finally, a clean copy will show the results of editing and demonstrate its value.

The Language of Good Relationships

The words you choose, your non-verbal expressions, and the way you listen all affect how well a writer receives your messages. Your language can communicate a sense of collaboration, or it can communicate power and arrogance.

Active Listening

One purpose of a review conference is information gathering. Thus, an important method of communication is listening. Active listening means drawing out the writer and working to understand his or her point of view. It contrasts with the pretense of listening while mentally formulating your own statements. If writer and editor listen to each other only superficially, the chances of real communication taking place are not good.

One strategy of active listening is to repeat or paraphrase something a writer says.

"Are you saying, then, that . . . ?"

The echo of the writer's statement guards against misinterpretation. If your paraphrase is inaccurate, the writer can correct you. This strategy also forces you to listen carefully enough to repeat. An active listener also probes for more information when a writer expresses ideas incompletely. Probing statements encourage the writer to elaborate.

> "Please go on."
> "What do you mean by . . . ?"
> "How will readers use this information?"
> "How does this point relate to . . . ?"
> "Please explain how . . ."

These are not statements to challenge the writer's competence but to seek the information you need to complete the editing.

With active listening, you also encourage the writer to share and cooperate. You can show your interest with non-verbal signals (nodding your head, smiling, leaning forward) and with verbal signals ("I see," "uh-huh").

Positive Language

The words you choose in the conference can encourage partnership or they can provoke defensiveness. Writers will respond to goal-oriented language more positively than to criticism.

Critical Language	**Goal-Oriented Language**
wordy	"Condensing the text will increase the chance that a person in an emergency will find the necessary information quickly."
poorly organized	"The tasks are rearranged chronologically so that the crew will know when to do each one."

The goals (finding information in an emergency, knowing when to do the tasks) are explanations that show the editing to be purposeful rather than arbitrary. Don't model after teachers who write critical words such as "awkward" and "unclear" on student papers. An editor, like a teacher, must evaluate, but the evaluation is not an end in itself. It should suggest goals for revision based on reader needs.

Another way to emphasize document goals rather than resort to criticism is to use "I" statements when you may seem critical, but "you" statements when you are praising. "I" statements do not suggest that you are stupid but rather that you are assuming the role of readers who may not have enough information.

Bad news	*Good news*
"I" statement (preferable for bad news):	**"You" statement** (preferable for good news; avoid for bad news):
"I don't understand how this example explains the thesis."	"You haven't been clear here."
"I can't tell whether I should turn off the hard drive or the monitor first."	"Your instructions are incomplete."
	"You did a good job of interpreting the table."

Try to focus on neutral subjects—the document and the reader—rather than on the writer and editor.

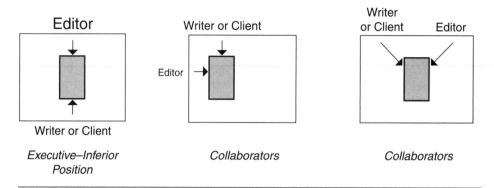

Figure 25.1   Chair Arrangement for a Conference with a Writer or Client

The Executive–Inferior arrangement makes it difficult to look at the document and invites confrontation between the editor and the writer or client.

One final caution: avoid words that suggest inappropriate editorial intervention, especially *change*. No writer wants to hear that you have "changed" his or her text. In fact, writers fear that editors will change the meaning. While polysyllabic euphemisms are generally undesirable stylistically, it may be better when dealing with a sensitive writer to substitute *emendation* for *change*.

Confidence

Your own confidence in your editing will encourage the writer's confidence. Wishy-washy comments can turn off writers as much as aggressive comments can. You should not apologize for your good suggestions and honest explorations, nor should you avoid making recommendations for fear of alienating a writer. Keep in mind that you and the writer are partners working on behalf of the project's readers.

Furniture Arrangement

Psychologists define non-verbal communication as the information we give by our posture and facial expressions. Similarly, a room's furniture arrangement suggests the relationship between the people in the room.[1] In Figure 25.1, the executive arrangement, with an executive protected behind an imposing desk, suggests a relationship of superior to inferior. Such a relationship is inconsistent with the partnership of editor with writer, as well as being inefficient when editor and writer look at a document together. In this arrangement, editor and writer look at each other, perhaps confrontationally.

The collaborators' arrangements, by contrast, place editor and writer in more equal positions, as partners. These arrangements allow editor and writer to view

---

[1]Rosemarie Arbur. 1977. Student-Teacher conference. *College Composition and Communication* 28: 338–342.

a document together and therefore encourage the conversation's focus to be on the document rather than on the personalities. A conference room may be a better place to meet than the office of editor or writer. Using a psychology term, it is more *neutral territory*. Likewise, a company table may be more appropriate than a desk that belongs to either the editor or the writer.

Presenting the Project Orally

In a professional setting, you may be asked to present the results of your work orally, either to a client (individual or committee) or to co-workers. The presentation to the client accompanies the handoff, the point at which you declare the objectives met and explain what you have accomplished. This presentation is an opportunity to showcase the value you have added to the project. It also helps the client understand the work of an editor and the standards of expert writing. You may pave the way for future assignments and increased responsibilities. Coworkers may review the work of others on related projects partly to understand what everyone is doing but also to stimulate thinking about ways to improve the processes of publications development.

In a classroom, presentations by class members can give the other students vicarious (secondhand) editing experiences. No one's project is quite like any other, and the class will see a variety of editorial situations and solutions to problems that will increase their wisdom about editing. Preparing a presentation for class also encourages you to reflect on the significance of work you have done.

Like a conference with a writer or client, a presentation is a form of communication. It is organized to reveal its purpose. Visual and verbal signals help the audience keep track of how the presentation is progressing. The style and tone are professional.

Content, Organization, and Illustrations

Your presentation will work best if you plan it and organize it to emphasize key points. A good strategy for professional presentations is to focus on problems and solutions. What were the problems you identified in the original, and how did you solve them in the edited copy? What were the constraints, if any, on what you could accomplish?

You don't have to tell the audience everything you did (that information is likely to be apparent from the text you submit) nor about every step you took in editing. Rather, audiences will appreciate your efforts to select and focus. Topics like these can be engaging:

- organizing an apparently chaotic list of items: alternatives and reasons for your choice
- making style readable/adjusting style to audience
- accommodating different browsers
- designing for development of the site
- using space on the screen effectively

- designing to accommodate people with disabilities
- acknowledging that readers will probably throw away the document after skimming and yet need to learn the material
- accommodating the likelihood of frequent revisions to the document
- compromising between design principles and cost constraints

Pick a topic like one or two of these that made your work interesting. Other topics that might interest your classmates could relate to planning and management strategies, such as your system of version control, or your negotiation with a client when you disagreed about some feature of visual design.

If you organize the whole presentation by problem and solution, some illustrations of the original compared with the edited version will let the audience see the impact of the change. Again, you can be selective: find your best examples that illustrate the problem you are foregrounding. Unless you are speaking in a room that lacks equipment, these examples should be available for screen projection so that everyone can see.

Presentation

If you have planned and organized your content and prepared some illustrations, you have done the hard work of a presentation. The quality of your presentation depends on good content that is organized to support the points you would like to make.

The remaining goal is to enable the audience to understand your points and follow your presentation. For these goals, you will need to consider how the people in the audience will listen and absorb information. One basic principle is that reinforcing oral information with visual information helps people understand and remember. That is why you should try to prepare some information for screen projection that shows the structure of the presentation, not just the examples. You can use presentation software, or you can project text files.

You have been in audiences often enough to know that people get distracted, even when the speaker is good. If you were reading a book and your mind wandered, you could find the place and begin reading again, but if you are in an audience, the presentation goes on without you. You need help reconnecting. When you read, you also depend on signals about where you are in relation to what is coming next. These signals include the order of chapters, page numbers, headings, and more. A presenter provides analogous signals.

Good presenters anticipate the needs of audience members to locate their place in the presentation. Here are some ways to do so:

- Provide an overview slide, listing your major topics. Explain the structure orally. The oral and verbal information will reinforce each other and provide a map of what is to come.
- When you move from one major point to another, use transition words: *next . . . , the second point . . . , in contrast to. . . .* These signals explain concepts but they also identify the progress of the presentation.
- Provide signals in a slide to indicate the topic. Figure 25.2 is a sample slide from a presentation on five types of ethical considerations in technical

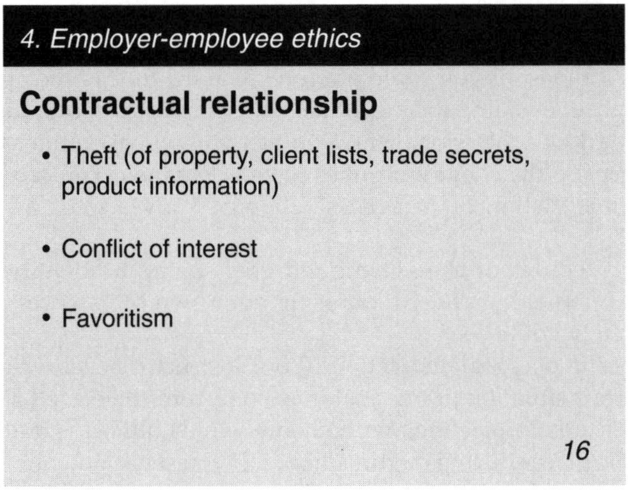

> ### 4. Employer-employee ethics
>
> ## Contractual relationship
>
> - Theft (of property, client lists, trade secrets, product information)
>
> - Conflict of interest
>
> - Favoritism
>
> *16*

Figure 25.2 Sample Slide

communication. The italic title above the slide title indicates that this slide relates to the fourth topic, employer-employee ethics. The number at the bottom is the slide number.

Keep your slides uncluttered. Minimize the number of words, and avoid patterns in the background. Use a light background with dark text. Don't be tempted by the possibility of inserting sound effects.

Many speakers are uncomfortable facing an audience, but planning prepares you to do well. What remains is focusing on the audience: literally, by looking at them, by projecting your voice toward them, by thinking of what they need to know about your subject. You may have to practice speaking when there is a slide behind you. Instinctively, you will want to turn around and look at it, but your audience doesn't want to look at your back. When you direct your attention to your audience and your subject, you will probably communicate well.

Finally, when you are in the audience for a classmate's presentation, help that person do well by being an attentive listener. You can affect the quality of the presentation by the way you respond—by facial expression and body language as well as by the questions you ask. Acting interested will bring out the best in the speaker.

Professionalism

A client project requires the highest standards of professionalism from you. Even though you are an apprentice, the quality of your work and your behavior represent the profession you hope to enter, and you affect client attitudes about other students, your instructor, and the college or university.

Your project plan outlines what you will do and when you will do it. An obvious mark of professionalism is getting your work done on time. If you will have to miss due dates for reasons you could not foresee at the time of the plan, contact the client in advance, explain, and negotiate different dates. Communication is also an obvious mark of professionalism. Don't let your client wonder where you are, and don't surprise the client with unexpected demands. You can communicate by phone, email, letter, or in person. Chapter 3 gives some advice about correspondence.

Follow the procedures of assessment and goal setting to identify good reasons for editorial intervention. Resist imposing your own preferences on a document that is not really yours.

Use good sense in personal interactions. Your instructor never wishes to send you into a dangerous situation or one that may compromise your ethics. This advice especially pertains to meetings with clients, which almost certainly should take place in public places during daytime hours. If you sense potential problems, ask your instructor to help you figure out solutions.

Treat your client and the work the client represents with respect. Editing has an inherent component of evaluation, and you would not be asked to edit if the document were already perfect. You may at times be tempted to laugh. If you are, enjoy the moment, but remind yourself that you are part of a team of people with different kinds of expertise working together to enable a document to do its work.

When you submit the final version of your work, thank the client for working with you. An oral thanks is appreciated, but a written note is even more gracious.

Using Your Knowledge

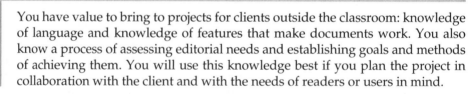

You have value to bring to projects for clients outside the classroom: knowledge of language and knowledge of features that make documents work. You also know a process of assessing editorial needs and establishing goals and methods of achieving them. You will use this knowledge best if you plan the project in collaboration with the client and with the needs of readers or users in mind.

Discussion and Application

1. Conduct editor–writer conferences in class and then answer the accompanying questions. Each student should present a document he or she has written to a classmate for editing. The document should be in a technical rather than a creative genre. In a brief planning session, the writer explains to the editor the purpose and readers of the document and any particular editing needs that he or she has identified. The editor, in turn, may query the writer. If the document is substantial, the editing will take place out of class. After editing is complete, each student editor meets the writer for a review conference. After the conference, the writer evaluates

the editing and the editing conference and makes suggestions for future conferences. These questions can be used for evaluation:

a. *Consider the editing.* What needs did the editor discuss? Did the editor overlook any editing needs? Comment on accuracy, completeness, and objectivity. Was the editor too aggressive or too passive on some issues? Did the editor seem to be a collaborator? Comment.

b. *Consider the conference.* How did the editor organize it? Did the editor clarify his or her plan for the conference at the beginning? Was the language goal oriented? Was the editor confident? How did you feel about being edited by this editor? Comment.

c. *Suggestions:* What should this editor work on to increase effectiveness in editing and in communicating with writers?

You should edit the work of a classmate other than the one who edits your work.

Glossary

abstract Summary or description of document contents; part of the body of a report or book.

accessibility Online site design that makes information available to people with disabilities. Includes customizable display choices and text alternatives for illustrations.

acknowledgments Credits to persons who have helped with the development of a publication; often the concluding part of a preface.

adjective Part of speech whose function is to modify a noun.

adverb Part of speech whose function is to modify a verb, adjective, or other adverb.

appositive Noun or noun phrase placed with another as equivalents; e.g., Joe, *the new writer, . . .*

ascender Part of the letter that rises above the x-height, such as the top part of the letters *b, d,* and *h.*

back matter Parts of a book following the body, including the appendix, glossary, and index.

baseline Imaginary horizontal line at the base of capital letters of type. Used to measure space between lines.

basic copyediting Editing for correctness and consistency but not for content and organization. Distinguished from comprehensive editing.

basis weight Weight of a ream of paper in its uncut size; e.g., a ream of 20-pound bond paper weighs 20 pounds in its uncut size of 17×22 inches.

blueline In offset lithography, a proof following the making of printing plates. Shows page sequence and folding, not just the individual pages. Printed on light-sensitive paper with blue characters.

body type Type for the text of a document as opposed to its headings and titles.

boilerplate Text that is standard for various documents and can be inserted, with minimal or no revision, in a new document.

browser Software that enables searching the World Wide Web; person who searches the Web or print pursuing particular interests rather than following the structure of the document.

bullet Heavy dot (•) used to mark items in a list. Sometimes set as a square, an unfilled circle, or another shape.

callout Words placed outside an illustration but referring to a part of the illustration.

camera ready Copy that may be photographed for printing without a change in type or format or special processes, such as screens; production ready.

caption Brief explanation or description of an illustration printed beneath the title.

case Form of a noun or pronoun that shows its relationship to other words in the sentence, whether subjective, objective, or possessive.

casting off   Process of calculating the document length according to the number of characters in a typescript and the specifications for typeface and type size, line length, and page depth of the final document.

clause   Group of words that contains a subject and verb.

    dependent/subordinate   Group of words with a subject and verb plus a subordinating conjunction or a relative pronoun. Cannot be punctuated alone as a sentence.

    independent/main   Group of words with a subject and verb but no subordinating conjunction or relative pronoun to make it dependent; may be punctuated as a complete sentence.

    nonrestrictive   Modifying clause beginning with a relative pronoun (*who, which*) that gives additional information about the subject but that is not essential to identify what the subject is. A comma separates this clause from the subject: "Nita Perez, *who is president this year of the users' group, . . .*"

    restrictive   Modifying clause beginning with a relative pronoun (*who, that*) that restricts the meaning of the subject (gives essential identifying information); no punctuation separates the modifier from the subject it modifies: "The officer *who sat at the end of the head table . . . .*"

clip art   Pictures of common objects that can be pasted into documents. Available on paper and electronically.

CMYK   Color model for printing with cyan, magenta, yellow, and black inks.

color separation   Photographic or laser process of creating the four primary printing colors from a full-color original. See **CMYK**.

complement   Words used to complete the sense of the verb. A subject complement completes a linking verb, while a direct object completes a transitive verb.

complex sentence   Sentence structured with a dependent as well as an independent clause.

compositor   (1) Person who uses typesetting equipment to key in text for a typeset document; a typesetter. (2) Person who assembles type and illustrations into pages (composes the pages).

compound sentence   Sentence composed of at least two independent clauses.

compound-complex sentence   Sentence consisting of two independent clauses plus a dependent clause.

comprehensive editing   Editing for the full range of document qualities, including content, organization, and design as well as grammar and punctuation, with the goal of making a document more usable, suitable for its purpose and readers, and comprehensible. Distinguished from **basic copyediting**.

configuration management   Software for file storage and management.

conjunction   Word that joins words (nouns, verbs, modifiers) or clauses in a series.

    coordinating   Joins items of equal value, including two independent clauses: *and, but, or, for, yet, nor, so.*

    subordinating   Joins items of unequal value, especially a dependent to an independent clause. Makes a clause dependent: *although, because, since, while, . . .*

conjunctive adverb   Modifier of a clause that shows its relationship to the content of another clause: *however, nevertheless, therefore, thus.*

content management   Creation, modification, and use of content throughout an organization, using single sourcing.

contextual inquiry   Research about document users and purposes at the site of use and with users as advisors.

contract   Agreement for a person or company to perform work for another company. Instead of being employed by the company for which they write and edit, technical communicators may work independently on contract or work for a contract company, one that places its employees in short-term assignments for other companies.

controlled language   Assigning a single definition to a term.

copy   Typescript or graphics used in preparing a document for publication.

copyediting   Editing a document to ensure its correctness, consistency, accuracy, and completeness. Distinguished from **comprehensive editing.**

copyfitting   Process of fitting copy into a prescribed space. May be done mathematically with the character count and document specifications, or electronically, with a page layout program.

copyright   Protection provided by law to authors of literary, dramatic, musical, artistic, and certain other intellectual property, both published and unpublished. The law gives the owner of copyright the exclusive right to reproduce, distribute, perform, or display the copyrighted work and to prepare derivative works.

copyright holder   Person or organization that owns the rights to the copyrighted work.

copyright page   Page in a document identifying the copyright; usually the back of the title page.

cropping   Cutting one or more edges from an illustration to remove irrelevant material and to center and emphasize the essential material.

data ink   Ink in an illustration or website used for the content as opposed to marking spaces with lines, decorations, or unnecessary callouts.

dead copy   Version of a document during production that has been superseded by a later version. The typescript becomes dead copy once page layout has produced page proofs.

deliverable   Specific outcome of a project. Deliverables in publications include printed manuals, online help, brochures, presentations, policy statements, or other such products using verbal and visual information.

descender   The part of a letter that descends below the baseline, such as the tail on the letters *g* and *y*.

design, document   The plan for a document and all its features (content, organization, format, style, typography, paper, binding) to make it useful and readable.

design, graphic   The visual features of the document, including typeface, size, and style, page size, line length and depth, paper, and binding.

desktop publishing   The preparation of final copy using office equipment and thus bypassing the professional typesetter. Printing and binding could then be done professionally or in house.

digital photography   Images recorded in dots per inch, perhaps in different colors.

digital printing   Imaging process for text and graphics from digital files. Bypasses the photographic processes and prepress labor of offset lithography. Uses electric charges on the image areas of a toner drum. Digital files can draw from databases, enabling customized individual pages. Popular for small print jobs and quick turnaround.

display type   The titles and headings of a document as opposed to the body copy or the type on a poster.

document set   A group of related documents, such as all the manuals for a piece of equipment or all the manuals published by a particular organization.

DPI   Dots per inch. Measures how fine the print will be. More DPI means a finer quality of print because each dot is smaller. *See* **LPI.**

dummy   A graphic designer's sketch of pages as they are to be printed, showing line length, margins, and placement of headings and illustrations.

editing, comprehensive   Development and revision of a document in the context of its uses and purposes, considering readers, content, organization, visual design, style, and use of illustrations.

em   Linear measure about equal to the width of a capital letter *M* in any given typeface and size; the square of a type size. The normal paragraph indent in typeset copy.

emendation   Term used to describe editorial intervention; alternative to "change." Its use may discourage the sense that an editor has the privilege to "change" a document, possibly for the worse.

en   Half an em. Used in ranges, as in ranges of numbers.

end focus   Structuring sentences to emphasize important information in the predicate.

FAQs   Frequently asked questions; a type of user support.

figure   Illustration that is not tabular; e.g., line drawing, photograph, bar graph.

font   Typeface. Historically, a collection of characters for a typeface in one size, including roman and italic characters, capital and lowercase characters, and sometimes small caps.

foreword   Part of the front matter of a book; introductory remarks written by someone other than the writer or editor.

format   Placement of text and graphics on a page including the number and width of columns; the dimensions of margins, spacing, and type; and the form of the text (paragraphs, lists).

four-color printing   Printing of full-color reproductions from four ink colors: cyan, magenta, yellow, and black. See also **CMYK** color model.

front matter   Parts of a document that precede the body; e.g., title page, table of contents, preface.

galley proof   Copy of text after typesetting but before page breaks have been established. Printed on long, shiny paper. Used to check accuracy of typesetting.

genre   A class of documents that share purposes and conventions of structure; examples: instructions, reports.

gerund   Noun substitute formed from a verb plus the suffix *-ing;* see **verbal**.

globalization   Developing products, and their documentation, for international audiences. Strategies include minimization of cultural metaphors and illustrations and language that can be translated readily.

glossary   Short dictionary of key terms used in the document.

grammar   System of rules governing the relationships of words in sentences.

graphics   (a) Text with a strong visual component consisting of more than words arranged in paragraphs; e.g., tables, line drawings, graphs. (b) Visuals with mathematical content (as would be drawn on graph paper); e.g., graphs, architectural drawings.

gutter   Inner margin of a book, next to the binding.

half title   Page at the beginning of a book or division that names only the main title, not the subtitle or other identifying information.

halftone   Picture with shading (different tones of light and dark) created by dots of different density. Produced by photographing the subject through a screen or by digital photography.

handoff   The end of the publications development process, when the completed document is turned over to the project manager or client.

hanging indention   Paragraph or list form in which all the lines are indented except the first.

hard copy   Document or draft printed on paper as opposed to the digital version, (soft copy).

heuristic   Guide to investigation. The questions *who, what, when, where, why, how,* and *so what* are a heuristic that guides a writer in developing content.

home page   The first screen of a World Wide Web site that functions like a book cover and table of contents to identify contents and structure of the site.

house style   Mechanical style choices preferred by a publishing organization. Derives from the designation of any publishing organization as a *house*.

HTML   Hypertext Markup Language; code to create various type styles in hypertext publication on the World Wide Web.

hyperlink   Electronic connection from one file to another or to another location in the same file; clicking on the link activates the code that points to the new file.

hypermedia   Multimedia, such as text, sound, and moving images, with electronic links.

hypertext   Text with electronic connections (links) to other documents or to sections within the text.

icon   Visual representation of a process or concept; a visual symbol.

imperative mood   See **mood**.

imposition   The arrangement of page proofs in a form before platemaking so that they will appear in correct order when the printed sheet is folded.

in house   Adverb or adjective (in-house) indicating that work is completed or applies within the organization rather than without; e.g., an in-house style manual designates style choices preferred by that organization.

index   List at the end of a book or manual of key terms used in the document and the page number(s) where they are used.

indicative mood   See **mood**.

infinitive   Verbal consisting of the word *to* plus the verb; e.g., *to edit*. Used primarily as a noun.

inflection   Change in the form of a word to show a specific meaning or a grammatical relationship to another word. The verb *edit* would be inflected as follows: *edit, edits, edited*. The adjective *good* inflected is *good, better, best*.

inspection   Formal review of a document by a group that may include writers, editors, managers, subject matter experts, and legal experts. An inspection meeting brings them together to determine revisions needed.

intellectual property   Property, including books, photographs, computer programs, and customer lists, that can be protected by federal law.

interface   Visual and textual features on the computer screen that enable users to use the website or software.

Internet   Global network of regional networks; the cables and computers that form the network. The World Wide Web uses the Internet to transmit information.

intransitive verb   Verb that does not "carry over" to a complement. The predicate ends with an intransitive verb.

introduction   Substantive beginning section for a document; in a book, generally the first chapter rather than part of the front matter, where a preface or foreword would go.

ISBN   International Standard Book Number; 10-digit number that identifies the book, including its publisher and country of publication.

ISO   International Organization for Standardization; sets standards for many businesses and technologies. "ISO 9000" is a standard for a quality management system for organizations doing business in the European Union; the standard requires a manual documenting the system.

italics   Style of printing type with the letters slanted to the right.

iteration   Version of the typescript as it moves through various editorial passes. A new iteration incorporates some editorial emendations.

justification   Adjustment of lines of text to align margins; alignment.

landscape orientation   Position of lines of type on pages parallel with the long side of the page to create pages wider than they are tall. See **portrait orientation**.

layout   (a) Spread and juxtaposition of printed matter. (b) Dummy or sketch for matter to be printed.

leading   (pronounced *ledding*) Space between lines of type.

legacy document   Existing document that must be adapted and converted for a new publication medium, new software, or new version of a product or organization.

legend   Explanation of symbols, shading, or type styles used in a graph.

legibility   The ease with which type or illustrations can be read. Legibility refers to recognition, while readability refers to comprehension.

line drawing   Drawing created with black lines, without shading. It may be photographed for printing without a halftone.

lingua franca   Single language, understood by all cultures in which the document is used.

linking verb   Verb that connects a subject with a predicate adjective or predicate nominative rather than with a direct object.

list of references   List of works cited in a document, with publication data.

localization   Adapting a document for a specific area. Includes translation but also use of cultural values of the readers' country.

LPI   Lines per inch used in printing. A higher number (such as 133) means higher resolution.

macro   Shortcut in word processing that encodes repeated procedures; can be user defined.

manuscript   Unpublished version of a document. Because the term literally suggests handwriting, it is often replaced by *typescript.*

markup   Process of marking a typescript, on paper or electronically, for graphic design or editing. Procedural markup directs changes in the text or graphic design; structural markup identifies structural parts that will later be formatted by a style sheet.

milestone   Defined point of achievement in project development.

mood   Verb form indicating the writer's attitude toward the factuality of action or condition expressed.

   imperative   Verb form used to express commands: "Turn on the computer."

   indicative   Verb form used for factual statements: "The computer is turned on."

   subjunctive   Verb form used to indicate doubt or a hypothetical situation: "If the server were turned off every night, browsers could not access our website."

multimedia   Information expressed in more than one medium, such as print, sound, video, and graphics.

noise   Distracting material in a document, such as errors, excess words, or an inappropriate voice, that interferes with the reader's attention to the content.

nominalization   Noun formed from a verb root, usually by the addition of a suffix; e.g., consideration, agreement.

nonfinite verb   See **verbal**.

noun   Part of speech representing a person, place, thing, or idea.

object   Noun or noun substitute that is governed by a transitive active verb, a nonfinite verb, or a preposition. A direct object tells what or who. An indirect object tells to whom or what or for whom or what.

offset lithography   Common printing method. The design or print is photographically reproduced on a plate, which is placed on a revolving cylinder of the printing press; the print is transferred to, or *offset* on, the paper by means of a rubber blanket that runs over another cylinder.

online   Electronic file on a computer rather than in print. *Online* instructions appear on the computer screen rather than in a printed manual.

page proof   Copy of typeset text. Used to check the accuracy of corrections and the logic of page breaks.

parallel structure; parallelism   Use of the same form (e.g., noun, participle) to express related ideas in a series.

participle   Modifier formed from a verb with the addition of the suffix *-ing* or *-ed;* e.g., *dripping* pipe, *misplaced* cap.

pass, editorial pass   One-time review of the document. Comprehensive editing may require multiple editorial passes.

PDF   Portable Document Format; a file format developed by Adobe and used for documents that are distributed electronically on different platforms.

perfect binding   Binding for paperback books in which the cover is attached to the pages with adhesive.

persona   Character or personality of the writer as projected in a document by his or her style.

phrase   Group of related words that function as a grammatical unit; does not contain both a subject and verb.

> **infinitive phrase**   Includes the infinitive form of a verb plus modifiers; e.g., "to write well."
>
> **noun phrase**   Consists of a noun and its modifiers; e.g., "stainless steel."
>
> **participial phrase**   Includes a participle plus modifiers; e.g., "diffusing quickly."
>
> **prepositional phrase**   Begins with a preposition; e.g., "above the switch."

pica   Unit of linear measure used by graphic designers and printers; roughly one-sixth of an inch. Used to describe both vertical and horizontal measures.

pica ruler, pica stick   Measuring device marked with increments of both picas and inches.

plate   Light-sensitive sheet of metal upon which a photographic image can be recorded. When inked, will produce printed matter in offset lithography.

point   Unit of linear measure used by graphic designers and printers especially in describing type size; one-twelfth of a pica.

portrait orientation   Position of lines of type on pages parallel with the short side of the page to create pages taller than they are wide. See **landscape orientation**.

predicate   Division of a sentence that tells what is said about the subject. Always includes a verb; may also include a complement of the verb and modifiers.

preface   Part of the front matter of a document stating the purposes, readers, scope, and assumptions about the document. Often includes acknowledgments as well.

preposition   Part of speech that links a noun with another part of the sentence.

printer   Person who reproduces or who supervises the reproduction in multiple copies of a document.

production   Process of developing a document from manuscript to distribution. Requires scheduling and coordination of services such as editing, graphic design, printing, binding, coding, and media reproduction.

pronoun   Part of speech that takes the position and function of a noun. May be personal (*I, we, you, they*), relative (*who, whose, which, that*), indefinite (*each, someone, all*), intensive (*myself*), reflexive (reflecting on the grammatical subject: she chided *herself*), demonstrative (*this, that, these*), or interrogative (*who?*).

prose   Words in sentence form, as opposed to verse.

prototype   An early example of the document or part of it that can be tested and modified before the rest of the document is developed.

publisher   Person or organization that funds the publication and its distribution. Usually separate from the printer.

query   Question to the writer posed by the editor requesting information that is necessary for completing the editing correctly.

query slip   Piece of paper on which a query is written; attached to the typescript; may be replicated online with "comments."

ragged right   Irregular right margin. Characters are not spaced to create lines of equal length.

readability   As applied to formulas, a quantifiable measure of the ease with which a text can be read. Based on counts of sentence and word features such as number of syllables and number of words per sentence. More broadly, the ease with which a reader can read and understand a document, based on content, level of technicality, organization, style, and format.

recto   In a book, the righthand page, numbered with an odd number. The *verso* is on the back.

redundant   Duplicate information; unnecessary repetition.

register   Alignment of printing plates one on top of the other to reproduce colored prints accurately.

relative pronoun   Pronoun that introduces a relative clause and has reference to an antecedent; *who, which, that.*

resolution   Number of dots of ink per inch (DPI) used to form characters. Low resolution (e.g., 300 DPI) can produce coarse or bumpy strokes in print, but low resolution means smaller files and faster downloads online.

RFP   Request for proposals; document that identifies a need for research, a service, or a product and invites competitive proposals to provide it.

river   White space running through a paragraph that forms a distracting diagonal or vertical line.

roman type   Type style characterized by straight vertical lines in characters rather than the slanted lines that characterize italic type.

running head   Title repeated at the top of each page of a book. May be the book title, chapter title, or author's name. May vary on recto and verso pages.

saddle stitching   Binding for a booklet with staples through the fold in the middle.

sans serif   Type style characterized by absence of *serifs* or short horizontal or vertical lines at the ends of the strokes in letters. Also called gothic. Common sans serif typefaces are Helvetica and Gothic.

schema   (plural *schemata*) Structured representation of a concept in memory.

screen   Glass plate marked with crossing lines through which continuous-tone art may be photographed for halftone reproduction.

script   Outline of visual and verbal elements of a document or module.

semantics   Study of meanings.

serif   Small horizontal or vertical line at the end of a stroke in a letter; also a category of typefaces characterized by the use of serifs, such as Times Roman and Bookman.

server   Computer that sends World Wide Web pages to client computer programs requesting the information.

SGML   Standard Generalized Markup Language; code for marking structural parts of documents to facilitate transfer among computer platforms, archiving, and databases.

side stitching   Binding in which staples are forced through the edge of the book.

signals   In text, the visual, verbal, and structural cues about structure and relationships of ideas. Examples: headings, color, terms such as *however.*

signature   Group of pages in a book folded from a single sheet of paper. Typically includes 16 pages, but may include 8, 32, or even 64 pages depending on the size of the pages and the number of folds.

simple sentence   Sentence consisting of one independent clause.

single-source documentation   Product description and instructions in various forms that draw on a source of information coded in a database.

SME   Subject matter expert, often the writer but primarily an engineer, scientist, or programmer.

soft copy   Digital or electronic copy.

solidus   Slanted line (/) used in math to show division and in prose to mean *per* or to separate lines of poetry that are run together.

specifications   Written, detailed description of a product's functions, capabilities, and physical characteristics. Establishes a plan for product development and enables documentation before the product is complete.

stakeholders   People with an interest in a product or organization: clients, customers, employers, co-workers.

standard American English   Widely accepted practice in North America in spelling, grammar, and pronunciation; the speech and writing patterns of educated persons in America; edited American English.

storyboard   Poster-sized visual and verbal outline of the structure and contents of a document; used in document planning.

stripping   In printing, the arrangement and taping of negatives from text and illustrations in a flat before platemaking; also the cutting and pasting of a corrected version in place of an incorrect one.

style   Choices about diction and sentence structure that affect comprehension and emphasis as well as projecting a *voice* or *persona*.

style, electronic   Typography and spacing specifications for a document part such as a heading; once defined, a style enables application of all the specifications in one command.

style, mechanical   Choices about capitalization, spelling, punctuation, abbreviations, numbers, etc., when more than one option exists.

style, typographic   Choices about appearance of type, such as bold, italic, or condensed.

style manual   Collection identifying preferred choices on matters of mechanical style including capitalization, abbreviations, and documentation.

style sheet   List of choices for spelling, mechanics, and documentation for a specific document. Helps the editor make consistent choices throughout the document.

subjunctive mood   See **mood**.

syllable   Unit of a word spoken as a single uninterrupted sound. Includes a vowel or a syllabic consonant.

syntax   Structure of phrases, clauses, and sentences.

table   Text or numbers arranged in rows and columns.

table of contents   List in the front matter of a document of the major divisions, such as chapters, and the page numbers on which the divisions begin.

tag   Code in markup languages and styles specifying the type of text.

template   In word processing or publishing software, a collection of styles (formatting directions) that define a document. See **style, electronic**.

tense   Form of the verb that indicates time of the action as well as continuance or completion.

   past tense   Indicates time in the past. With regular verbs, formed with the addition of the suffix *-ed*.

   present tense   Indicates current time.

   future tense   Indicates action that will occur in the future. Usually formed with a helping verb, *shall* or *will*.

terminology management   Knowing which terms have been used before and using the same terms consistently.

title page   Page in the front matter of a document identifying the title. May include other information, such as the name of the writer or editor, date of publication, and publisher.

tone   Sound that the voice of the writer projects—serious, angry, flippant, concerned, silly, etc.

type size   Height (and proportionate width) of a letter, expressed in points. For example, 12-point type will almost fill a 12-point (1-pica) line and is good for body copy, while 72-point type will almost fill a 6-pica line and is so large that its use would be restricted to banners and announcements.

type style   Shape of letters as determined by the slant and thickness of the lines and the presence or absence of serifs. *Roman* style uses vertical lines while *italic* style uses slanted ones; *serif* style uses serifs while *gothic* or *sans serif* style does not. *Bold* or regular weight and *condensed* or *expanded* may also define type styles. Also denotes classes of type as *body type,* or the body of the text, as compared with *display type,* or titles and headings.

typeface   Type design with distinguishing character shapes and a name; e.g., Times Roman, Helvetica. Equivalent of *font.*

typescript   Draft of a document, before it is typed in final form or typeset; the copy on which an editor works; the parallel of *manuscript,* when documents were written first in longhand.

typeset   Adjective describing text that has been prepared by photoelectronic typesetting equipment rather than by a typewriter or desktop publishing equipment.

typesetter   Person whose job is to prepare typeset copy in fullscale printing. May refer to the owner of a typesetting business or to the compositor, the person who keyboards the documents.

typesetting   Process of keying text into photoelectronic typesetting equipment in order to produce typeset galleys or proofs. Formerly done manually or mechanically, with lead characters.

usability   Ease with which a document, such as a manual or website, can be used.

usability test   Test with representative users to determine whether they can readily use the document for its intended purposes, such as completing a task accurately or finding reference material.

usage   Accepted practice in the use of words and phrases.

verb   Part of speech that denotes action, occurrence, or existence. Characterized by *tense, mood,* and *voice.*

   intransitive   Verb that does not carry over to a complement.

   linking   Verb that links the subject to a subject complement.

   to be   "Is" or a variant of *is*; links the subject to a subject complement.

   transitive   Verb that requires a direct object to complete its meaning.

verbal   Verb used as a noun, adjective, or adverb. Verbals may be *participles* (modifiers formed from a verb plus the suffix *-ing* or *-ed*), *gerunds* (noun substitutes formed from a verb plus the suffix *-ing*), and *infinitives* (verbs plus *to,* used chiefly as nouns). Also called *nonfinite* verbs.

version control   Process of ensuring that all people who work on a document use the latest version.

verso   In a book, the page on the left side as the book lies open, numbered with an even number; the back side of a *recto* page.

voice   Form of a verb that indicates the relation between the subject and the action expressed by the verb.

   active voice   Verb form indicating that the subject performs the action expressed by the verb.

   passive voice   Verb form indicating that the subject of the sentence receives the action expressed by the verb; always identified by a *to be* verb plus a past participle.

white space   Graphic design concept: blank space on the page that functions to draw attention to certain parts of the page, to provide eye relief, to signal a new section, or to provide aesthetic balance.

wordspacing   Amount of space between words. Manipulated in order to achieve right justification and to eliminate rivers.

World Wide Web; the Web; WWW   Multimedia, graphic international database of information stored in multiple sites that are linked by the Internet. Accessed electronically through a browser.

x-height   Size of a letter without its descender or ascender, or the equivalent to the x in the alphabet.

XML   Extensible Markup Language; system for coding the structural parts of documents for display on the Internet and for storing the information in databases. Derives from SGML and is related to HTML.

Index